Annual Fishes

Life History Strategy, Diversity, and Evolution

Annual Fishes

Life History Strategy, Diversity, and Evolution

EDITED BY
NIBIA BEROIS • GRACIELA GARCÍA • RAFAEL O. DE SÁ

CRC Press
Taylor & Francis Group
Boca Raton London New York

CRC Press is an imprint of the
Taylor & Francis Group, an **informa** business

CRC Press
Taylor & Francis Group
6000 Broken Sound Parkway NW, Suite 300
Boca Raton, FL 33487-2742

First issued in paperback 2020

Version Date: 20150513

ISBN 13: 978-0-367-57531-1 (pbk)
ISBN 13: 978-1-4822-9971-7 (hbk)

Library of Congress Cataloging-in-Publication Data

Annual fishes : life history strategy, diversity, and evolution / editors, Nibia Berois, Graciela Garcia, and Rafael O. de Sa.
     pages cm
"A CRC title."
Includes bibliographical references and index.
ISBN 978-1-4822-9971-7
     1. Fishes--Seasonal distribution. 2. Fishes--Migration. 3. Fishes--Life cycles. I. Berois, Nibia, 1946- editor. II. Garcia, Graciela, 1953- editor. III. De S?, Rafael O., editor.

QL639.5.A56 2015
597.15'6--dc23

2015017780

Visit the Taylor & Francis Web site at
http://www.taylorandfrancis.com

and the CRC Press Web site at
http://www.crcpress.com

Contents

Part I
Diversity, Life History, and Reproductive Biology

Chapter 1

Marcelo Loureiro and Rafael O. de Sá

Chapter 2

Nibia Berois, María José Arezo, Nicolás G. Papa, and Cora Chalar

Chapter 3

**Magela Rodao, Jimena Montagne, Graciela A. Clivio, Nicolás G. Papa,
and Gabriela Casanova Larrosa**

Chapter 4

Jason E. Podrabsky, Amie L. Romney, and Kristin M. Culpepper

Chapter 5

**María José Arezo, Nicolás G. Papa, Nibia Berois, Graciela García,
Verónica Gutiérrez, and Marcelo Antonelli**

Chapter 6

Alessandro Cellerino

Chapter 7

Nicolás G. Papa, Graciela A. Clivio, and Jimena Montagne

Preface

Modern fishes are the basal and most diverse group of extant vertebrates, comprising more than 28,000 species (Nelson, 2006; Fishbase, www.fishbase.org) and exhibiting multiple levels of diversity, including development, morphology, ecology, behavior, genetics, reproduction, and life history strategy. Furthermore, they are an important model species in basic and applied research, as wells as being food resources and environmental biomonitors. Among fishes, some teleosts have emerged as leading models in genetics and developmental studies due to their easy maintenance and reproduction in laboratories. Zebrafish (*Danio rerio*) and medaka (*Oryzias latipes*) in particular have become commonly used in basic and technological approaches as the representative teleosts.

Annual fishes (Cyprinodontiformes, Aplocheiloidei) stand as an attractive fish model that offers exceptional advantages for studies in genetics and developmental studies, as well as in comparative and evolutionary biology due to their unique life cycles. They inhabit extremely variable freshwater environments, have a short lifespan, and exhibit a set of unique and fascinating survival strategies. South American and African annual fishes inhabit ponds that completely disappear during the annual dry season. During the rainy season, the adults engage in complex mating courtships that result in burrowing in the substrate, mating, and laying desiccation-resistant eggs. Developing eggs/embryos remain within the substrate, whereas the entire adult population dies during the dry season. Thus, the survival of the population and the species is entirely dependent on the success of buried embryos when they begin to hatch in the next rainy season, once the ponds are flooded. Consequently, annual fishes are among the most remarkable extremophile vertebrates.

Existence in such an unfavorable environment is quite challenging, and annual fishes display unique physiological, ecological, and behavioral evolutionary traits, among others. In addition, their annual life cycle shows unique developmental pathways and mechanisms related to the survival of the developing embryos, for example, epiboly detached from organogenesis and embryonic diapauses as reversible arrests. It is not surprising that annual fishes show high species diversity as well as intraspecific morphological, chromosomal, and molecular diversity. Part of this variation can be related to high plasticity in relation to environmental instability. Furthermore, speciation in this group is the result of the interaction of evolutionary mechanisms at the individual, population, and ecological levels. Taken together, the characteristics of annual fishes make them a challenging model for understanding the mechanisms involved in their adaptive attributes. Therefore, they are truly an extraordinary vertebrate model for evolutionary, ecological, behavioral, and developmental studies.

Currently, there are a few international and interdisciplinary research groups working on annual fishes. The aims of these groups involve a wide range of topics, such as behavior, sexual selection, molecular evolution, morphological evolution, speciation, population dynamics, competition, reproductive strategies, regulation of developmental mechanisms, evo-devo studies, and senescence. This research will contribute not only to understanding the complexity of annual fishes but also to understanding the general vertebrate model, particularly in areas such as speciation and senescence, where the life cycle of annual fishes is undoubtedly advantageous over other vertebrate models.

Until now, there has not been a single book that summarizes the current status of our understanding about the many facets of annual fish biology. The goal of this book is to bring together colleagues working in different areas of the biology of annual fishes to provide an overview of the current knowledge. In addition, we aim to stimulate current and future students to engage in understanding the biology of these fishes. The book is organized in three areas with the following contents. Under "Diversity, Life History, and Reproductive Biology," seven chapters explore the topics of diversity, reproduction and general development, gamete interactions, alternative developmental patterns, sex determination and differentiation, aging phenotypes, and maintenance of annual fishes under laboratory conditions. In "Ecology and Conservation," four chapters discuss studies on annual fishes in the field of evolutionary ecology of African annual fishes, tolerance of environmental stress, and diversity and conservation of annual fishes from Brazil. Finally, under "Evolution," seven chapters focus on reproductive behavior and sexual selection, brain anatomy and neurogenesis in adults, transposable elements, chromosome evolution and speciation, meiotic pattern, hybridization among species, and parallel evolution in New and Old World aplocheloids.

In summary, this book provides an insight into the different topics that are being investigated in the biology of annual fishes. Also, it indirectly notes gaps in our knowledge that will challenge future researchers. We are hopeful and enthusiastic about this book serving to attract undergraduate and graduate students as they are considering careers in different areas of biology. Furthermore, as annual fishes are common aquarium organisms, we hope this book will be of interest both to professional researchers and the large and important aquarist community because our joint efforts may ensure the future conservation of this fascinating group of fishes.

Editors

Nibia Berois is professor of cell biology and developmental biology at the Facultad de Ciencias, Universidad de la República, Montevideo, Uruguay. She earned her PhD in biology from the same university in 1998. She is a Grade 4 researcher at Programa de Desarrollo de Ciencias Básicas (PEDECIBA) and Level 2 Researcher at Sistema Nacional de Investigadores (SNI). Her expertise focuses on biology of reproduction and development, with emphasis on neotropical annual fishes. She is the author of more than 40 scientific papers in her research area. She has been an advisor for 42 student theses (bachelor's to doctoral level).

Graciela García is professor of genetics at the Evolutionary Genetics Section, Facultad de Ciencias, Universidad de la República, Montevideo, Uruguay. She is a Grade 4 researcher at PEDECIBA and Level 2 researcher at the SNI. She is the author of more than 40 scientific papers, mainly concerning neotropical annual fishes. Dr. García earned her PhD in biology from the Universidad de la República, Montevideo, Uruguay, in 1996. She is in general an animal geneticist with special interests in population genetics and evolution of fishes, phylogeography, and speciation in different fish models.

Rafael O. de Sá is professor of biology at the University of Richmond, Richmond, Virginia. Dr. de Sá earned his BS from the Universidad de la República, Montevideo, Uruguay; his MS in ecology and evolution from the University of Kansas; and his PhD in zoology from the University of Texas at Austin in 1991. In 2013–2014, he held a Fulbright Scholar Distinguished chair at the University of São Paulo, Brazil. He is a Grade 4 researcher at PEDECIBA and associate researcher at the SNI. His general training has been as an evolutionary biologist with expertise in molecular and morphological phylogenies, systematics, and taxonomy. He is the author of over 100 papers.

Contributors

Marcelo Antonelli
Instituto de Ciencias Biomédicas
Universidad de Chile
Santiago, Chile

María José Arezo
Sección Biología Celular
Universidad de la República
Montevideo, Uruguay

Matías Arim
Departamento de Ecología y Evolución
Universidad de la República
Montevideo, Uruguay

and

Centro Universitario Regional Este
Universidad de la República
Maldonado, Uruguay

Nibia Berois
Sección Biología Celular
Universidad de la República
Montevideo, Uruguay

Ana Borthagaray
Departamento de Ecología y Evolución
Universidad de la República
Montevideo, Uruguay

and

Centro Universitario Regional Este
Universidad de la República
Maldonado, Uruguay

Alessandro Cellerino
Scuola Normale Superiore
Pisa, Italy

and

Fritz-Lipmann Institute for Age Research
Leibniz Institute
Jena, Germany

Cora Chalar
Sección Bioquímica
Universidad de la República
Montevideo, Uruguay

Graciela A. Clivio
Sección Biología Celular
Universidad de la República
Montevideo, Uruguay

Kristin M. Culpepper
Department of Biology
Portland State University
Portland, Oregon

Rafael O. de Sá
Department of Biology
University of Richmond
Richmond, Virginia

Alejandro Duarte
Departamento de Ecología y
  Evolución
Universidad de la República
Montevideo, Uruguay

Anabel S. Fernández
Instituto de Investigaciones Biológicas
  Clemente Estable and Unidad
  Neuroanatomia Comparada
Universidad de la República
Montevideo, Uruguay

Graciela García
Sección Genética Evolutiva
Universidad de la República
Montevideo, Uruguay

Ândrio Cardozo Gonçalves
Laboratório de Ictiologia
Instituto Pró-Pampa
Rio Grande do Sul, Brazil

Demetrio Luis Guadagnin
Departamento de Ecologia
Universidade Federal do Rio Grande do Sul
Porto Alegre, Brazil

Verónica Gutiérrez
Sección Genética Evolutiva
Universidad de la República
Montevideo, Uruguay

Daniel Hernández
Centro Universitario Regional Este
Universidad de la República
Maldonado, Uruguay

Luis Esteban Krause Lanés
Laboratório de Ecologia e conservação
    de Ecossistemas Aquáticos
Universidade do Vale do Rio dos Sinos
and
Laboratório de Ictiologia
Instituto Pró-Pampa
Rio Grande do Sul, Brazil

Gabriela Casanova Larrosa
Unidad de Microscopía Electrónica
Universidad de la República
Montevideo, Uruguay

Marcelo Loureiro
Departamento de Ecología y Evolución
Universidad de la República
and
Departamento de Ictiología
Museo Nacional de Historia Natural
Montevideo, Uruguay

Jimena Montagne
Sección Biología Celular
Universidad de la República
Montevideo, Uruguay

Silvia Olivera-Bravo
Departamento Neurobiologia Celular
    y Molecular
Instituto de Investigaciones Biológicas
    Clemente Estable
Montevideo, Uruguay

Sebastián Oviedo
Sección Genética Evolutiva
Universidad de la República
Montevideo, Uruguay

Nicolás G. Papa
Sección Biología Celular
Universidad de la República
Montevideo, Uruguay

Carlos Passos
Sección Etología
Universidad de la República
Montevideo, Uruguay

Verónica Pinelli
Departamento de Ecología y Evolución
Universidad de la República
Montevideo, Uruguay

and

Centro Universitario Regional Este
Universidad de la República
Maldonado, Uruguay

Jason E. Podrabsky
Department of Biology
Portland State University
Portland, Oregon

Martin Reichard
Institute of Vertebrate Biology
Academy of Sciences of the
    Czech Republic
Brno, Czech Republic

Claire L. Riggs
Department of Biology
Portland State University
Portland, Oregon

Néstor Ríos
Sección Genética Evolutiva
Universidad de la República
Montevideo, Uruguay

Magela Rodao
Unidad de Microscopia Electrónica
Universidad de la República
Montevideo, Uruguay

Amie L. Romney
Department of Biology
Portland State University
Portland, Oregon

Gil G. Rosenthal
Department of Biology
Texas A&M University
College Station, Texas

and

Centro de Investigaciones Científicas
    de las Huastecas "Aguazarca"
Hidalgo, Mexico

Juan Carlos Rosillo
Instituto de Investigaciones Biológicas
   Clemente Estable
Universidad de la República
Montevideo, Uruguay

Mareney Rovira
Sección Genética Evolutiva
Universidad de la República
Montevideo, Uruguay

Bettina Tassino
Sección Genética Evolutiva
Universidad de la República
Montevideo, Uruguay

Matheus Vieira Volcan
Laboratório de Ictiologia
Instituto Pró-Pampa
Rio Grande do Sul, Brazil

and

Programa de Pós Graduação em
   Biodiversidade Animal
Universidade Federal de Santa Maria
Rio Grande do Sul, Brazil

Josiah T. Wagner
Department of Biology
Portland State University
Portland, Oregon

Diversity, Life History, and Reproductive Biology

CHAPTER 1

Diversity of Aplocheiloidei

Marcelo Loureiro and Rafael O. de Sá

CONTENTS

1.1 INTRODUCTION

Annual fishes are a fascinating group of freshwater fishes due to their particular developmental, ecological, physiological, and evolutionary adaptations. They possess an uncharacteristically short life cycle among vertebrates, that is, less than a year long. This unique life cycle is correlated to the seasonal pools of water they inhabit both in the African and Neotropical realms (Figure 1.1). The term "annual fishes" was coined by Myers (1952) to specifically refer to these short-lived vertebrates. Recent papers have reviewed the distinct developmental and growth pathways (Berois et al. 2012; Blažek et al. 2013) and overall biology (Berois et al. 2014) of annual fishes. This chapter will review hypothesized phylogenetic relationships from higher taxa to genus levels.

Annual fishes belong to the diverse Order Cyprinodontiformes (Series Atherinomorpha). Compared with the other freshwater diverse groups (e.g., Cypriniformes, Characiformes, Siluriformes) whose species diversity is accompanied by an amazing diversity in morphology and trophic habits, Aplocheiloidei killifishes are morphologically and trophically constrained; except for some alternate reproductive strategies (Costa 1998a), most of the variation they exhibit is in their coloration patterns.

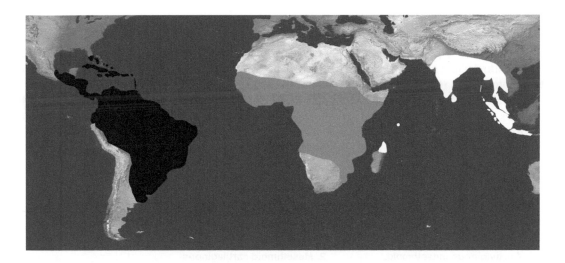

Figure 1.1 Distribution of Aplocheiloidei. White shade = Aplocheilidae; Black shade = Rivulidae; Gray shade = Notobranchiidae. (source: http://naturalearth.springercarto.com.)

Cyprinodontiformes exhibit a variety of reproductive modes, including viviparous species (e.g., Poecilidae, Anablepidae), species with delayed embryonic development (e.g., Fundulidae), and even a parthenogenetic species, *Poecilia formosa*. However, Aplocheiloidei exhibit several unique reproductive characteristics. The most well-known examples are self-mating in two species of *Kryptolebias* (Tatarenkov et al. 2009) and the amazing annual life cycles of Notobranchiidae and many Rivulidae. The life cycle involves complex developmental, ecological, and behavioral adaptations. Furthermore, some annuals also have internal insemination, for example, *Campelollebias* and *Cynopoecilus* (Costa 1998a; Ferrer et al. 2014).

One well-known characteristic of Aplocheiloidei is the striking sexual dimorphism that many species exhibit, wherein males are usually slightly larger than females and exhibit brighter coloration patterns (Sedlácek et al. 2014). Among Neotropical annual species, breeding males are colorful and show various patterns of bright vertical bands and/or dots on body, fins, and head (particularly over the operculum), whereas females are mostly brown with scattered darker spots on fins and body (Berois et al. 2014). A recent study demonstrated that intra and intersexual selection on male body size favors larger size in males of the Neotropical rivulid *Austrolebias charrua* (Passos et al. 2013). Among the African nothobranchids, females are duller than males, with overall silvery or brownish coloration and spots ranging from barely visible to darker dots (e.g., *N. melanospilus* species group, Wildekamp et al. 2009). Males exhibit striking coloration, with brighter blues, yellow, orange, and red colorations than are found among Neotropical species. Despite the general patterns described here, both males and females of Africa and the Neotropical show a great deal of variation among species (Sedlácek et al. 2014). Sexual selection through female choice is responsible for the chromatic dimorphism in the African annual *Notobranchius* (Haas 1976). A recent study documented that, at least in some lineages, annual killifish coloration is androgen dependent, and brighter coloration can be induced in females with treatment with exogenous androgens; furthermore, the authors suggest that the striking sexual dimorphism evolved by the disappearance of bright coloration in females (Sedlácek et al. 2014).

On the other hand, size and general morphology in Aplocheiloids seems to be constrained, at least in comparison with other diverse clades of freshwater fishes such as Characiformes, Siluriformes, and Cichliformes. Most aplocheiloid species do not surpass 10 cm in body length and present an overall fusiform body shape. Only the annual rivulid tribe Cynolebisini and, to a lesser extent, the annual notobranchid *Notobranchius* evolved relatively compressed bodies. Diet also seems constrained in Aplocheiloidei, with all species analyzed so far shown to be generalist predators, with some piscivorous specialization in larger species of *Austrolebias* (Costa 2009a) and *Notobranchius* (see Chapter 10).

Parenti (1981) provided the first phylogenetic analysis of Cyprinodontiformes and recovered Aplocheiloidei as a monophyletic group based on nine synapomorphies, mostly osteological characters. Later, Costa (1998b) proposed 18 additional synapomorphies, one of them confirmed recently by Costa's (2012) analysis of the caudal skeleton of Cyprinodontiformes (Table 1.1). Costa's (2004), analysis of 18 terminal taxa (mostly Rivulidae) identified additional morphological synapomorphies for Aplocheiloidei and modified some of his previously proposed synapomorphies (Costa 1998b).

Parenti (1981) proposed two families for the suborder Aplocheiloidei: Aplocheilidae, for the Old World species, and Rivulidae, for the New World taxa. Subsequently, this taxonomic

Table 1.1   Parenti (1981) and Costa (1998b) Synapomorphies for Aplocheiloidei

Parenti (1981)	Costa (1998b)
1. **Attached orbital rim** (1)	1. **Orbital rim attached** (1)
2. **Cartilaginous mesethmoid**	2. **Mesethmoid cartilaginous**
3. **Broad anterior end of basihyal**	3. **Anterior portion of basihyal widened**
4. **Tubular anterior naris**	4. **Anterior narus opening at the tip of a distinctively cylindrical structure**
5. **Males more elaborately pigmented than females**	5. **Males with color pattern much more elaborate than females**
6. Narrow and twisted lacrimal[a](2)	6. Lacrimal slightly twisted with posterior rim reduced and bone formed mainly by canal[a](2)
7. Reduced cephalic sensory pores[a]	7. Supraorbital canals open with neuromasts exposed externally[a]
8. Close-set pelvic girdles[b]	8. Caudal accessory cartilaginous absent[c]
9. Posterior extension of vomer dorsal to anterior arm of parasphenoid[b]	9. Dorsal process of maxilla short, anteriorly directed, not parallel to ventral process
	10. Coronoid process of anguloarticular reduced
	11. Lateral flange of hyomandibula expanded posterodorsally
	12. A distinct anteromedial process of second hypobranchial directed toward second basibranchial
	13. Anterior cartilage head of fifth ceratobranchial narrow and reduced
	14. Membranous aperture between fourth and fifth ceratobranchial reduced
	15. A distinct posterior process on fourth epibranchial
	16. Vomerine teeth present
	17. A wide process on the anterior portion of lateral ethmoid
	18. Neurocranium flattened
	19. Demosphenotic short
	20. Medial process of pelvic girdle short
	21. Distal radial of anal fin with an expanded posteroventral rim
	22. Proximal end of the parhypural laminar, without dorsal paired process and not contacting preural centrum
	23. Anterior portion of dorsal fin with one long ray attached to two proximal radials and preceded by one or two short rays
	24. 20 to 25 radii on anterior abdominal scales
	25. Dentition on second pharyngobranchial reduced

Note: Bold = Character coincident to both authors.
[a] Character modified by Costa (1998b).
[b] Character not mentioned after Parenti (1981).
[c] Character confirmed by Costa (2013); (1) = character modified by Costa (2004): orbital rim ventrally attached as synapomorphic for Aplocheiloidei, and completely attached, synapomorphic for Notobranchoidea; (2) = in Costa (2004) synapomorphic for Notobranchoidea.

arrangement was supported by the first molecular analysis (using nuclear genes) of aplocheloids (Meyer and Lydeard 1993) and morphological data (Costa 1998b). The combined analysis of molecular and nonmolecular data of Cyprinodontiformes recovered a trichotomy consisting of Rivulidae (three species included), Aplocheilidae (only *Nothobranchius melanospilus* included), and Cyprinodontoidei (18 species included; Parker 1997). Alternatively, Murphy and Collier (1997) recovered a clade consisting of species from Madagascar, the Seychelles, and Indo-Malaysia as the sister taxa to a clade formed by the continental African species and Rivulidae based on mitochondrial sequences. Next, Costa's (2004) morphological analysis agreed with Murphy and Collier's (1997) results and proposed two superfamilies: Aplocheilidea (*Aplocheilus* and *Pachypanchax*, from noncontinental African and Asian taxa) and Nothobranchioidea, divided into two families, Notobranchiidae and Rivulidae, for continental Africa and New World taxa, respectively.

A study of cranial soft tissues (Hertwig 2008) identified 78 characters; these were combined with Costa's (1998b) morphological data set. The phylogenetic analysis differs from the previous phylogeny by recovering the New World Rivulidae as the sister group to a clade consisting of the Aplocheilidae and Nothobranchiidae (Hertwig 2008). In the analysis in which only soft-tissue characters were used, *Aplocheilus* was recovered as the sister taxa to all other Cyprinodontiformes (Hertwig 2008). This author noted the agreement of these results with a functional morphological study in which the jaw apparatus of *Aplocheilus* possess numerous plesiomorphic characters shared with *Atherina* (Atheriniformes; Alexander 1967).

The most recent phylogenetic analysis of Aplocheiloidei (Costa 2013) included mitochondrial sequences (cytochrome b, 12S rRNA, and 16S rRNA), 199 nonmolecular characters for 61 aplocheiloid taxa and six outgroups, including caudal osteological characters from Costa (2012). This latest study disagrees with Hertwig's (2008) results and provides additional support to Murphy and Collier's (1997) and Costa's (2004) phylogenies. Also, the study noted that the split of Aplocheiloidea from Notobranchoidea is congruent with the vicariant event associated with the breakup of the Somali basin (Costa 2013), which resulted in the separation of Madagascar and India from a northern South America–Africa block (ca. 121 mya; Sanmartín and Ronquist 2004). Interestingly, a similar pattern is found among Cichliformes, where species from branches of the basal nodes are distributed in India and Madagascar, while cichlids from continental Africa and the Neotropics showed a sister clade relationship (Smith et al. 2008).

We summarize here proposed relationships of Aplocheiloidei, building consensus topologies based on different authors and data types: Murphy and Collier (1997, 1999); Murphy et al. (1999a); Van Der Zee et al. (2007); Sonnenberg and Busch (2009); Collier et al. (2009); Huber (2000); and Costa (2013). Summaries are provided in Figure 1.2 and Table 1.1.

1.2 SUPERFAMILY APLOCHEILOIDEA

The single family Aplocheilidae (distributed in Madagascar, India, Pakistan, and Vietnam) is diagnosed by a black blotch on the dorsal fin of females (Costa 2004). The family consists of 14 currently valid species (Eschmeyer 2014) grouped in two genera, *Aplocheilus* and *Pachypanchax*, which may have been originated by the vicariant event associated with the breakup of the Mascarene basin (ca 84 mya; Sanmartín and Ronquist 2004; Costa 2013).

1.2.1 *Aplocheilus*

Seven small- to medium-sized nonannual species (50–100 mm SL) distributed from Pakistan to Vietnam (Figure 1.3a). Parenti (1981) defined *Aplocheilus* by an attenuated lower jaw and premaxillary ascending processes that are medially expanded and overlapping.

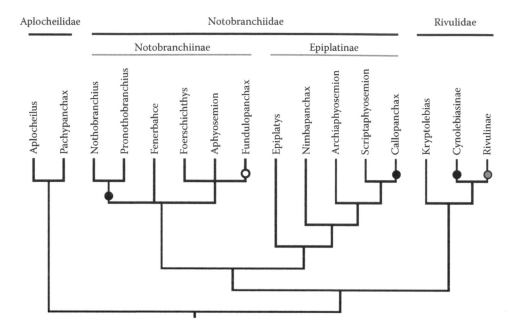

Figure 1.2 Summary of proposed relationships of Aplocheiloidei with emphasis in Notobranchiidae. Consensus topology was manually reconstructed on *Mesquite: A Modular System for Evolutionary Analysis*, release version: 3.01. (Based on Costa, W. E. J. M. 2013. *Vertebrate Zoology* 63:139–154; Murphy, W. J. and G. E. Collier. 1997. *Molecular Biology and Evolution* 14:790–799; Murphy, W. J. and G. E. Collier. 1999. *Molecular Phylogenetics and Evolution* 11:351–360; Murphy, W. J. et al. 1999a. *Molecular Phylogenetics and Evolution* 11:343–350; Van Der Zee, J. R., T. Woeltjes, and R. H. Wildekamp. 2007. *The Fresh and Brackish Water Fishes of Lower Guinea, West-Central Africa*, Vol. 2, Tervuren: Musée Royal de l'Afrique Centrale; Sonnenberg R. and E. Busch. 2009. *Zootaxa* 2294:1–22; Collier, G. E., W. J. Murphy, and M. Espinoza. 2009. *Molecular Phylogenetics and Evolution* 50:190–196; Huber, J. H. 2000. *Killi-Data 2000. Updated Checklist of Taxonomic Names, Collecting Localities and Bibliographic References of Oviparous Cyprinodont Fishes (Cyprinodontiformes)*. Paris: Cybium Société Française d'Ichtyologie.) Symbols: black circle = annual life cycle; white circle = semiannual or facultative annual life cycle; gray circle = clade presents annual and nonannual species.

1.2.2 *Pachypanchax*

Seven small- to medium-sized nonannual species (52–100 mm SL) distributed in Madagascar and the Seychelles. The genus is diagnosed by maxillary relatively immobile, attached at its posterior end to the preorbital by a skin fold; ascending processes of premaxillae flat, broad, tapered posteriorly, not overlapping at the midline; one pair of tubular nares; reflective pineal spot absent; E-type frontal squamation with prominent H scales; absence of papillae on scales of fin rays; haemal arches not expanded and lacking pleural ribs on haemal spines; hypural plates fused forming a hypural fan in adults; caudal fin rounded to rounded-truncate with middle rays not extended; proximal caudal fin heavily scaled; caudal fin lacking a median lobe; pigmentation pattern without cross bars on body; basal dorsal-fin spot absent in males (Loiselle 2006).

1.3 SUPERFAMILY NOTOBRANCHOIDEA

This clade is supported by eight synapomorphies: twisted and reduced lacrimal; distinctive neural spine on first vertebra; shortened laminar proximal end of parahypural; narrowed haemal

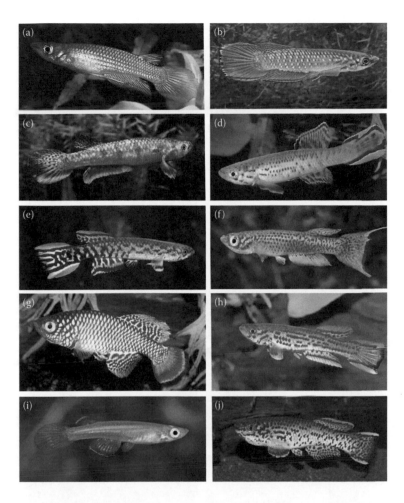

Figure 1.3 Live specimens of Aplocheilidae and Notobranchiidae: (a) *Aplocheilus lineatus* (photo by Frans Vermeulen); (b) *Epiplatys fasciolatus* (photo by Tony Terceira); (c) *Nimbapanchax petersi* (photo by Tony Terceira); (d) *Callopanchax monroviae* (photo by Frans Vermeulen); (e) *Scriptaphyosemion geryi* (photo by Tony Terceira); (f) *Fenerbahce devosi* (photo by Tony Terceira); (g) *Nothobranchius rachovii* (photo by Frans Vermeulen); (h) *Aphyosemion hannelorae* (photo by Frans Vermeulen); (i) *Foerschichthys flavipinnis* (photo by Tony Terceira); (j) *Fundulopanchax gardneri* (photo by Frans Vermeulen). (With permission.)

spine of preural centrum 2; one or two short rudimentary rays at the beginning of dorsal and anal fins; pectoral fin insertion ventrolateral; completely attached orbital rim; and superior mouth (Costa 2004). The superfamily consists of two families, Nothobranchiidae and Rivulidae, and their split is congruent with the opening of the northern South Atlantic Ocean (ca.110 mya; Sanmartín and Ronquist 2004; Costa 2013).

The continental Africa family Nothobranchiidae is diagnosed by the presence of bifid epipleural ribs; a prominent keel-shaped lateral process on the middle part of the compound caudal centrum (Costa 2012); and three oblique red stripes on the postorbital region of males (Costa 2004). Notobranchid species have been diagnosed based on coloration patterns (most commonly based on male colorations, e.g., Sonnenberg 2007). However, in *Scriptaphyosemion*, female coloration patterns, as well as neuromast arrangements, are also diagnostic for species or groups of species. Species identities have subsequently been supported by molecular data (e.g., Legros and Zentz 2007; Agnèse et al. 2009; Shidlovskiy et al. 2010; Sonnenberg et al. 2011; Ng'oma

Figure 1.4 Live specimens of Rivulidae: (a) *Kryptolebias marmoratus* (photo by J.P. Cicéron); (b) *Papiliolebias bitteri* (photo by P. Calviño); (c) *Melanorivulus punctatus* (photo by D. Nielsen); (d) *Neofundulus paraguayensis* (photo by P. Calviño); (e) *Trigonectes aplocheiloides* (photo by F. Alonso); (f) *Pterolebias longipinnis* (photo by F. Alonso); (g) *Hypsolebias magnificus* (photo by Amer Four); (h) *Spectrolebias semiocellatus* (photo by Amer Four); (i) *Cynopoecilus melanotaenia* (photo by M. Loureiro); (j) *Austrolebias luteoflammulatus* (photo by M. Loureiro).

et al. 2013; Dorn et al. 2014). Two clades are recognized within Notobranchiidae: Epiplatinae and Notobranchiinae (Figure 1.4). The two subfamilies, as well as some clades within them, are currently supported by molecular (Murphy and Collier 1997) and morphological data (Van Der Zee et al. 2007; Costa 2013).

1.4 SUBFAMILY EPIPLATINAE

A molecular phylogenetic analysis recovered a monophyletic Epiplatinae (Murphy and Collier 1997); however, only six species (of the current 60 described species) were included in the study.

The subfamily is diagnosed by preopercular canal tubular with five pores; anterior part of lower jaw broad; and a downward obliquely oriented series of scales on the gill cover (Van Der Zee et al. 2007).

The clade consisting of *Nimbapanchax*, *Archiaphyosemion*, *Callopanchax*, and *Scriptaphyosemion* (tribe Callopanchacini) is considered the sister group of *Epiplatys* (Murphy et al. 1999a; Sonnenberg and Busch 2009; Costa 2015). It is diagnosed by six synapomorphies recently proposed by Costa (2015): posterior portion of mandibular channel shaped as a single open groove; basihyal pentagonal; urohyal dorsal process absent or rudimentary; wide bony flap next to proximal portion of fourth ceratobranchial; broad bony flap next to proximal portion of fifth ceratobranchial; haemal prezygapophysis of pre-ural vertebra number 2 directed ventrally.

The ((*Callopanchax* + *Scriptaphyosemion*) (*Archiaphyoseminon*; formerly known as *Roloffia*)) clade shares a derived condition of an abrupt widening of the posterior portion of the parasphenoid. This clade is also supported by molecular data (Sonnenberg and Busch 2009; Collier et al. 2009) and the recently outlined molecular and morphological phylogeny (Costa 2015). The sister relationship between *Callopanchax* and *Scriptaphyosemion* was recovered on molecular studies (Murphy and Collier 1997; Murphy and Collier 1999; Collier et al. 2009; Sonnenberg and Busch 2009) and combined analysis (Costa 2015) and is supported by five morphological synapomorphies (Costa 2015): urohyal drop-shaped; second pharyngobranchial teeth absent; parietal short and elliptical; first vertebra with an anteromedian laminar process on the neural process.

1.4.1 *Epiplatys*

The most diverse genus in the subfamily with 37 small- to medium-sized recognized nonannual species (40–80 mm SL); grouped into two reciprocally monophyletic clades (Collier et al. 2009) and distributed throughout East and West Africa (Figure 1.3b). It is diagnosed by a postemporal bone with a lower ossified branch; orbital rim completely attached; hypobranchials without teeth; and lack of darkened caudal marginal dorsal *ocellus* (Parenti 1981). *Epiplatys* has a unique extensive distribution, largely overlapping with the distributions of other genera of Epiplatinae (Western Africa) and with some species distributed in the eastern part of the continent (Collier et al. 2009).

1.4.2 *Nimbapanchax*

Consists of five valid small nonannual species (max. 70 mm SL) endemic to the Upper Guinean region (Liberia, Guinea, and western Ivory Coast in the upper region of coastal rivers) (Sonnenberg and Busch 2009) (Figure 1.3c). This genus is diagnosed by a single synapomorphy (Costa 2015): second vertebra with a fan-shaped neural process.

1.4.3 *Archiaphyosemion*

A monotypic and nonannual genus distributed in Guinea, Sierra Leone, and the upper Niger and its affluents in Guinea, *A. guineense* is a small species (max. 70 mm SL) defined by nine autapomorphies (Costa 2015): frontal squamation E-patterned; dorsal process of maxilla broad; dorsal process of autopalatine prominent; anterior dorsal condyle of hyomandibula longer than posterior dorsal condyle; cartilaginous portion of basihyal equal in length to osseous portion; bifid proximal portion of first epibranchial; second and third epibranchials longer than half the length of adjacent ceratobranchials; bony flap adjacent to the articular head of fourth ceratobranchial in a subproximal position; short fifth ceratobranchial with a row of medial broad teeth.

1.4.4 *Callopanchax*

Consists of four currently recognized small- to medium-sized annual species (60–85 mm SL) inhabiting the coastal plains from southwestern Guinea and Sierra Leone to western Liberia (Costa 2009b; Sonnenberg and Busch 2010) (Figure 1.3d). The monophyly of *Callopanchax* is supported by four synapomorphies (Costa 2015): dorsal portion of autopalatine twisted and narrow; antero-ventral margin of hyomandibula convex; neural processes of first vertebra medially fused; neural prezygapophyses of anterior caudal vertebrae long.

1.4.5 *Scriptaphyosemion*

Is a relatively species-rich nonannual genus (Figure 1.3f), compared with closely related genera, currently consisting of 13 small recognized species (Eschmeyer 2014). Species diversity is highest in Sierra Leone, Liberia, and Guinea, and the distribution range includes Burkina Faso, Gambia, Guinea-Bissau, Ivory Coast, Mali, and Senegal (Huber 2007). This genus has been diagnosed by a red coloration pattern with yellow/red margins of the unpaired fins; chromosome numbers $n = 20$ or 21; and standard length less than 40 mm (Radda and Pürzl 1987). Coloration patterns of females have been used to diagnose species or groups of species. However, most species exhibit interspecific variance in male color pattern that, coupled with intraspecific variance among and within populations, makes species identification problematic and not clear. Furthermore, some of the most variable species may consist of complexes of cryptic species (Sonnenberg and Busch 2012). Recently, Costa (2015) proposed two synapomorphies for the genus: vomerine teeth absent; neural and haemal spines of preural vertebra 2 curved.

1.5 NOTOBRANCHIINAE

This subfamily, together with Rivulinae, is probably the most controversial clade of Aplocheiloidei regarding phylogenetic relationships and recognition of genera. Three diagnostic characters were proposed for the subfamily: preopercular canal tubular with six pores; anterior part of lower jaw less broad and somewhat rounded; and an upward, obliquely oriented scale row on the gill cover (Van Der Zee et al. 2007). The clade consisting of *Aphyosemion* and *Fundulopanchax*, two genera from central-western Africa, has been supported by molecular (Murphy and Collier 1999) and morphological (Costa 2009b) analysis; the latter work proposed expanded hypobranchials as a synapomorphy for this clade. In addition, Murphy and Collier (1999) also proposed a sister group relationship of (*Aphyosemion* + *Fundulopanchax*) to (*Nothobranchius* + *Pronothobranchius*). A recent phylogenetic analysis of *Notobranchius* suggested *N. thierryi* as the sister taxon to *Pronothobranchius* + *Nothobranchius* (Dorn et al. 2014) in agreement with previous results (Murphy and Collier 1999). The position of *Foerschichthys* and *Fenerbahce* remains provisional, since they have not been included in any recent phylogenetic analysis. Parenti (1982) included *Foerschichthys* in a clade that included Epiplatinae species and Aplocheilidae, and placed *Fenerbahce* (a senior synonym of *Adamas*) as closely related to *Aphyosemion*. However, taxon sampling (considering the current number of valid species) was very low. More recently, *Foerschichthys* was placed in the subtribe Aphyosemina (together with *Aphyosemion* and *Fundulopanchax*) and *Fenerbahce* was given its own subtribe; both of these generic arrangements lack supporting synapomorphies (Huber 2000).

1.5.1 *Fenerbahce*

Consists of two small nonannual species, *F. devosi* and *F. formosus* (Figure 1.3f) endemic to the southeastern Congo Basin (Sonnenberg et al. 2011). The genus is characterized by the following combination of characters: maximum total length less than 40 mm; reflective scales dorsally on the head; pre, post, supraorbitals; opercular neuromast system without tubular structures; and small dorsal fin (Sonnenberg et al. 2011).

1.5.2 *Pronothobranchius*

Contains four currently recognized annual species (Eschmeyer 2014) with a combined distribution that includes Gambia, Mali, Niger, Nigeria, Burkina Faso, Chad, Cameroon, and Ghana. The species are small (65 mm total length), and occur in temporary brooks and swamps (Wildekamp et al. 1986). They are bottom spawners and embryonic development lasts 1.5 months (Huber 1996). The genus can be differentiated from other nothobranchins by the presence of a fused fan-shaped hypural (Valdesalici 2013).

1.5.3 *Nothobranchius*

Is a large annual genus (Figure 1.3g) consisting of 62 currently valid small (30–70 mm SL) species (except *N. ocellatus*, which may reach 150 mm SL), most of them distributed in temporary savannah pools in Eastern Africa and two species in central Africa (Dorn et al. 2014). Most derived *Nothobranchius* species are found in the lowland plains, although a significant number of species occur at higher-altitude plains in Kenya, Tanzania, and Zimbabwe (Dorn et al. 2014). Synapomorphies of the genus were proposed by Parenti (1981): preopercular canal open, not represented by pores; attachment of the interarcual cartilage directly to the second pharyngobranchial; and oval eggs. Male coloration in the genus is sexually selected (Haas 1976), and males of many species occur in two color morphs that can occur allopatrically or sympatrically (Reichard and Polacik 2010).

1.5.4 *Foerschichthys*

A monotypic small-size nonannual genus (*F. flavipinnis*; max. 25 mm SL) distributed in southeastern Ghana, southern Togo, southern Benin, and southwestern Nigeria to the Niger delta (Figure 1.3i). It is distinguished from all other aplocheiloid killifishes by the derived placement of the dorsal fin, which is placed entirely posterior to the anal fin (Parenti 1981).

1.5.5 *Fundulopanchax*

Consist of about 40 described small- to large semi-annual species (40–130 mm SL, most less than 60 mm SL; Huber 2000) distributed from southwest Nigeria to southeast Cameroon (Figure 1.3j). However, some species have been reported from small temporal ponds of the Ecucu River drainage system of equatorial Guinea (Malumbres and Castelo 2001). The genus monophyly is supported by molecular analysis (Murphy and Collier 1999). However, the genus has been hard to define morphologically. Two characters distinguished it from other genera: having 16 or more circumpeduncular scales and unique otolith morphology (Van Der Zee and Wildekamp 1994). However, not all species present the first character; furthermore, there is not a clear consensus about which species belong to this genus or which to *Aphyosemion*. A distinct difference between the two genera, although this has not been verified for all species, would be that *Fundulopanchax* species have an annual life cycle (Murphy and Collier 1999).

1.5.6 *Aphyosemion*

Is the most speciose genus (Figure 1.3h) in the subfamily, consisting of more than 80 nonannual small species (max. 45 mm SL; Valdesalici and Eberl 2013). The species in this genus inhabit small streams within the Congo basin from Togo to Angola. The taxonomy of the genus is still unstable, and the species are grouped in several subgenera that have alternatively been considered separate genera or subgenera (Sonnenberg 2000; Agnèse et al. 2009; Van Der Zee and Sonnenberg 2011); although a molecular phylogeny has recovered some proposed subgeneric groups as monophyletic (Agnèse et al. 2006). The taxonomical problems of *Aphyosemion* arise, in part, because the older-type specimens have lost key characters, for example, male coloration patterns, and original descriptions lack detailed information about live or preserved coloration. Furthermore, many species exhibit variation in color pattern within and between populations, which confounds identifications, diagnoses, and species descriptions (Van Der Zee and Sonnenberg 2011). Also, phylogenetic relationships and species boundaries may be problematic to assess due to genetic introgression among species in the genus (Sonnenberg and Van Der Zee 2012). No synapomorphic characteristics have been proposed for the genus.

1.6 RIVULIDAE

The first synapomorphies for Rivulidae (Figures 1.4 and 1.5) derive from a phylogenetic analysis of Cyprinodontiformes (Parenti 1981; Table 1.2). The 1980s was a period of discovery of the amazing diversity in this clade, which also provided new derived characters for the family or modified some of Parenti's synapomorphies (Costa 1990, 1998a, 2004; Table 1.2). More recently, additional characters derived from soft tissue analysis have been proposed (Hertwig 2008; Table 1.2). Also, the

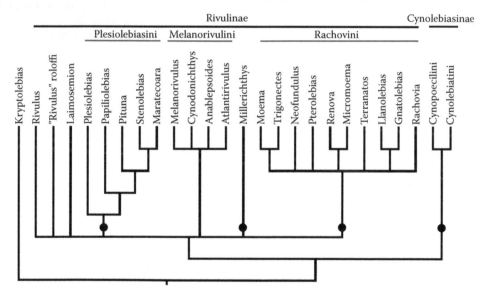

Figure 1.5 Summary of proposed relationships of Rivulidae with emphasis on Rivulinae. Consensus topology was manually reconstructed on *Mesquite: A Modular System for Evolutionary Analysis*, release version: 3.01. Conflicting clades with no synapomorphies are shown as polytomies. Symbols: black circle = annual life cycle. (Based on Costa, W. J. E. M. 1998a. *Journal Comparative Biology* 3:33–94; Costa, W. E. J. M. 2004. *Ichthyological Exploration Freshwaters* 15:105–120; Costa, W. J. E. M. 2011a. *Ichthyological Exploration of Freshwaters* 22:233–249; Costa, W. J. E. M. 2011b. *Zoological Journal of the Linnean Society* 162:131–148; Costa, W. E. J. M. 2013. *Vertebrate Zoology* 63:139–154; Costa, W. E. J. M. 2014a. *Vertebrate Zoology* 64:177–192; Murphy, W. J. and G. E. Collier. 1997. *Molecular Biology and Evolution* 14:790–799; Murphy, W. J. and G. E. Collier. 1996. *Molecular Biology and Evolution* 13:642–649; Murphy, W. J., J. E. Thomerson, and G. E. Collier. 1999b. *Molecular Phylogenetics and Evolution* 13:289–301; Hrbek, T. and A. Larson. 1999. *Evolution* 53:1200–1216; Hrbek, T., C. P. Deus, and I. P. Farias. 2004. *Copeia* 2004:569–576; Hrbek, T. and D. C. Taphorn. 2008. *Zootaxa* 1734:27–42.)

Table 1.2 Proposed Synapomorphies for Rivulidae

Parenti 1981	Costa 1990	Costa 1998	Costa 2004	Hertzwig 2008
First postcleithrum absent	First postcleithrum absent	Absence of the first postcleithrum	First postcleithrum absent	A1 *Adductor mandibulae* with a completely separated medial portion (character 5)
Branchiostegal and opercular membranes united	Membranes over opercular and branchiostegal rays fused	Branchiostegal and opercular membranes united	Branchiostegal and opercular membranes united	A1 *Adductor mandibulae* inserts caudally on the maxillary (character 14)
Obsolescent preopercular and lacrimal canals	Mandibular and lacrimal canals absent	Absence of mandibular canal and reduction of the preopercular bone	Infraorbital, preopercular, and mandibular canals open	A1 *Adductor mandibulae* origin expanded posteriorly and ventrally at the preopercle (character 23)
Lacrimal, preopercular, and mandibular canals represented by neuromasts	Reduced preopercular canal Anteromedian process of lateral ethmoid curved	**Well-developed process on the anterior border of lateral ethmoid**	**Pronounced retrorse process of lateral ethmoid**	Origin of the *m. intermandibularis* expanded posteriorly (character 56)
Medial process of lateral ethmoid expanded	**Expanded lateral ethmoid surface pointing toward parasphenoid (Note: subsequently modified by Costa [1998])**	Triangular ventral process of maxilla (Note: modified from Parenti [1981])[a]	Frontal scales circularly arranged around scale A[a]	Presence of a separate *foramen mandibulare* in the dorsal part of the hyomandibula (character 63)
Head scales arranged in circular pattern (Note: Costa [1990] disagrees; Costa [2004] proposed again but modified)[a]	Preopercle elongated (with internal curvature reduced)	Thin, C-shaped preopercle	Ventral process of maxilla bent and directed posteriorly (Note: modified from Costa [1998])[a]	
Process on ventral arm of maxilla (Note: Costa [1990] disagrees; Costa [1998] modified; Costa [2004] modified again)[a]	Neural spine of first vertebrae present	Neural spine of first vertebra (Note: synapomorphy for Notobranchoidea; Costa 2004)	Thin C-shaped preopercle	
	Cartilaginous interhyal (Note: deleted by Costa [2004], since bony condition in *Kryptolebias*)	Cartilaginous interhyal (Note: deleted by Costa [2004], since bony condition in *Kryptolebias*)	Posterior distal border of distal premaxilla straight	
	Diving into substrate to lay eggs (Note: Costa [1998] disagrees)	Straight posterior border of the distal tip of premaxilla	Coronoid process of dentary narrow	
		Reduced coronoid process of dentary	Ventral tip of autopalatine shortened	
		Short ventral portion of autopalatine	Uncinate process of third epibranchial median	
		Reduced uncinate process of the third epibranchial	Shortened lateral rim of frontal lacrimal very twisted and narrow	
		Reduced lateral border of frontal	Minute dermosphenotic	
		Extremely twisted and narrowed lacrimal	Posterior tip of ascending process of premaxilla curved medially	
		Minute dermosphenotic	Ventral process of angulo-articular median	
		Reduced basihyal cartilage (Note: homoplasic; Costa 2004)	Proximal edge of first hypobranchial bifid	
		Absence of a posterior flange on cleithrum (Note: present in *Kryptolebias*; Costa 2004)	Supracleithrum and postcleithrum separated	
		Hypurals ankylosed to form a single hypural plate (Note: not discussed in Costa [2004])		
		Transverse dark bar crossing the eye (Note: only in Cynolebiasinae; Costa 2004)		
		Annual life cycle (Note: nonannual in Costa [2004])		
		Reduced mesopterygoid, short snout, and elongate rostral cartilage (Note: not discussed in Costa [2004])		

Note: Bold = Character proposed by Costa (1998), ater modified by Costa (1998, 2004); underlined = Character proposed by Costa (1990), later deleted by Costa (2004).
[a] Character first proposed by Parenti (1981), later modified by Costa (1998, 2004).

monophyly of Rivulidae has been supported by molecular phylogenies (Murphy and Collier 1997; Murphy et al. 1999b; Hrbek and Larson 1999; Costa 2013). Furthermore, the monophyly of Rivulidae and its relationship with other Aplocheiloidei, suggests its origins as a vicariant event as a result of the opening of the southern Atlantic Ocean (Costa 2013). However, as fossils of Rivulidae have not been found yet, this question remains open. The family consists of three clades: Kryptolebiatinae, Rivulinae, and Cynolebiasinae; with Kryptolebiatinae (with only the genus *Kryptolebias*) as the sister group to a clade consisting of (Rivulinae + Cynolebiasinae) (Costa 2013). A basal *Kryptolebias* was previously recovered in the Hrbek and Larson (1999; *K. caudomarginatus* was then included in the genus *Rivulus*) and Costa (2004) studies, although nodes support was weak. *Kryptolebias* also was recovered monophyletic in a clade with Rivulinae with Cynolebiasinae as their sister taxon (Murphy and Collier et al. 1999); this topology was also recovered with low support by Costa (2011a) and statistically supported by Hrbek et al. (2004). Parenti's (1981) results suggested a paraphyletic Rivulinae relative to Cynolebiasinae (*Cynolebias* in that article); however, poor sampling of low known diversity at the time probably biased those results. Costa's (1998a) phylogenetic analysis of Rivulidae recovered *Kryptolebias* nested within *Rivulus*. A close relationship of *Cynopoecilus melanotaenia* and *Leptolebias fractifasciatus* (Cynolebiasinae) to *Rivulus* (Rivulinae), supported by having two narrow and distantly attaching dorsal insertion sites of the muscle protractor hyoidei, has also been recovered (Hertwig 2008), conflicting with previous hypotheses of relationships.

1.7 KRYPTOLEBIASINAE

Costa (2004) recognized the subfamily Kryptolebiasinae provided synapomorphies for *Kryptolebias*, and described the diversity of the genus; unfortunately the phylogenetic analysis (as well as Costa [2013]) did not report branches support.

1.7.1 *Kryptolebias*

Eight nonannual species (Figure 1.4a), inhabiting Atlantic coastal basins of North, Middle, and South America (Costa 2013). Two synapomorphies support the genus: dorsally expanded flange on dorsal tip of the autopalatine and vestigial or absent anterodorsal process of urohyal (Costa 2004). *Kryptolebias* is also differentiated from other rivulids by the presence of four plesiomorphic character states: ossified interhyal, well-developed posterior flange of cleithrum, distinct uncinate process of third epibranchial, and narrow distal articulated face of first hypobranchial (Costa 2004).

1.8 RIVULINAE + CYNOLEBIASINAE

The sister relationship of Rivulinae and Cynolebiasinae was supported by eight synapomorphies (Costa 2004): anterodorsal process of urohyal long; cartilaginous or absent interhyal; dorsal process of distal portion of second epibranchial absent; uncinate process of third epibranchial shortened; distal articulate face of first hypobranchial expanded; neural spine of first vertebra narrow; vestigial or absent posterior flange of cleithrum; and E-scales not overlapped.

1.9 RIVULINAE

Based on a combined molecular and morphological analysis, eight synapomorphies were proposed for the subfamily: ventral process of the angulo-articular narrow; interhyal bone rudimentary or absent; first epibranchial twisted; subdistal process of the second epibranchial absent; uncinate process of the third epibranchial forming an angle of about 45–60°; anteriorly directed paired process on the first vertebra; membrane between preopercular laterosensory series and isthmus continuous; E-scales not overlapped (Costa 2011a). The nonoverlapping of E-scales was previously proposed as a synapomorphy for Rivulidae (Costa 2004); this discrepancy was not addressed and remains in the literature. Costa (2004) restricted *Rivulus* to two species and subsequently (Costa 2011a) elevated to genera former subgeneric categories within it (e.g., *Laimosemion*, *Anablepsoides*, etc.). A tree topology supporting this hypothesis was previously reported based on a smaller species sampling (Murphy and Collier 1996). In addition, a molecular analysis placed *Rivulus* as the sister group of the remaining Rivulinae (Hrbek et al. 2004). Subsequent molecular and nonmolecular analyses placed *Rivulus* or *Laimosemion* alternatively as the sister clade to the rest of all other Rivulinae (Costa 2011a,b, 2013). In these analyses, Melanorivulini and Rachovinni (tribes created by Costa 2011a) nodes were not statistically supported. Furthermore, a molecular analysis (mitochondrial sequences) by Hrbek et al. (2004), which placed *Rivulus* as the sister genus of the remaining Rivulinae, considered Melanorivulini as polyphyletic. Costa (2011a) considered *Rivulus roloffi* as *insertae sedis* due to lack of specimens to analyze morphological characters. According to Murphy and Collier (1999); Hrbek and Larson (1999); and Hrbek et al. (2004), this species is the sister taxon of all remaining Rivulinae except *Rivulus sensu stricto*.

1.9.1 *Rivulus*

Two (one described and one undescribed) medium-sized nonannual species (max. 50 mm SL) found in rivers and swamps along coastal plains and mountains of Cuba (Ponce de León et al. 2014). The genus is distinguished from other nonannual rivulines by the following apomorphies: all hypurals fused (single plate); neural prezygapophysis of caudal vertebrae rudimentary; fourth ceratobranchial teeth absent; about 50% of anterior half of caudal fin covered by scales; four neuromasts on the anterior supraorbital series; and black round spot with white margin on the dorsoposterior portion of caudal peduncle in females (Costa 2011a).

1.9.2 *Laimosemion*

A diverse nonannual clade consisting of approximately 25 small- to medium-sized species (20–40 mm SL) with great diversity of coloration patterns, formerly assigned to *Rivulus*. They are widely distributed in northwestern South America, central and eastern Brazilian Amazon, Guianas, upper Orinoco River basin in Venezuela and Colombia, Rio Negro basin in Brazil, and lower Peruvian Amazon. Although its monophyly is well supported (Hrbek et al. 2004; Costa 2011a, 2013), no morphological synapomorphies have been identified, and the genus can only be distinguished from other Rivulinae by a combination of characters. Most species of *Laimosemion* differ from all other rivulines by the intense yellow or orange pigmentation of the anal fin in adult females (vs. hyaline or pale pink; Costa 2011a).

The clade ((Plesiolabiasine) + (Melanorivuline + Rachovine)) is supported by one synapomorphy: presence of a cartilage extending along the lateral part of the first hypobranchial (Costa 2011a).

1.10 PLESIOLEBIASINI

This clade is diagnosed by the following synapomorphies: absence of an anterior expansion on the ventral process of the maxilla; anterior portion of the entopterygoid not overlapping the autopalatine; anterior retrorse process of lateral ethmoid contacting the main axis of the bone; first two proximal radials of anal fin fused; pelvic bones medially overlapped; increase in number of neuromasts (between five and seven in the anterior, and four or more in the posterior sections) of supraorbital series; and anal fin with basal region dark gray to black with a row of white to yellow spots (reversed in species of *Maratecoara* and in *Stenolebias damascenoi*) (Costa 2011b).

1.10.1 *Plesiolebias*

Eleven small annual species (18–50 mm SL) distributed in the Araguaia, Canabrava, Tocantins, and Xingu rivers in Brazil and eastern Paraguay. The genus is diagnosed by the following unambig-uous synapomorphies: posterior margin of the autopalatine with a deep concavity; metapterygoid narrow; ventral process of the maxilla shorter than dorsal process; symplectic slender, twice as long as deep; interarcual cartilage absent; basibranchials and hypobranchials thin; parasphenoid with dorsal process; pupil with anterior pointed extension; and anterior portion of the anal fin longer than posterior portion (Costa 2011b).

The clade (*Papiliolebias* + ((*Pituna*) + (*Stenolebias* + *Maratecoara*))) is supported by: dark orange jaws; upper and lower sections of the preopercular neuromast series continuous; and anterior and posterior sections of the mandibular neuromast series continuous (Costa 2011b). Hrbek et al. (2004) did not recover this topology; instead, the clade consisting of *Papiliolebias* + *Pituna* is sister taxa of *Maratecoara*; however, *Stenolebias* was not included in the analysis.

1.10.2 *Papiliolebias*

Four small annual species (32–50 mm SL) distributed in the Paraguay River basin, Bermejo River, upper Rio Mamoré basin, and Rio Madeira river basin (Eschmeyer 2014). The genus (Figure 1.4b) is supported by four synapomorphies: nine pelvic-fin rays; dark blue unpaired fins of males; white line along distal margin of anal fin of male; and humeral metallic green spot (Costa 1998a).

The clade ((*Pituna*) + (*Stenolebias* + *Maratecoara*)) is supported by pointed dorsal and anal fins, a condition that also occurs in several other annual rivulids, and the presence of red lines on the dis-tal region of the dorsal-fin rays in males (reversed in *Stenolebias bellus*; Costa 2011b). Whereas the sister relationship between *Stenolebias* and *Maratecoara* is supported by an expanded dorsomedial process of the autopalatine; a ventrally directed ventral process of the fourth ceratobranchial; and the presence of intense light blue iridescence on the flank in males (Costa 2011b).

1.10.3 *Maratecoara*

Four small annual species (32–50 mm SL) distributed in the Tocantins, Xingu, and Canabrava basins. The genus is diagnosed by posterior supraoccipital processes medially fused; supracleithrum shorter than posttemporal half length; branchiostegal membrane expanded to overlap pectoral-fin base; frontal scales arranged transversely; dorsal and anal fins long in males, tip surpassing poste-rior margin of caudal fin; caudal fin lanceolate in males, tip terminating in a filamentous ray; lower section of preopercular series with numerous neuromasts (13–15); anterodorsal region of flank with three horizontal rows of orange spots in males; iris green in males (also in *Plesiolebias*); and bran-chiostegal membrane bright blue in males (Costa 2011b).

1.10.4 *Stenolebias*

Two small annual species (max. 30 mm SL) distributed in the Paraguay River basin; diagnosed by posterior tip of entopterygoid not reaching the metapterygoid (also in *Plesiolebias* and *Pituna brevirostrata*) and a single anterior branchiostegal ray (also in *Plesiolebias*) (Costa 2011b).

1.10.5 *Pituna*

Six medium-sized annual species (30–40 mm SL) distributed in the Araguaia area, Tocantins basin, Paranaíba area, and the Canabrava area. Monophyly of *Pituna* is supported by a wide, triangular basihyal (anterior width more than 50% of its length); caudal-fin squamation extending over 50% of fin length in males (also in *Rachovia maculipinnis* and *Trigonectes rubromarginatus*); flank with small yellow spots arranged in oblique rows in males; flank with oblique rows of dark spots in females (also in *Plesiolebias*) (Costa 2011b).

1.11 MELANORIVULINI

Melanorivulus, *Cynodonichthys*, *Anablepsoides*, and *Atlantirivulus* form a weakly supported clade diagnosed by the presence of a small rounded black spot on the dorsal portion of the caudal-fin base in females, with a white dot on the anterior margin (Costa 2011a); as mentioned before, Hrbek et al. (2004) found this clade to be polyphyletic.

1.11.1 *Melanorivulus*

Includes 38 recognized small nonannual species (20–40 mm SL) (Figure 1.4c). They are distributed in the Paraguay and Uruguay river basins in Bolivia, Paraguay, Brazil, and Argentina (Costa 2011a). Species are found in shallow swampy parts of stream borders. The genus is distinguished from all other rivulines by the presence of black pigmentation concentrated on the ventral margin of the caudal fin and on the distal margin of dorsal and anal fins in females. The genus possesses some apomorphic character states that are homoplastic with other Rivulinae (Costa 2011a).

1.11.2 *Cynodonichthys*

Twenty-seven small- to medium-sized nonannual species (50–80 mm SL) distributed in Central America and northwest South America, including Pacific and Caribbean drainages of Central America (southern Mexico, Guatemala, Honduras, Belize, Nicaragua, Costa Rica, and Panama), river basins west of the Cordillera Occidental, and hills between the Magdalena and Maracaibo basins in Colombia (Costa 2011a). They live in streams from sea level to about 1500 m. asl. Although most species may be recognized by the presence of a yellow stripe with an upper red border on the ventral margin of caudal fin of males, no synapomorphies have been identified for the genus, and it can be distinguished from other melanorivulines only by a combination of characteristics (Costa 2011a).

1.11.3 *Anablepsoides*

Forty-nine small- to large-sized nonannual species (30–120 mm SL) distributed in Southern Lesser Antilles, Orinoco River basin in Venezuela, river basins of Guianas and adjacent parts of northeastern Venezuela and northern Brazil; Amazon River basin in Colombia, Ecuador, Peru, Bolivia, and Brazil; and small isolated river basins of northeastern Brazil (Costa 2011a). It inhabits the shallowest parts of streams and swamps inside forest or open savannah, from lowlands to mountain regions; some species may be found along shores of lakes and larger rivers (see references in Costa [2011a]). It differs from all other rivulines by the ventral surface of the head, which is entirely covered with scales (Costa 2011a).

1.11.4 *Atlantirivulus*

Eleven small nonannual killifishes (25–45 mm SL) distributed in coastal river drainages of eastern Brazil, between 10°S and 31°S (Costa 2011a). Species in this genus inhabit the shallowest parts of swamps, within dense forests or open waters. Two characters distinguish this genus from all other rivulids: the infraorbital series of neuromasts forming a zigzag row and a curved ventral process of the angulo-articular (Costa 2008a).

1.12 RACHOVINI

All species of this clade are seasonal fishes diagnosed by the presence of a prominent urogenital papilla in adult males (Costa 1998a); monophily of Rachovine is well supported (Hrbek et al. 2004; Hrbek and Taphorn 2008; Costa 2011a). However, these analyses recovered different topologies, which may be the result of differences in taxon sampling in the analyses.

1.12.1 *Moema*

Nineteen small- to large-sized annual species (40–130 mm SL) distributed in the rainforest of the Amazon River basin and the Pantanal of the Paraguay River basin. Recently, the genus *Aphyolebias* was placed in the synonym of *Moema* (Costa 2014a). *Moema* (including *Aphyolebias*) is diagnosable by the unique opercle shape (the ventroposterior part of the opercle is expanded and its dorsal portion is distinctively longer than the ventral portion) and by a pectoral fin with distal portion narrow, forming a long pointed tip (Costa 2014a).

1.12.2 *Trigonectes*

Six large-sized annual species (max. 90 mm SL) distributed in the Paraguay, Guaporé, Mamoré, Araguaria, and Tocantins river basins (Costa 2014a). The genus (Figure 1.4e) is diagnosed by an elongate premaxilla and dentarium; molariform teeth on the third pharyngobranchial and fifth ceratobranchial; elongate pelvic fins in males; and increased overall size (Costa 1998a).

1.12.3 *Neofundulus*

Six small- to medium-sized annual species (30–70 mm SL) distributed in the Paraguay, Mamoré, Guaporé, and Das Velhas river basins (Figure 1.4d). It is diagnosed by a fourth pectoral radial not expanded ventrally; subbasal longitudinal white to yellow stripe on the anal fin of males; transverse black bars on the caudal fin of females; and black spots on the pectoral fin of males (Costa 1998a).

1.12.4 *Pterolebias*

Three medium-to-large annual species (70–120 mm SL) distributed in the eastern and southern Amazon River basin and the Paraná-Paraguay river system (Costa 1998a). *Pterolebias* is diagnosed by possessing a narrow and pointed ventral process of the angulo-articular (Figure 1.4f); an expanded medial flap on second pharyngobranchial; a small humeral metallic orange spots in males; and dark gray bars on pectoral fin in males (Costa 2005).

The clade (*Renova* + *Micromoema*) is supported by the relative position of the first two proximal radials of the dorsal fin, which strongly diverge ventrally from each other, and the presence of a yellowish-white or pale yellow stripe with a broad, dark-red upper margin on the ventral portion of the caudal fin in males (Costa 2014a). *Renova* was recovered as the sister genus to *Pterolebias* (Hrbek et al. 2004; Hrbek and Taphorn 2008).

1.12.5 *Renova*

Monotypic, *R. oscari*, small-sized species (max. 47 mm SL) distributed in the Orinoco River basin in Venezuela. Diagnosed by the absence of two well-defined oblique reddish-brown stripes on the preopercular region of males and an ocellate caudal black spot in females (Costa 1998a).

1.12.6 *Micromoema*

Monotypic, *M. xiphophora*, medium-sized annual species (max. 60 mm SL) distributed in the Orinoco River basin in Venezuela. Diagnosed by an anteriorly expanded ventral tip of autopalatine; reduced number of vertebrae (31); absence of ventral process of posttemporal; and elongate posterior extension on the lower border of caudal fin of males (Costa 1998a).

1.12.7 *Terranatos*

Monotypic, *T. dolichopterus*, small-sized annual species (max. 48 mm SL) endemic to the Orinoco River basin in Venezuela. Diagnosed by anteriorly expanded quadrate; short anterior portion of urohyal; median laminar process between supraoccipital processes; dorsal fin anteriorly positioned; second dorsal proximal radial between neural spines of vertebrae 9 and 10; anal proximal radials strongly widened; widened dorsal portion of cleithrum; short ventral tip of cleithrum; long pectoral fin in males; elongate dorsal and anal fins of males; elongate and pointed anal fin of males; transverse arrangement of frontal scales; reduced caudal-fin squamation; slightly posteriorly directed, black suborbital bar; and absence of alternate black and white spots on the basal and posterior parts of anal fin of males (Costa 1998a).

Molecular phylogenies support a sister group relationship between species of the genera currently known as *Gnatholebias* and *Llanolebias* (Hrbek and Larson 1999; Hrbek et al. 2004; Hrbek and Taphorn 2008; Costa 2014a). The clade (*Gnatholebias* + *Llanolebias*) is also supported by the following morphological synapomorphies: a prominent projection on the lateral border of the middle portion of parasphenoid; middle portion of coracoid with a circular groove; scapula dorsally projected; tip of pelvic fin in males with a single filamentous ray; pelvic fins medially fused; and a white stripe with ventral narrow black margin on the ventral portion of pectoral fin in males (Costa 2014a).

1.12.8 *Gnatholebias*

Two large-sized annual species (max. 100 mm SL), distributed in the Orinoco River basin. Diagnosed by long premaxilla and dentary; hypurals forming a single plate; proximal anal radials curved posteriorly; elongate filamentous rays on dorsal and anal fins of males; numerous anal-fin rays (22–26); horizontal reddish-brown stripe close to dorsal border of caudal fin of males; narrow black margin on ventral border of pectoral fin of males; and white to light green border on lower edge of pectoral fin of males (Costa 1998a).

1.12.9 *Llanolebias*

Monotypic, *L. stellifer* (previously assigned to *Rachovia*), medium-sized annual species (max. 75 mm SL) distributed in Los Llanos of the Orinoco River basin (Venezuela and Colombia).

The species allocation to its own genus is based on molecular analysis (Hrbek and Taphorn 2008); however, it possesses no unique morphological characters that unambiguously distinguish it from all other rivulid species. In addition, the combination of 13 characters distinguishes it from the genus *Rachovia*, and 33 characters distinguish it from their putative sister genus *Gnatholebias* (Hrbek and Taphorn 2008).

1.12.10 *Rachovia*

Four medium-sized annual species (max. 60 mm SL) distributed in the Magdalena River basin, Orinoco River basin in Colombia and Venezuela, and the Lago Maracaibo basin. It is diagnosed by the presence of a foramen on the hyomandibula, near the limit with the metapterygoid; an adipose ridge on the predorsal region in old males; and a distinctive projection in the posterolateral region of the telencephalon, which is medially coalesced (Costa 1998a). Recently, *Austrofundulus* was considered a junior synonym of *Rachovia* (Costa 2014a).

1.12.11 *Millerichthys*

Monotypic (*M. robustus*), small annual species (30 mm SL); the annual life cycle of this species was recently reported (Domínguez-Castanedo et al. 2013). The species is diagnosed by the following apomorphies: reduced mesopterygoid; conical and elongate ventral process of posttemporal; long posterior series of supraorbital neuromasts; absence of transverse dark bar crossing the eye; three transverse yellow stripes on the anal fin of males; and black spots along the caudal-fin base of females (Costa 1998a). Alternative relationships have been proposed for the group as a sister taxon to Rachovine supported by three synapomorphies (i.e., dorsally pointed preopercle; reduced neural prezygapophyses of caudal vertebrae; and reduced dark pigmentation in females; Costa 1998a) or as the sister taxa to Cynolebiasinae supported also by three synapomorphies (i.e., small posterior tip of mesopterygoid not contacting metapterygoid; hypurals completely ankylosed, forming a single plate; and an arrangement of frontal scales in transverse pattern without a central scale; Costa 2004). The species has not been included in any phylogenetic analysis; consequently, its position within Rivulidae remains unclear. Interestingly, Hrbek and Larson (1999) proposed that Rachovine and Cynolebiatine were sister taxa, raising the interest of including *Millerichthys* in future phylogenies.

1.13 CYNOLEBIASINAE

This subfamily is diagnosed by numerous morphological synapomorphies: dorsally positioned dorsomedial process of autopalatine; reduced mesopterygoid (Figure 1.6); thin C-shaped opercle with reduced dorsal tip, median rim, and sensory canal; elongate epibranchials and interarcual cartilage; rounded second hypobranchial; absence of teeth in the fourth ceratobranchial; absence of a posteroventral process on the anterior portion of the fourth ceratobranchial; absence of teeth in vomer; thin lacrimal, with narrow lower portion and vestigial sensory canal; absence of dermosphenotic; absence of anterior processes of first vertebra; vestigial or absent neural prezygapophysis of caudal vertebrae; narrow and anteriorly curved epural and parahypural; anteriorly placed dorsal fin origin; reduced first proximal anal-fin radial; reduced ossification of medial anal-fin radials; vestigial or absent ventral process of posttemporal; elongate supracleithrum; scale-like pectoral radials; cylindrical urogenital papilla in males; prominent pocket-like urogenital papilla in females; reduced supraorbital squamation; reduced caudal-fin squamation; continuous supraorbital series of

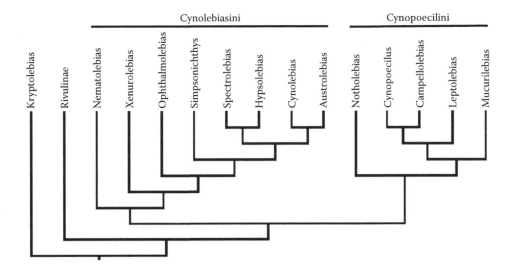

Figure 1.6 Summary of proposed relationships of Rivulidae with emphasis on Cynolebiasinae. Consensus topology was manually reconstructed on *Mesquite: A Modular System for Evolutionary Analysis*, release version: 3.01. (Based on Costa, W. J. E. M. 1998a. *Journal Comparative Biology* 3:33–94; Costa, W. J. E. M. 2007. *Zootaxa* 1660:1–134. Costa, W. J. E. M. 2008. *Zoological Journal of the Linnean Society* 153:147–160; Costa, W. E. J. M. 2010. *Journal of Biogeography* 37:1995–2004; Costa, W. E. J. M. 2013. *Vertebrate Zoology* 63:139–154; Costa, W. E. J. M. 2014b. *Vertebrate Zoology* 64:23–33; Murphy, W. J., J. E. Thomerson, and G. E. Collier. 1999b. *Molecular Phylogenetics and Evolution* 13:289–301; Hrbek, T. and A. Larson. 1999. *Evolution* 53:1200–1216; Hrbek, T., C. P. Deus, and I. P. Farias. 2004. *Copeia* 2004:569–576; Ferrer, J., J. M. Wingert, and L. R. Malabarba. 2014. *Zoological Journal of the Linnean Society* 172:846–866.)

neuromasts; long posterior series of supraorbital neuromasts; minute neuromast on dermosphenotic; and vertical bars on body sides of juveniles (Costa 1998a). Cynolebiasinae consists of two clades, Cynolebiasini and Cynopoecilini (Costa 1998a; Murphy et al. 1999b; Hrbek and Larson 1999; Hrbek et al. 2004; Costa 2013).

1.14 CYNOLEBIASINI

The following synapomorphies support this clade: a broad posteroventral process of dentary; concave anterodorsal margin of opercle; anterior displacement of anal-fin origin in males; elongate filamentous rays on dorsal and anal fins of males; sexual dimorphism in number of dorsal- and anal-fin rays; numerous anal-fin rays in males (19–32); elongate dorsal portion of cleithrum; absence of posteroventral flange of cleithrum; ventrally directed ventral tip of cleithrum; broad anteroventral tip of coracoid; numerous supraorbital neuromasts (12–25); and spawning within substrate (Costa 1998a).

1.14.1 *Nematolebias*

Three medium-sized annual species (max. 80 mm SL) distributed in coastal basin of southeastern Brazil. Diagnosed by the presence of two male characteristics: hypertrophied papillae on the pectoral fin and the presence of a broad subdistal orange stripe with overlapped golden lines on the anal fin (Costa 2002a, 2006a, 2010).

1.14.2 *Xenurolebias*

Four small-sized annual species (max. 30 mm SL) endemic to the Atlantic forest in southeastern Brazil. Three uniquely shared synapomorphies diagnose the genus: presence of filamentous rays on the posterior margin of the anal fin in females; caudal fin sublanceolate to lanceolate with an expansion of the ventral portion of the fin in males; and a dark reddish bar on the posterior portion of the caudal peduncle extending over the dorsal and ventral margins of the caudal fin in males (Costa 2006a, 2007; Costa and Amorim 2014).

1.14.3 *Ophthalmolebias*

Five small-sized annual species (65 mm SL) distributed in eastern Brazil (Costa 2010). The genus is distinguished by the following synapomorphies: medial crest with a prominent ventral expansion on palatine; eyes positioned laterally on the head; anterior and posterior sections of supraorbital series of neuromasts separated by an interspace; metallic blue bars on the flank in females; and anal fin pink in females (Costa 2006a, 2007).

1.14.4 *Simpsonichthys*

Eight small-sized annual species (55 mm SL) distributed in the central Brazil plateau (Costa 2010). The genus is distinguished by males having red flanks with bright blue bars on the anterior portion and vertical rows of blue dots on the posterior portion; however, one species, *S. cholopteryx*, presents blue bars on the entire flank (Costa 2006a, 2007).

1.14.5 *Spectrolebias*

Eight small-sized annual species (Figure 1.4h) (32 mm SL) distributed in the Paraguay, Guapore, Xingu, Araguaia, and Tocantins river basins (Costa 2010). Distinguished by two synapomorphies: long hyomandibula and narrow proximal tip of fourth ceratobranchial (Costa 2006a, 2007).

1.14.6 *Hypsolebias*

Thirty-one medium-sized annual species (80 mm SL) distributed in Jequitinhonha, Sao Francisco, and Urucuia river basins (Costa 2010). The genus (Figure 1.4g) is distinguished by a second pharyngobranchial longer than it is wide (Costa 2006a, 2007).

The clade (*Cynolebias* + *Austrolebias*) was recovered as monophyletic (Costa 1998a). However, genera validity and supporting synapomorphies have varied over the years. The genus *Megalebias* first proposed by Costa (1998a) was subsequently placed in the synonymy of *Austrolebias* (Costa 2006b). On the other hand, a morphological analysis recovered a topology (Loureiro 2004) wherein some members of *Austrolebias*, that is, *Megalebias* species (Costa 1998a, 2002a,b,c) would be more closely related to *Cynolebias* than to *Austrolebias* as currently defined. In this study, the clade consisting of the former (*Megalebias* + *Cynolebias*) is supported by eight morphological synapomorphies: presence of a robust sphenotic process (also in Costa 1998a); denticles over gill rakers in ceratobranchial 1; reduced basihyal; presence of ventral process of posttemporal in 80%–89% of individuals; an average of 30 caudal fin rays; posteriorly placed dorsal fin; more than 40 scales on lateral line series; and five or more vertebrae supporting the caudal skeleton. The most recent revision of *Austrolebias* proposed the following synapomorphies for the genus: tip of pterotic posterolateral rim thickened and directed laterally; presence of a process on the dorsomedial portion of the palatine; dorsal portion of opercle as long as the ventral portion; medial and distal radials of dorsal and anal fins cartilaginous or poorly ossified; cartilaginous space of shoulder girdle broad; radials discoid, thin, the uppermost usually absent or extremely reduced (Costa 2006b). However, this recent work did not discuss systematic changes done in previous publications (Costa 1998a, 2002b).

1.14.7 *Cynolebias*

Twenty medium- to large-sized annual species (50–120 mm SL) distributed in the Sao Francisco and Urucuia river basins. The genus is distinguished by: rostral cartilage widened; coronoid process of angulo-articular vertically elongated; three to four lateral supratemporal neuromasts; double row of neuromasts on lateral mandibular series; a deep and sharp ventral expansion of infraorbital line of neuromast; large and rounded supraorbital spot; dark brown to black notches scattered on dorsoposterior portion of the head; and light blue lines among rays on distal portion of the anal fin of males (Costa 2001).

1.14.8 *Austrolebias*

Forty small- to large-sized annual species (35–150 mm SL) distributed in the La Plata and Patos-Merín basins (Figure 1.4j). *Austrolebias* synapomorphies are: absence of scales between corner of mouth and anterior portion of preopercular region and ventral portion of opercular region; deep urohyal; and dark gray to black infraorbital bar and supraorbital spot (Costa 2006b). Conflicting phylogenetic hypotheses have been proposed for this genus using morphological (Costa 2002b, 2006b, 2010; Loureiro 2004) and molecular (García et al. 2014) data analyses.

1.15 CYNOPOECILINI

Monophyly of this clade has been supported by several phylogenetic studies (Costa 1998a, 2008, 2013, 2014b; Hrbek and Larson 1999; Murphy et al. 1999b; Ferrer et al. 2014). Fourteen synapomorphies support the monophyly of this group: vomerine teeth absent; ventral process of angulo-articular vestigial or absent; sympletic elongate; muscular fibers circularly arranged around the base of the urogenital papilla to form an ejaculatory pump; supraorbital scales absent; minute contact organs along margin of flank scales in males; iris green (blue to yellowish-green in males, in Ferrer et al. 2014); egg chorion surface reticulate with mushroom-like projections; and dorsal and anal fins twisted in males during courtship behavior (Costa 2008; Ferrer et al. 2014). Additional synapomorphies not mentioned in Ferrer et al. (2014) include: wide posterior process of quadrate; short mesopterygoid; supraorbital series of neuromasts anteriorly interrupted; and between four and seven neuromasts on caudal-fin base (Costa 2008).

1.15.1 *Notholebias*

Three small-sized annual species (max. 30 mm SL) distributed in coastal basins of southeastern Brazil. The genus is distinguished from the other genera of the tribe by the presence of a dermosphenotic; basihyal narrow; well-developed contact organs on most inner surfaces of pectoral fin in males; opercular region with red bars in males; and caudal fin of males with iridescent bars or vertical lines (Costa 2008a). However, a recent study recovered only the narrow basihyal as a synapomorphy for the genus (Ferrer et al. 2014).

1.15.2 *Mucurilebias*

Monotypic, *M. leitaoi*, small-sized annual species (30 mm SL) distributed in coastal basin of southeastern Brazil (Costa 2014b). Diagnosed by seven synapomorphies: eye large, occupying a great part of the head side; low number of caudal-fin rays (21); distal portion of epural broader than distal portion of parahypural; oblique red bar through opercle; isthmus bright red in males; white stripe on distal margin of dorsal fin in males; and red stripe on distal margin of anal fin in males (Costa 2014b).

A clade ((*Leptolebias*) + (*Campellolebias* + *Cynopoecilus*)) diagnosed by two synapomorphies: dermosphenotic absent and opercular region in males with a reticulate color pattern (Costa 2008). Ferrer et al. (2014) only recovered the first character as a synapomorphy of the clade.

1.15.3 *Leptolebias*

Six small-sized annual species (max. 29 mm SL) distributed in coastal basins of southeastern Brazil. Distinguished by having caudal fin longer than deep; single anterior supraorbital neuromast; and distal third of dorsal fin in males without dark pigmentation (Costa 2008). However, a recent study listed only two synapomorphies: single anterior supraorbital neuromast and distal third of dorsal fin in males without dark pigmentation (Ferrer et al. 2014).

According to Costa (1998a), this clade (*Campellolebias* + *Cynopoecilus*) is supported by an elongate posterior process of vomer; narrow lateral wings of vomer; rod-shaped proximal anal radials; elongated first proximal anal radial; elongated and pointed anal fin of males; a putative ejaculatory pump; absence of vertical bars in juveniles; internal fertilization; and midwater spawning. The addition of a new species of *Cynopoecilus* forced the three synapomorphies listed for *Campellolebias* to be considered as synapomorphies of this entire clade (Ferrer et al. 2014): posterobasal portion of the last neural spine with a small process bearing a thin ligament attached to the anterior tip of epural (distinctively bent); dark ventral stripe between lower jaw and pseudogonopodium; and dark lateroventral stripe on head.

1.15.4 *Campellolebias*

Four small-sized annual species (max. 32 mm SL) distributed in coastal basins of southeastern Brazil. Distinguished by the following three synapomorphies: presence of a pseudogonopodium; a derived behavior pattern ("coiled retrorse motion" of Costa 1998a) (Costa 2006c; Ferrer et al. 2014); and long and narrow first proximal radials of the anal fin (Costa 2006c).

1.15.5 *Cynopoecilus*

Six small-sized annual species (max. 40 mm SL) distributed in the coastal plains and Jacui River basin of the Pato-Merin drainage basin in Brazil and Uruguay (Figure 1.4i). It is diagnosed by: minute caudal accessory cartilages; anteriorly curved sixth and seventh anal-fin rays of males; numerous anal-fin rays (23–29); five minute rays at the anterior portion of the anal fin; a minute pelvic fin; a pocket-like structure around the urogenital region of males; a broad longitudinal black stripe from the tip of lower jaw to the caudal-fin base; an oblique longitudinal black stripe from the pectoral-fin base to the posterior part of the anal-fin base; yellow to yellowish-brown iris in males; and a zigzag display during male courtship behavior (Costa 2002c; Ferrer et al. 2014).

ACKNOWLEDGMENTS

We thank the following individuals who provided live photos of fishes: J. P. Cicéron, Pablo Calviño, Dalton Nielsen, Felipe Alonso, Amer Faour, Frans Vermeulen, and Tony Terceira. ML thanks the editors for their invitation to contribute to this book and acknowledges support from PEDECIBA (Programa de Desarrollo de Ciencias Básicas), CSIC (Comisión Sectorial de Investigación Científica, Universidad de la República), SNI (Sistema Nacional de Investigadores), and ANII (Agencia Nacional de Investigación e Innovación). RdS was supported by award NSF–DEB 1144692.

REFERENCES

Agnèse, J. F., R. Brummet, P. Caminade, J. Catalan, and E. Kornobis. 2009. Genetic characterization of the *Aphyosemion calliurum* species group and description of a new species from this assemblage: *A. campomaanse* (Cyprinodontiformes: Aplocheiloidei: Nothobranchiidae) from Southern Cameroon. *Zootaxa* 2045:43–59.

Agnèse, J. F, F. Zentz, O. Legros, and D. Sellos. 2006. Phylogenetic relationships and phylogeography of the Killifish species of the subgenus *Chromaphyosemion* (Radda, 1971) in West Africa, inferred from mitochondrial DNA sequences. *Molecular Phylogenetics and Evolution* 40:332–346. doi: 10.1016/j.ympev.2006.03.018

Alexander, R. M. 1967. Mechanisms of the jaws of some atheriniform fish. *Journal of Zoology* 151:233–255. doi: 10.1111/j.1469-7998.1967.tb02876.x

Berois, N., J. M. Arezo, and R. O. de Sá. 2014. The Neotropical genus *Austrolebias*: An emerging model of annual killifishes. *Cell and Developmental Biology* 3:1–9. doi: 10.1146/annurev.cellbio.3.1.1

Berois N., M. J. Arezo, N. G. Papa, and G. Clivio. 2012. Annual fish: Developmental adaptation for an extreme environment. *Wire Developmental Biology* 1:595–602. doi: 10.1002/wdev.39

Blažek, R., M. Polačik, and M. Reichard. 2013. Rapid growth, early maturation and short generation time in African annual fishes. *EvoDevo* 4:24. doi: 10.1186/2041-9139-4-24

Collier, G. E., W. J. Murphy, and M. Espinoza. 2009. Phylogeography of the genus *Epiplatys* (Aplocheiloidea: Cyprinodontiformes). *Molecular Phylogenetics and Evolution* 50:190–196. doi: 10.1016/j.ympev.2008.10.006

Costa, W. J. E. M. 1990. Análise filogenética da família Rivulidae (Cyprinodontiformes, Aplocheiloidei). *Revista Brasileira Biologia* 50:65–82.

Costa, W. J. E. M. 1998a. Phylogeny and classification of Rivulidae revisited: Origin and evolution of annualism and miniaturization in rivulid fishes (Cyprinodontiformes: Aplocheiloidei). *Journal Comparative Biology* 3:33–94.

Costa, W. J. E. M. 1998b. Phylogeny and classification of the Cyprinodontiformes (Euteleostei: Atherinomorpha): A reappraisal. In *Phylogeny and Classification of Neotropical Fishes*, eds. L. R. Malabarba, R. E. Reis, R. P. Vari, Z. M. S. Lucena, and C. A. S. Lucena. Porto Alegre: EDIPUCRS, 537–560.

Costa, W. J. E. M. 2001. The Neotropical annual fish genus *Cynolebias* (Cyprinodontiformes: Rivulidae): Phylogenetic relationships, taxonomic revision and biogeography. *Ichthyological Exploration of Freshwaters* 12:333–383.

Costa, W. J. E. M. 2002a. The Neotropical seasonal fish genus *Nematolebias* (Cyprinodontiformes: Rivulidae: Cynolebiatinae): Taxonomic revision, with description of a new species. *Ichthyological Exploration of Freshwaters* 13:41–52.

Costa, W. J. E. M. 2002b. Monophyly and phylogenetic relationships of the Neotropical annual fish genera *Austrolebias* and *Megalebias* (Cyprinodontiformes: Rivulidae). *Copeia* 2002(4):916–927. doi: org/10.1643/0045-8511

Costa, W. J. E. M. 2002c. The annual fish genus *Cynopoecilus* (Cyprinodontiformes: Rivulidae): Taxonomic revision, with descriptions of four new species. *Ichthyol. Explor. Freshwaters* 13:11–24.

Costa, W. E. J. M. 2004. Relationships and redescription of *Fundulus brasiliensis* (Cyprinodontiformes: Rivulidae), with description of a new genus and notes on the classification of the Aplocheiloidei. *Ichthyological Exploration of Freshwaters* 15:105–120.

Costa, W. J. E. M. 2005. The Neotropical annual killifish genus *Pterolebias* Garman (Teleostei: Cyprinodontiformes: Rivulidae): Phylogenetic relationships, descriptive morphology, and taxonomic revision. *Zootaxa* 1067:1–36.

Costa, W. J. E. M. 2006a. Descriptive morphology and phylogenetic relationships among species of the Neotropical annual killifish genera *Nematolebias* and *Simpsonichthys* (Cyprinodontiformes: Aplocheiloidei: Rivulidae). *Neotropical Ichthyology* 4:1–26. doi: org/10.1590/S1679-6225

Costa, W. J. E. M. 2006b. The South American annual killifish genus *Austrolebias* (Teleostei: Cyprinodontiformes: Rivulidae): Phylogenetic relationships, descriptive morphology and taxonomic revisión. *Zootaxa* 1213:1–162.

Costa, W. J. E. M. 2006c. Taxonomy and phylogenetic relationships among species of the seasonal, internally inseminating, South American killifish genus *Campellolebias* (Teleostei: Cyprinodontiformes: Rivulidae), with the description of a new species. *Zootaxa* 1227:31–55.

Costa, W. J. E. M. 2007. Taxonomic revision of the seasonal South American killifish genus *Simpsonichthys* (Teleostei: Cyprinodontiformes: Aplocheiloidei). *Zootaxa* 1669:1–134.

Costa, W. J. E. M. 2008. Monophyly and taxonomy of the Neotropical seasonal killifish genus *Leptolebias* (Teleostei: Aplocheiloidei: Rivulidae), with the description of a new genus. *Zoological Journal of the Linnean Society* 153:147–160. doi: 10.1111/j.1096-3642.2008.00380.x

Costa, W. E. J. M. 2009a. Trophic radiation in the South American annual killifish genus *Austrolebias* (Cyprinodontiformes: Rivulidae). *Ichthyological Exploration of Freshwaters* 20:179–191.

Costa, W. E. J. M. 2009b. Osteology of the African annual killifish genus *Callopanchax* (Teleostei: Cyprinodontiformes: Nothobranchiidae) and phylogenetic implications. *Vertebrate Zoology* 59:31–40.

Costa, W. E. J. M. 2010. Historical biogeography of cynolebiasine annual killifishes inferred from dispersal–vicariance analysis. *Journal of Biogeography* 37:1995–2004. doi: 10.1111/j.1365-2699. 2010.02339.x

Costa, W. J. E. M. 2011a. Phylogenetic position and taxonomic status of *Anablepsoides*, *Atlantirivulus*, *Cynodonichthys*, *Laimosemion* and *Melanorivulus* (Cyprinodontiformes: Rivulidae). *Ichthyological Exploration of Freshwaters* 22:233–249.

Costa, W. J. E. M. 2011b. Comparative morphology, phylogenetic relationships, and historical biogeography of plesiolebiasine seasonal killifishes (Teleostei: Cyprinodontiformes: Rivulidae). *Zoological Journal of the Linnean Society* 162:131–148. doi: 10.1111/j.1096-3642.2010.00672.x

Costa, W. E. J. M. 2012. The caudal skeleton of extant and fossil cyprinodontiform fishes (Teleostei: Atherinomorpha): Comparative morphology and delimitation of phylogenetic characters. *Vertebrate Zoology* 62:161–180.

Costa, W. E. J. M. 2013. Historical biogeography of aplocheiloid killifishes (Teleostei: Cyprinodontiformes). *Vertebrate Zoology* 63:139–154.

Costa, W. E. J. M. 2014a. Phylogeny and evolutionary radiation in seasonal rachovine killifishes: Biogeographical and taxonomical implications. *Vertebrate Zoology* 64:177–192.

Costa, W. E. J. M. 2014b. A new genus of miniature cynolebiasine from the Atlantic Forest and alternative biogeographical explanations for seasonal killifish distribution patterns in South America (Cyprinodontiformes: Rivulidae). *Vertebrate Zoology* 64:23–33.

Costa, W. E. J. M. 2015. Comparative morphology, phylogeny, and classification of West African callopanchacine killifishes (Teleostei: Cyprinodontiformes: Nothobranchiidae). *Zoological Journal of the Linnean Zociety.* doi: 10.1111/zoj.12270.

Costa, W. J. E. M. and P. F. Amorim. 2014. Integrative taxonomy and conservation of seasonal killifishes, *Xenurolebias* (Teleostei: Rivulidae), and the Brazilian Atlantic Forest. *Systematics and Biodiversity* 12:350–365. doi: 10.1080/14772000.2014.918062

Domínguez-Castanedo, O., M. Á. Mosqueda-Cabrera, and S. Valdesalici. 2013. First observations of annualism in *Millerichthys robustus* (Cyprinodontiformes: Rivulidae). *Ichthyological Exploration of Freshwaters* 24:15–20.

Dorn, A., Z. Musilová, M. Platzer, K. Reichwald, and A. Cellerino. 2014. The strange case of East African annual fish: Aridification correlates with diversification for a savannah aquatic group? *BMC Evolutionary Biology* 14:210. doi: 10.1186/s12862-014-0210-3

Eschmeyer, W. N. ed. 2014. *Catalog of Fishes: Genera, Species, References.* http://researcharchive.calacademy.org/research/ichthyology/catalog/fishcatmain.asp. Electronic version accessed 30 11 2014.

Ferrer, J., J. M. Wingert, and L. R. Malabarba. 2014. Description of a new species and phylogenetic analysis of the subtribe Cynopoecilina, including continuous characters without discretization (Cyprinodontiformes: Rivulidae). *Zoological Journal of the Linnean Society* 172:846–866. doi: 10.1111/zoj.12190

García, G., V. Gutiérrez, N. Ríos et al. 2014. Burst speciation processes and genomic expansion in the neotropical annual killifish genus *Austrolebias* (Cyprinodontiformes, Rivulidae). *Genetica* 142:87–98. doi: 10.1007/s10709-014-9756-7

Haas, R. 1976. Sexual selection in *Nothobranchius guentheri* (Pisces: Cyprinodontidae). *Evolution* 30: 614–622. doi.org/10.2307/2407584

Hertwig, T. 2008. Phylogeny of the Cyprinodontiformes (Teleostei, Atherinomorpha): The contribution of cranial soft tissue characters. *Zoologica Scripta* 37:141–174. doi: 10.1111/j.1463-6409.2007.00314.x

Hrbek, T., C. P. Deus, and I. P. Farias. 2004. *Rivulus duckensis* (Teleostei; Cyprinodontiformes): New species from the Tarumã basin of Manaus, Amazonas, Brazil, and its relationships to other neotropical Rivulidae. *Copeia* 2004:569–576. doi: 10.1643/CI-03-277R1

Hrbek, T. and A. Larson. 1999. The evolution of diapause in the killifish family Rivulidae (Atherinomorpha Atherinomorpha, Cyprinodontiformes): A molecular phylogenetic and biogeographic perspective. *Evolution* 53:1200–1216. doi.org/10.2307/2640823

Hrbek, T. and D. C. Taphorn. 2008. Description of a new annual rivulid killifish genus from Venezuela. *Zootaxa* 1734:27–42.

Huber, J. H. 1996. *Killi-Data 1996. Updated Checklist of Taxonomic Names, Collecting Localities and Bibliographic References of Oviparous Cyprinodont Fishes (Atherinomorpha, Pisces)*. Paris: Société Française d'Ichtyologie, Muséum National d'Histoire Naturelle.

Huber, J. H. 2000. *Killi-Data 2000. Updated Checklist of Taxonomic Names, Collecting Localities and Bibliographic References of Oviparous Cyprinodont Fishes (Cyprinodontiformes)*. Paris: Cybium Société Française d'Ichtyologie.

Huber, J. H. 2007. *Killi-Data 2007*. Paris: Killi-Data Editions.

Legros, O. and F. Zentz. 2007. *Aphyosemion malumbresi* n. sp., un nouveau *Chromaphyosemion* (Teleostei: Aplocheilidae) originaire de Guinée Equatoriale continentale. *Association Killiphile Francophone de Belgique, Killi-Contact* 35:1–28.

Loiselle, P. V. 2006. A review of the Malagasy *Pachypanchax* (Teleostei: Cyprinodontiformes, Aplocheilidae), with descriptions of four new species. *Zootaxa* 1366:1–44.

Loureiro, M. 2004. *Sistemática y Biogeografía de los Peces Anuales de la Subtribu Cynolebiatina (Cyprinodontiformes: Rivulidae: Cynolebiatinae)*. PhD thesis, Montevideo, Uruguay, 'PEDECIBA.

Malumbres, F. and R. Castelo. 2001. Descripción de una nueva especie del género *Fundulopanchax* Myers, 1924 (Cyprinodontiformes, Aplocheilidae), para la Ictiofauna Continental de Guinea Ecuatorial. *Graellsia* 57:175–180.

Meyer, A. and C. Lydeard. 1993. The evolution of copulatory organs, internal fertilization, placentae and viviparity in killi—fishes (Cyprinodontiformes) inferred from a DNA phylogeny of the tyrosine kinase gene X-src. *Proceedings of the Royal Society of London B* 254:153–162.

Murphy, W. J. and G. E. Collier. 1996. Phylogenetic relationships within the aplocheiloid fish genus *Rivulus* (Cyprinodontiformes, Rivulidae): Implications for Caribbean and Central American biogeography. *Molecular Biology and Evolution* 13:642–649. doi: 10.1093/oxfordjournals.molbev.a025624

Murphy, W. J. and G. E. Collier. 1997. A molecular phylogeny for aplocheiloid fishes (Atherinomorpha, Cyprinodontiformes): The role of vicariance and the origins of annualism. *Molecular Biology and Evolution* 14:790–799. doi: 10.1093/oxfordjournals.molbev.a025819

Murphy, W. J. and G. E. Collier. 1999. Phylogenetic relationships of African killifishes in the genera *Aphyosemion* and *Fundulopanchax* inferred from mitochondrial DNA sequences. *Molecular Phylogenetics and Evolution* 11:351–360.

Murphy, W. J., T. N. P. Nguyen, E. B. Taylor, and G. E. Collier. 1999a. Mitochondrial DNA phylogeny of West African aplocheiloid killifishes (Cyprinodontiformes, Aplocheilidae). *Molecular Phylogenetics and Evolution* 11:343–350. doi: 10.1006/mpev.1998.0567

Murphy, W. J., J. E. Thomerson, and G. E. Collier. 1999b. Phylogeny of the neotropical killifish family Rivulidae (Cyprinodontiformes, Aplocheiloidei) inferred from mitochondrial DNA sequences. *Molecular Phylogenetics and Evolution* 13:289–301. doi: 10.1006/mpev.1999.0656

Myers, G. S. 1952. Annual fishes. *Aquarium Journal* 23:125–141.

Ng'oma, E., S. Valdesalici, and R. K. Cellerino A. 2013. Genetic and morphological studies of *Nothobranchius* (Cyprinodontiformes) from Malawi with description of *Nothobranchius wattersi* sp. nov. *Journal of Fish Biology* 82:165–188. doi: 10.1111/jfb.12001

Parenti, L. R. 1981. A phylogenetic and biogeographic analysis of cyprinodontiform fishes (Teleostei: Atherinomorpha). *Bulletin of the American Museum of Natural History* 168:341–557.

Parenti, L. R. 1982. Relationships of the African killifish genus *Foerschichthys* (Teleostei: Cyprinodontiformes: Aplocheilidae). *Proceedings of Biological Society of Washington* 95:451–457.

Parker, A. 1997. Morphological data in fish systematics: Examples from the Cyprinodontiformes. In *Molecular Systematics of Fishes*, eds. T. D. Kocher and C. A. Stepien. San Diego: Academic Press, 163–188.

Passos, C., B. Tassino, M. Loureiro, G. G. Rosenthal. 2013. Intra- and intersexual selection on male body size in the annual killifish *Austrolebias charrua*. *Behavioral. Processes* 96:20–26. doi: 10.1016/j.beproc.2013.01.008

Ponce de León, J. L., G. León, R. Rodríguez et al. 2014. Phylogeography of Cuban *Rivulus*: Evidence for allopatric speciation and secondary dispersal across a marine barrier. *Molecular Phylogenetics and Evolution* 79:404–414. doi: 10.1016/j.ympev.2014.07.007

Radda, A. C. and E. Pürzl. 1987. *Colour Atlas of Cyprinodonts of the Rain Forests of Tropical Africa*. Wien: O. Hofmann-Verlag.

Reichard, M. and M. Polacik. 2010. Reproductive isolating barriers between colour-differentiated populations of an African annual killifish, *Nothobranchius korthausae* (Cyprinodontiformes). *Biological Journal of the Linnean Society* 100:62–72. doi: 10.1111/j.1095-8312.2010.01406.x

Sanmartín, I. and F. Ronquist. 2004. Southern Hemisphere biogeography inferred by event-based models: Plant versus animal patterns. *Systematic Biology* 53:216–243. doi: 10.1080/10635150490423430

Sedlácek, O., B. Baciaková, and L. Kratochvíl. 2014. Evolution of body colouration in killifishes (Cyprinodontiformes: Aplocheilidae, Nothobranchiidae, Rivulidae): Is male ornamentation constrained by intersexual genetic correlation? *Zoologischer Anzeiger. A Journal of Comparative Zoology* 253:207–215. doi: 10.1016/j.jcz.2013.12.004

Shidlovskiy, K. M., B. R. Watters, and R. H. Wildekamp. 2010. Notes on the annual killifish species *Nothobranchius rachovii* (Cyprinodontiformes; Nothobranchiidae) with the description of two new species. *Zootaxa* 2724:37–57.

Smith, Wm. L., P. Chakrabarty, and J. S. Sparks. 2008. Phylogeny, taxonomy, and evolution of Neotropical cichlids (Teleostei: Cichlidae: Cichlinae). *Cladistics* 24:625–641. doi: 10.1111/j.1096-0031.2008.00210.x

Sonnenberg, R. 2000. The distribution of *Chromaphyosemion* Radda, 1971 (Teleostei: Cyprinodontiformes) on the coastal plains of West and Central Africa. In *Isolated Vertebrate Communities in the Tropics Proceedings 4th International Symposium Bonn*, ed. G. Rheinwald. *Bonn Zool. Monogr.* 46.

Sonnenberg, R. 2007. Description of three new species of the genus *Chromaphyosemion* Radda, 1971 (Cyprinodontiformes: Nothobranchiidae) from the coastal plains of Cameroon with a preliminary review of the *Chromaphyosemion* splendopleure complex. *Zootaxa* 1591:1–38.

Sonnenberg, R. and E. Busch. 2009. Description of a new genus and two new species of killifish (Cyprinodontiformes: Nothobranchiidae) from West Africa, with a discussion of the taxonomic status of *Aphyosemion maeseni* Poll, 1941. *Zootaxa* 2294:1–22.

Sonnenberg, R. and E. Busch. 2010. Description of *Callopanchax sidibei* (Nothobranchiidae: Epiplateinae) a new species of killifish from southwestern Guinea, West Africa. *Bonn Zoological Bulletin* 57:3–14.

Sonnenberg, R. and E. Busch. 2012. Description of *Scriptaphyosemion wieseae* (Cyprinodontiformes: Nothobranchiidae), a new species from northern Sierra Leone. *Bonn Zoological Bulletin* 61:13–28.

Sonnenberg, R. and J. R. Van Der Zee. 2012. *Aphyosemion pseudoelegans* (Cyprinodontiformes: Nothobranchiidae), a new killifish species from the Cuvette centrale in the Congo Basin (Democratic Republic of Congo). *Bonn Zoological Bulletin* 61:3–12.

Sonnenberg, R., T. Woeltjes, and J. R. Van Der Zee. 2011. Description of *Fenerbahce devosi* (Cyprinodontiformes: Nothobranchiidae), a new species of dwarf killifish from the eastern Congo Basin in the Democratic Republic of Congo. *Zootaxa* 2966:1–12.

Tatarenkov, A., S. M. Q. Lima, D. S. Taylor, and J. C. Avise. 2009. Long-term retention of self-fertilization in a fish clade. *Proceedings of the National Academy of Sciences of the United States of America* 106:14456–14459. doi: 10.1073/pnas.0907852106

Valdesalici, S. 2013. *Pronothobranchius chirioi* n. sp. a new annual killifish species from the Niger River drainage, with descriptions of *P. kiyawensis*, *P. gambiensis* and *P. seymouri* (Cyprinodontiformes: Nothobranchiidae). *Killi-Data Series* 2013:21–41.

Valdesalici, S. and W. Eberl. 2013. *Aphyosemion grelli* (Cyprinodontiformes: Nothobranchiidae), a new species from the Massif du Chaillu, southern Gabon. *Vertebrate Zoology* 63:155–160.

Van Der Zee, J. R. and R. Sonnenberg. 2011. *Aphyosemion musafirii* (Cyprinodontiformes: Nothobranchiidae), a new species from the Tshopo Province in the Democratic Republic of Congo, with some notes on the *Aphyosemion* of the Congo Basin. *Bonn Zoological Bulletin* 1:73–87.

Van Der Zee, J. R. and R. H. Wildekamp. 1994. Description of a new *Fundulopanchax* species (Cyprinodontiformes: Aplocheilidae) from the Niger delta, with a redefinition of the genus *Fundulopanchax*. *Journal of African Zoology* 108:417–434.

Van Der Zee, J. R., T. Woeltjes, and R. H. Wildekamp. 2007. Aplocheilidae Bleeker, in *The Fresh and Brackish Water Fishes of Lower Guinea, West-Central Africa*, Vol. 2, eds. M. Stiassny, G. G. Teugels, and C. D. Hopkins. Tervuren: Musée Royal de l'Afrique Centrale.

Wildekamp, R. H., R. Romand, and J. J. Scheel. 1986. Cyprinodontidae. In *Check-list of the Freshwater Fishes of Africa (CLOFFA)*, eds. J. Daget, J. P. Gosse, and D.F.E. Thys van den Audenaerde, ISNB, Brussels, MRAC; Tervuren; and ORSTOM, Paris. 2, 165–276.

Wildekamp, R. H., K. M Shidlovsky, and B. R. Watters. 2009. Systematics of the *Nothobranchius melanospilus* species group (Cyrpinodontiformes: Nothobranchiidae) with description of two new species from Tanzania and Mozambique. *Ichtyological Exploration of Freshwaters* 20:237–254.

CHAPTER 2

Life Cycle, Reproduction, and Development in Annual Fishes
Cellular and Molecular Aspects

Nibia Berois, María José Arezo, Nicolás G. Papa, and Cora Chalar

CONTENTS

2.1 INTRODUCTION

Annual fishes or annual killifishes (Cyprinodontiformes; Aplocheiloidei) are freshwater teleosts exposed to an extremely variable environment that have a short lifespan. They inhabit temporary ponds in South America and Africa that dry out during the dry season, leading to the death of the entire adult population. During the rainy season the adults reproduce and generate desiccation-resistant embryos that remain buried in the muddy bottom of the dry ponds and hatch in the next rainy season once the ponds are flooded. The resulting juveniles reach sexual maturity in a few weeks, and a new reproductive cycle begins (Wourms, 1964, 1967; Arezo et al., 2005). Therefore, the survival of the species becomes entirely dependent upon buried embryos (Figure 2.1). The annual condition is defined by two characteristics: (1) a special and unique stage during early development, the dispersion–reaggregation of deep blastomeres that separates epiboly to embryonic axis formation and (2) the possibility to enter into developmental arrests or diapauses before hatching (Myers, 1952; Wourms, 1972a,b,c). The lifespan of annual fishes and the time they reach sexual maturity is variable among species. Among the African annual genus *Nothobranchius*, two species (*N. kadleci* and *N. furzeri*) have a median lifespan of 9 weeks, and juveniles attain sexual maturity at 17–19 days posthatching under laboratory conditions. Both species were reported as having the most rapid sexual maturation and minimum generation time of any reported vertebrate species

Figure 2.1 Life cycle of annual fish (*Austrolebias*) and temporary ponds they inhabit: (a) Flooded pond during the wet season (from fall to the end of spring) when the adults reproduce. *A. charrua* male (left) and female (right) are shown. (b) The same pond during the dry season (summer): adults have died and the embryos remain in a prehatching stage of developmental arrest. An embryo is shown.

(Blažek et al., 2013). Among Neotropical species the genus *Austrolebias* has a longer lifespan, about 9 months, and sexual maturity is attained 8–12 weeks after spawning under laboratory conditions (Vaz-Ferreira et al., 1964; Volcan et al., 2013). Furthermore, adults of *Austrofundulus limnaeus*, an annual killifish endemic to northern South America (Taphorn and Thomerson, 1978), remain fertile and produce viable embryos for more than 2 years in captivity (Podrabsky, 1999).

It was pointed out that, in contrast to other teleosts, annual fishes exhibit a unique developmental pattern (Myers, 1952). Epiboly is temporally and spatially detached from embryonic axis formation, and embryos can undergo one or more reversible arrests (diapauses) at three different stages: diapause I during epiboly, diapause II at the middle somite stage, and diapause III at the prehatching stage (Wourms, 1972a,b,c; Arezo et al., 2005). However, annual fish embryos can follow another alternative developmental trajectory, called "escape embryos," in which the embryos develop directly and escape to one or more diapauses (Wourms, 1972c; Podrabsky et al., 2010a; see Chapter 4). These developmental adaptations are closely related to the life cycle, and because of the exclusive features depicted above, annual fishes are outstanding organisms for comparative developmental approaches and for understanding some aspects about vertebrate evo-devo (Berois et al., 2014).

In this chapter, we consider the sexual strategy, gametogenesis, fertilization, and early development of annual fishes, especially at the cellular and molecular levels. Most of the data in these research areas derive from work on the genus *Austrolebias*. The genus consists of approximately 38 recognized species distributed in the Paraná-Plata basin and the Patos-Merín coastal lagoon system (Costa, 2006). Furthermore, the most studied species is *Austrolebias charrua* (Costa and Cheffe, 2001), distributed from southern Brazil (Patos-Merín lagoon) to eastern Uruguay (Rocha Department).

2.2 SEXUAL STRATEGY

Teleost fishes have different sexual strategies ranging from gonochorism (testes and ovaries in separate individuals) to hermaphroditism (functional male and female tissues in the same individual (Devlin and Nagahama, 2002). The only available data on sexual strategy in annual fishes were reported for *A. charrua* (Arezo et al., 2007). Under Yamamoto's (1969) classification, the sexual strategy corresponds to the "differentiated gonochoric" pattern wherein the early gonads develop directly into an ovary or a testis without intermediate stages (Arezo et al., 2007).

Austrolebias species have a high reproductive potential. After reaching sexual maturity, females spawn daily until senescence under laboratory conditions (Vaz-Ferreira et al., 1964; Wourms, 1967). Similar data were reported for *Cynopoecilus melanotaenia* based on macroscopic observation of the female cycle (Arenzon et al., 1999) and for *Austrolebias nigrofasciatus* in a reproductive study of the species husbandry under different temperatures (Volcan et al., 2013). The only histological study describing the pattern of sex differentiation and gametogenesis in a Neotropical annual fish is that of *A. charrua* (Arezo et al., 2007). This study showed that histological sex differentiation occurs at prehatching stages and begins earlier in females. This is a remarkable difference from nonannual gonochoric fish species for which sex differentiation has been reported at different post-hatching stages, also beginning earlier in females (Strüssmann and Nakamura, 2002).

2.3 GONADS: MORPHOLOGY AND FUNCTION

The female reproductive system of *A. charrua* consists of two elongated ovaries located ventrally to the swim bladder and dorsally to the gut. Both ovaries are connected with the genital papilla by a common duct. The ovarian parenchyma contains follicles at different stages of development. From sexual maturation to senescence four kinds of female germ cells are found in the follicles, described according to their size, nuclear cytoplasm, and vitelline envelope morphology: oogonia, previtellogenic, vitellogenic, and fully grown or total vitellogenic oocytes (Arezo et al., 2007).

Oogonias are grouped in clusters at the periphery of the ovary and are the smallest germ cells. The previtellogenic oocytes are associated with follicle cells in the ovarian follicle, show a high nucleus/cytoplasm ratio, and by the end of this stage they have cortical alveoli beneath the cell membrane. Vitellogenic oocytes undergo successive intermediate stages during vitellogenesis. The early ones have large peripheral cytoplasmic vacuoles that appear empty with standard staining techniques. The nucleus shows several nucleoli close to the nuclear envelope. As oogenesis proceeds, a fine granular acidophilic material accumulates among the vacuoles and progressively fuses to form the yolk (Figure 2.2a). Between the oocyte cell membrane and the follicle cells, the oocyte envelope is visible under light microscopy. This envelope (chorion, oocyte or egg envelope, zona pellucida) of teleost oocytes, the only coat that interacts with the sperm during fertilization, is formed by deposition of successive layers at the same time that vitellogenesis is taking place. At the fully grown stage, the envelope reaches its maximum width and complexity, showing a trilaminar structure (Berois et al., 2007). In *Austrolebias*, the ultrastructure of the oocyte envelope shows: a homogeneous electron-dense layer, that is, the zona radiata externa; a more electron-dense middle layer; and the widest fibrillar layer, that is, the zona radiata interna which is the last to appear during

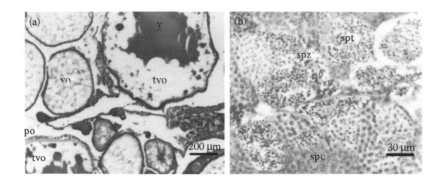

Figure 2.2 Histological organization of gonads in *A. charrua*. Hematoxylin and eosin stain. (a) Ovary with oocytes in different stages: po, previtellogenic oocyte; vo, vitellogenic oocyte; tvo, total vitellogenic oocyte; y, yolk. (b) Testis showing different cysts: spc, spermatocytes cyst; spt, spermatides cyst; spz, sperms.

the deposition process. The entire oocyte envelope is crossed by channels (Arezo et al., 2007). Finally, the fully grown oocytes, the largest germ cells, show the nucleus displaced to one pole and a large amount of acidophilic yolk fluid filling the cytoplasm. The overall yolk aspect agrees with the unique fluid yolk characteristic previously identified for Atherinomorpha (Parenti and Grier, 2004).

The male reproductive system of *A. charrua* consists of a pair of elongated testes, triangular in transverse section, and deferent ducts. Testes are located in the body cavity with the same anatomical relations as the ovaries. The parenchyma of the testis is organized in lobules limited by a basal membrane. The lobules extend from the external edge to the central region of the gonad, where they open in a network of deferent tubes. Among lobules there is a vascular interstitial connective tissue with Leydig-like cells. A main deferent duct connects both testes with genital papilla. Within the lobules, spermatogenesis occurs in cysts (Figure 2.2b).

Inside the lobules, there are cysts formed by clusters of germ cells in the same stage of spermatogenesis surrounded by Sertoli cells. According to germ cell size and nuclear and cytoplasmic features, five stages can be identified: spermatogonia, spermatocytes I and II, spermatids, and sperm. All five different cysts are observed in the testis from sexual maturation to senescence (Figure 2.2b).

Spermatogonia are the largest, rounded cells located at the blind external boundaries of the lobules. Their nuclei have fine granular chromatin. Spermatocytes show decreasing size as the cell proceeds through meiosis. Spermatids show a small amount of acidophilic cytoplasm and round nuclei with gradually condensing chromatin. Sperm are packed inside the cysts but can also be seen free in the lumen of the deferent ducts (Arezo et al., 2007).

The histological organization of the testes of *A. charrua* corresponds to the "restricted lobular pattern" that characterizes atherinomorph fishes, including Cyprinodontiformes. In the restricted pattern, spermatogonia are confined to the distal end of the lobules (Grier, 1981), whereas in the "unrestricted" pattern the spermatogonia occur along the testicular lobules, as seen in most teleosts. Parenti and Grier (2004) considered the presence of a restricted pattern to be phylogenetic informative; this testis pattern and a fluid-filled yolk in the oocyte appear to be uniquely derived characters for Atherinomorpha.

The organization pattern of adult gonads of annual fishes corresponds with an asynchronous spawning mode for females and a continuous spawning for males (Arezo et al., 2007).

2.4 GAMETES MORPHOLOGY

The oocytes of *Austrolebias* exhibit the general features of other telolecithal demersal eggs (i.e., large amount of yolk, nucleus at the animal pole just below the micropyle, and ornamented oocyte envelope or chorion). Teleost fishes, as well as insects and cephalopods, have a special structure, the micropyle, consisting of a narrow channel-like structure that perforates the chorion at the animal pole (Yanagimachi et al., 2013). The micropyle forms during deposition of the oocyte envelope, and its formation is related to the presence of a special cell in the animal pole that interrupts the deposition of the coat material at this place. In teleosts, the micropyle is the only entry site, across the chorion, for the male gamete. Furthermore, the micropyle has taxonomic, and potentially phylogenetic value, since it shows morphological differences among species and sperm–micropyle attraction would be species-specific (Yanagimachi et al., 2013).

The oocyte envelope of *Austrolebias* is tough and thick ($\chi = 6.0$ μm ±1.0). The hardness may play an important role in preventing dehydration during dry environmental periods. The surface of this envelope is ornamented with species-specific filaments whose ultrastructural pattern differs among species (Wourms and Sheldon, 1976; Loureiro and de Sá, 1996; Arezo et al., 2005, 2007; Fava and Toledo-Piza, 2007; García et al., 2009) (Figure 2.3a). The surface ornamentation of the oocyte envelope seems to be species specific for freshwater fishes in general (Johnson and Werner, 1986). A detailed description of ultrastructure of the oocyte envelope in several *Austrolebias* species is presented in Chapter 3.

The oocyte envelope is composed of glycoproteins (choriogenins, ZPs), the macromolecular complexity of which varies with the species (Spargo and Hope, 2003). Analyses of teleost fishes ZP gene sequences have shown two classes of genes derived from the duplication of an ancestral gene and further distinguished by their expression in the liver (estrogen regulated), in the ovary, or both, depending on the species (Conner and Hughes, 2003). In the annual fish *A. charrua*, two cDNAs, *achzpL* and *achzpH*, have been identified, both expressed in the liver (*achzpL* Genbank Accession number: KP083410; *achzpH*, KP083411). The deduced amino acid sequence of both expressed sequences showed identity values between 65% and 80% with ZPs from species belonging to diverse orders (e.g., *Fundulus heteroclitus*, *Oryzias latipes*, *Sparus aurata*, and *Danio rerio*). The expression pattern of both genes begins in the embryo before hatching and overlaps with the process of sex differentiation in this species (Arezo et al., 2007; N. Papa, personal communication).

Fish oogenesis is a sensitive biomarker to environmental pollutants and endocrine disrupting chemicals (EDCs) that cause noticeable changes in the synthesis of fish oocyte proteins, including envelope proteins and vitellogenin (Vtg) (Arukwe et al., 2000; Arukwe and Goksoyr, 2003). Using ZP proteins as environmental biomonitors seems to have a higher potential than using vitellogenin. Subtle changes in VTg would not threaten the survival of the offspring, whereas small changes in ZP expression might change the thickness and strength of the oocyte envelope, which can impact

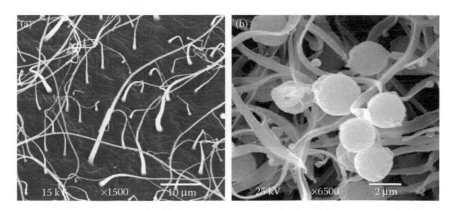

Figure 2.3 Ultrastructural aspects of *Austrolebias* gametes (SEM). (a) Oocyte envelope showing two types of filaments: thick and thin with an overall cone shaped morphology. (b) Sperms with round head, mitochondria in the midsection, and tails with lateral fins.

prevention of polyspermy at fertilization or protection of the embryo during development (Arukwe et al., 1997; Arukwe and Goksoyr, 2003).

As for other teleosts with external fertilization, annual fishes have typical uniflagellate anacrosomal aquasperm (Jamieson, 1991; see Chapter 3). Under SEM, the sperm exhibited a spherical head ($\chi - 3.8 + 0.5$ μm diameter), a midpiece with round mitochondria arranged in a single layer, and a flagellar tail. The tail had two, three, or four short lateral fins along almost its entire length (Figure 2.3b). A comparative sperm morphology study of *Austrolebias* suggested ultrastructural differences that, together with oocyte envelope patterns, could be complementary tools for phylogenetic and prezygotic isolation analyses among closely related species (García et al., 2009).

2.5 FERTILIZATION

Teleost fertilization exhibits different characteristics from those of other vertebrates and even other "fish" groups. Sperm–oocyte interaction only occurs at the micropyle. Since teleost sperm lack acrosomes, there is no acrosome reaction during fertilization. Once near the micropyle, fish spermatozoa show direct movement toward the micropyle opening (Hart, 1990). Furthermore, this attraction is mediated by a glycoprotein (lectin based) found on the surface of the oocyte's envelope, and its removal results in a drop of fertilization rate (Yanagimachi et al., 2013). The sperm fuses to the oocyte cell membrane underlying the micropyle (Coyne and Orr, 2004; Yanagimachi et al., 2013). Therefore, in the cell membrane of both gametes lies the key to species specificity in teleost fishes.

Annual fishes reproduce during the rainy season when ponds are flooded. They engage in elaborate courtship behaviors that result in the male and female partially burrowing into the substrate. The courtship pattern in *Austrolebias*, qualitatively described, is similar in different species of the genus (Vaz-Ferreira et al., 1964; Belote and Costa, 1998). More elaborate analyses of the sequence of behavioral units, as well as frequency and duration of male and female behavioral units, were reported for *A. reicherti* (García et al., 2008). Recent ethological studies provide additional data to elucidate the sexual barriers that could explain isolation among annual species with parapatric distribution; see Chapter 12 for further details.

In annual fishes both gametes are released during the courtship once the pair is buried in the bottom mud. The fertilized eggs develop buried in the mud and will survive the dry season (Figure 2.1b). In the subsequent rainy season, once the ponds are flooded, most eggs hatch. Under experimental conditions, pairs of *A. charrua* produce fertilized eggs for about 32 weeks with an average number of 20–30 eggs per day (Berois et al., 2014). Similar amounts of eggs were reported for the African annual genus *Nothobranchius* (Genade et al., 2005).

2.6 DEVELOPMENT

The onset of development is through meroblastic cleavage that follows the general teleost pattern, with the appearance of a discoidal blastoderm on the animal pole (Figure 2.4a). As development proceeds, three populations of cells can be distinguished from yolk to surface: (1) a yolk syncytial layer, where the cleavage nuclei share a common cytoplasm (Trinkaus, 1951, 1993); (2) a medial cell population, called "deep cells"; and (3) the enveloping multinuclear layer, the outmost layer, consisting of large and flat cells that surround and protect the embryo (Wourms, 1972a) (Figure 2.4b). Recent work comparing kinetics of cell division during early cleavage among African annual and nonannual species belonging to three phylogenetic clades reported that annual species showed cleavage times significantly longer than their nonannual sister taxa. In addition, it was demonstrated that in the annual *Nothobranchius furzeri*, the first five cleavage divisions are synchronous and do not present a G_1 phase. Cell cycle synchronization was lost after the fifth cleavage division, probably because of the activation of the very first zygotic genes (midblastula transition). The lengthening in duration of the annual fish cleavage would be an evolutionary advantage since it provides more time for embryos to face environmental conditions (Dolfi et al., 2014).

Teleost gastrulation begins when deep cells migrate and meet, forming the embryo at the same time that epiboly of the enveloping multinuclear layer and yolk syncytial layer takes place. Studies of early development in species considered representative of teleosts have shown the conservation of synchronicity of epiboly and convergence (presence of the germ ring) during gastrulation (Kimmel et al., 1995; Iwamatzu, 2004).

At this time, annual fish development shows the first unique trait: concurrent with epiboly of the other two cell populations, the deep cell layer disperses over the entire syncytial layer. This is known as the dispersion stage, and it represents, together with the following stage of reaggregation, one of the special features that define the annual condition (Wourms, 1972a,c; Arezo et al., 2005). During this stage the triangular or rhomboidal deep cells progressively migrate toward the vegetal pole (Figure 2.4c). The number of cells at early epiboly, compared with a typical teleost, is quite low in all species of annual fishes analyzed (Wourms, 1972c). For instance, *A. viarius* has about 100 migrating deep cells (Arezo et al., 2005), whereas in *Danio rerio* there are about 4000 migrating cells in this early stage (Kimmel et al., 1995).

After the deep cells have reached the vegetal pole, a second stage unique to annual fishes, described as the reaggregation stage, occurs (Wourms, 1972a,c; Arezo et al., 2005). At this time, deep cells that are surrounded by the enveloping layer on the top and by the syncytial layer on the bottom converge over the surface, forming a cell aggregate (Figure 2.4d). At the onset this cell aggregate is one cell thick, but the cells become progressively tightly packed, forming a disk-like plate several cells thick (Wourms, 1972a,b; Carter and Wourms, 1991; Arezo et al., 2005). This

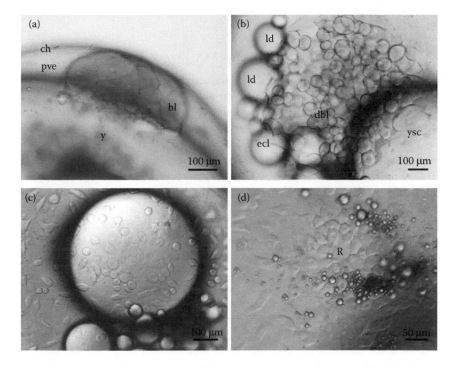

Figure 2.4 Early development in *A. charrua*. Differential interference contrast microscopy (DIC). (a) A 16-cell blastula: bl, blastomere; ch, chorion; pve, perivitelline space; y, yolk. (b) Blastula with the three types of cell population: ecl, enveloping cell layer; dbl, deep blastomere; ysc, yolk syncytial cell; ld, lipid droplet. (c) Dispersion stage. (d) Early reaggregate: R, reaggregation zone.

plate is evocative of the blastodisk of amniotes that forms at the end of cleavage; furthermore, in the middle of this structure is where the axis of the embryo appears, although the mechanisms of gastrulation involved in its formation remain to be determined.

A study of early gastrulation in typical teleosts (i.e., *Salmo, Salvelinus*; Ballard and Dodes, 1968) described a series of changes in the adhesiveness and motility of deep cells. At late blastula stage, these cells separate from one another and randomly migrate; this is referred to as the disengaged state. Thus, at the beginning of gastrulation, an incipient dispersed phase has been reported as a general teleost feature, and it could be considered a precursor of the dispersed stage of annual fishes (Berois et al., 2012, 2014). Furthermore, Ballard and Dodes (1968) established two overall patterns for teleost development that may contribute to understanding the dispersion–reaggregation phases in annual fishes: (1) embryogenesis solely involves the deep cells and (2) the first multicellular structure that forms during embryogenesis (the "nubbin"), considered to be equivalent to the prechordal plate, results from the localized "accumulation" (aggregation) of previously "disengaged" (dispersed) deep cells (Ballard and Dodes, 1968). Similar observations were reported for medaka (*Oryzias latipes*; Yokoya, 1966). Therefore, is the pattern of gastrulation in annual fishes different from the general teleost or is it a derived one? Based on these reports, the annual fish pattern at the onset of gastrulation with dispersed and reaggregation stages could be explained as an extension of the incipient processes previously described in typical teleosts (Berois et al., 2012, 2014).

The subsequent processes of axis formation, somitogenesis, and organogenesis in annual fish development are similar to those of other teleosts. However, in annual fishes, the overall duration of embryonic development has been documented to be longer, lasting between 40 and 320 days postfertilization (Wourms, 1972c). Taking into account that temperature influences the length of development, variation in this parameter could explain time differences in developmental rate. In the African *Nothobranchius guentheri*, an increase of 5°C resulted in a reduction of 10 days in development (Markofsky and Matias, 1977).

2.6.1 Diapauses

The developmental arrests or diapauses that characterize annual fish development are present in South American and African species and are of variable length. They could be facultative or obligatory depending on different genera. Moreover, embryos can sometimes bypass all diapauses and are then called "escape embryos" (see Chapters 4 and 9). A recent work suggests that the ability to produce diapausal eggs evolved independently at least six times within African and South American annual fishes (Furness et al., 2015). In *Austrolebias,* diapauses I and II are facultative, whereas diapause III is obligatory (Wourms, 1972c). This has been reported for *A. nigripinnis* (Peters, 1963), *A. bellotti* (Wourms, 1972c), and *A. viarius* (Arezo et al., 2005) in embryos cultured under standardized laboratory conditions. In *Austrofundulus*, diapause I is facultative, but diapauses II and III are obligatory (Wourms, 1972c; Podrabsky and Hand, 1999). In two African annual species (*N. furzeri* and *N. kadleci*) viable embryos can escape all diapauses and hatch in about 15 days. These species also have the most rapid sexual maturation and minimum generation time of any vertebrate species (Blažek et al., 2013).

Taking into account the possible combinations among the stages and durations, diapauses can mean different developmental pathways to the timing of an embryo trajectory. As previously noted, annual fish embryos can bypass diapauses when environmental conditions are favorable (Podrabsky et al., 2010a; Chapter 4). From an ecological perspective, these possibilities would allow annual fishes to fine-tune the duration of development in relation to a very unpredictable environment (Wourms, 1972c; Polačik et al., 2014). Entering diapause is accomplished by a dramatic reduction in metabolic rate and an increase in long-term embryo survival. Adaptation to the harsh aquatic environment could have happened through a strong selection during embryonic development (Furness et al., 2015).

Diapause I, which occurs during the dispersed stage of late blastula, has been interpreted as an early control mechanism exerted by the environment over the developmental pathway (Wourms, 1972b). A study in the African *Nothobranchius* reported that diapause I was induced by chemical signals from adult fishes (Inglima et al., 1981). The putative polar hydrophilic substance responsible for this developmental arrest remains unknown (Denucé, 1989). In laboratory conditions, embryos developing in aerated water without adults do not enter in diapause I in either *Austrolebias* or *Nothobranchius* (Arezo et al., 2005; Valenzano et al., 2011), suggesting similar diapause I–related mechanisms in Neotropical and African annual fishes.

Diapause II (at 35–40 somite stage) occurs in advanced embryos undergoing organogenesis (Wourms, 1972a,b). Most of the available data about diapause II are on the northern South American species *Austrofundulus limnaeus*. During diapause II, this species shows depressed metabolism, great tolerance to anoxia, and the presence of a heat shock protein; cells are in a G_0-like stage of the cell cycle (Podrabsky and Hand, 1999; Podrabsky et al., 2007; Meller et al., 2012). Furthermore, the species have two developmental trajectories, "escape" and diapause II, with respective differing morphological and physiological features. Young females and high incubation temperature were related to the "escape" trajectory (Podrabsky et al., 2010a). Other studies focused on the role of steroid hormones in *A. limnaeus* showed that treatment of embryos with exogenous E_2 induced the increase in the "escape embryos" during diapause II (Pri-Tal et al., 2011). This study also demonstrated that an age-related decrease in maternal E_2 is linked with a decline in the number of "escape embryos." Therefore, a steroid signal seems to be involved in the regulation of diapause II and progression of development in *A. limnaeus*.

Diapause III happens at the prehatching stage in embryos that are almost ready to hatch. This stage overlaps with the variable duration of the dry season. The metabolism of embryos in diapause III is lower when compared with prehatching embryos induced to hatch (Levels et al., 1986; Podrabsky and Hand, 1999). The capacity to reduce water loss reported in embryos of *A. limnaeus* during diapause III seems to be related to the resistance to desiccation (Podrabsky et al., 2001, 2010b).

Embryos of *A. charrua* developing in laboratory conditions can escape to one or more diapauses and hatch in about 40 days. Most of the resulting alevins are belly sliders, probably due to air bladder problems, but others are normal (M.J. Arezo, personal communication). This situation, shared with African *Nothobranchius* embryos (Valenzano et al., 2011; Blažek et al., 2013; Polačik et al., 2014), leads to the following hypothesis: in laboratory conditions (no adults, adequate temperature, water, or culture solution without desiccation) the embryos would not receive hazard signals from the environment, and the mechanisms that slow metabolism and prompt diapauses would not be activated. One could say that, in these conditions, embryos could develop with a "green light" until they hatch. It is necessary to note that the direct development has a cost in alevins' viability, because the belly sliders have a lower likelihood of survival, at least in nature. In laboratory conditions, they can reach adulthood, but their size and weight are lower than normal individuals (C. Passos, personal communication).

In brief, annual fish diapauses are an evolutionary strategy of innovation with survival function. They involve stages in which metabolic activities are reduced or suspended until the environmental conditions become favorable.

Some nonannual Cyprinodontiformes are able to delay hatching under special environmental conditions. *Fundulus heteroclitus* inhabits coastal marshes and usually has continuous development. In addition, spawning and hatching align with high-tide cycles (Taylor, 1999). However, developing embryos exposed to air for longer periods showed higher speed of development and reduced levels of an aquaporin protein, whereas embryos completely submerged under water experienced delayed hatching (Tingaud-Sequeira et al., 2009). More recently, it was reported that embryos of *F. heteroclitus* exposed to air showed 806 genes differentially expressed. These genes showed transcriptional features of "stress" response proteins (Tingaud-Sequeira et al., 2013). In this context, diapause III could be considered as the transition of the extended delayed hatching of some nonannual species to an attribute in annual fish development (Wourms, 1972c; Taylor, 1999).

2.7 CONCLUSIONS AND PERSPECTIVES

Living in a quite unfavorable and unpredictable environment, annual fishes have reproductive strategies that result in evolutionary advantages tightly related to their short lifespan. The main aspects related to reproduction are the early sexual differentiation and maturity, the elaborate species-specific courtship, and the asynchronous and continuous spawning from sexual maturity to death.

At the developmental level, the unique dispersion–reaggregation stages before gastrulation and the occurrence of diapauses of variable duration can generate embryos that develop and hatch at different times. This asynchrony could imply an advantage for the survival of annual fish species that face a harsh environment. This advantage in the early phase of life leads the embryo to be the key to the survival of the entire population.

The longer length and special features of early development in annual fishes offer the chance to study the migration of blastomeres and other special cell types, such as the primordial germ cells. Understanding the mechanisms that underlie the cell migration is a basic topic in organogenesis, but it could also be of value for the comprehension of aberrant cell migration in some disease conditions.

Also interesting and in need of additional studies is the gastrulation pattern. Its geometry in annual fishes is quite different from that reported in general for teleosts and resembles that of amniote embryos. This feature may be an evolutionary innovation or a derived character from the general teleost gastrulation pattern. Understanding the gastrulation pattern requires knowledge of the temporal and spatial expression patterns of genes involved in early gastrulation in other vertebrates, in addition to count on the transcriptome of annual fishes at this stage.

To understand the factors that control or underlie entry and exit of the reversible arrests in annual fishes is also of great interest. Studies are currently in progress to identify the differential expression of genes related to diapauses I and III in *A. charrua* using transcriptomic and proteomic approaches. The diapauses of the annual fish model is a valuable biological event for understanding the genetic and/or environmental traits that control tolerance to stress in vertebrates.

Currently, an interdisciplinary team is working to determine the main factors that generate the high diversity observed in Neotropical annual fishes and their adaptations in the genus *Austrolebias*. Topics related to evolutionary biology, ethology, reproductive strategies, regulation of development, and senescence are under consideration in the research community.

ACKNOWLEDGMENTS

The authors acknowledge financial support from Facultad de Ciencias (Universidad de la República), PEDECIBA (Programa de Desarrollo de Ciencias Básicas), CSIC (Comisión Sectorial de Investigación Científica, UdelaR), and ANII (Agencia Nacional de Investigación e Innovación).

REFERENCES

Arenzon, A., A. Carvalho, and M.B. Camino. 1999. Reproduction of the annual fish *Cynopoecilus melanotaenia* (Regan, 1912) in a temporary water body in Rio Grande do Sul, Brazil. (Cyprinodontiformes, Rivulidae). *Hydrobiologia* 411:65–70. doi: 10.1023/A:1003868711295

Arezo, M.J., S. D'Alessandro, N. Papa, R.O. de Sá, and N. Berois. 2007. Sex differentiation pattern in the annual fish *Austrolebias charrua* (Cyprinodontiformes: Rivulidae). *Tissue and Cell* 39:89–98. doi: 10.1111/j.0022-1112.2005.00688.x

Arezo, M.J., L. Pereiro, and N. Berois. 2005. Early development in the annual fish *Cynolebias viarius*. *Journal of Fish Biology* 66:1357–1370. doi: 10.1016/j.tice.2007.01.004

Arukwe, A., T. Celius, B.T. Walther, and A. Goksoyr. 2000. Effects of xenoestrogen treatment on zona radiata protein and vitellogenin expression in Atlantic salmon (*Salmo salar*). *Aquatic Toxicology* 49:159–170. doi: 10.1016/S0166-445X(99)00083-1

Arukwe, A. and A. Goksoyr. 2003. Eggshell and egg yolk proteins in fish: Hepatic proteins for the next generation: Oogenetic, population, and evolutionary implications of endocrine disruption. *Comparative Hepatology* 2:4. doi:10.1186/1476-5926-2-4

Arukwe, A., F.R. Knudsen, and A. Goksoyr. 1997. Fish zona radiata (egg shell) protein: A sensitive biomarker for environmental estrogens. *Environmental Health Perspectives* 105:418–422.

Ballard, W.W. and L.M. Dodes. 1968. The morphogenetic movements of the lower surface of the blastodisc in Salmonid embryos. *Journal of Experimental Zoology* 168:76–84. doi: 10.1002/jez.1401680107

Belote, D.F. and W.J.E.M. Costa. 1998. Reproductive behavior patterns in three species of the South American annual fish genus *Austrolebias* Costa, 1998 (Cyprinodontiformes, Rivulidae). *Boletim do Museu Nacional, Nova Serie, Rio de Janeiro* 514:1–7.

Berois, N., M.J. Arezo, C. Chalar, M.M. Brauer, and C. Barros. 2007. Ultrastructure and protein composition of the oocyte envelope in the whitemouth croaker (*Micropogonias furnieri*, Desmarest, 1823, Sciaenidae, Perciformes). *Journal of Applied Ichthyology* 23:34–39. doi: 10.1111/j.1439-0426.2006.00762.x

Berois, N., M.J. Arezo, and R.O. de Sá. 2014. The neotropical genus *Austrolebias*: An emerging model of annual killifishes. *Cell and Developmental Biology* 3:136. doi: 10.4172/2168-9296.1000136

Berois, N., M.J. Arezo, N.G. Papa, and G.A. Clivio. 2012. Annual fish: Developmental adaptations for an extreme environment. *WIREs Developmental Biology* 1:595–602. doi: 10.1002/wdev.39

Blažek, R., M. Polačik, and M. Reichard. 2013. Rapid growth, early maturation and short generation time in African annual fishes. *EvoDevo* 4:24. doi: 10.1186-9139-4-24

Carter, C.A. and J.P. Wourms. 1991. Cell behavior during early development in the South American annual fishes of the genus *Cynolebias*. *Journal of Morphology* 210:247–266. doi: 10.1002/jmor.1052100305

Conner, S.J. and D.C. Hughes. 2003. Analysis of fish ZP1/ZPB homologous genes-evidence for both genome duplication and species-specific amplification models of evolution. *Reproduction* 126:347–352. doi: 10.1093/humrep/deh835

Costa, W.E.J.M. 2006. The South American annual killifish genus *Austrolebias* (Teleostei: Cyprinodontiformes: Rivulidae): Phylogenetic relationships, descriptive morphology and taxonomic revision. *Zootaxa* 1213:1–162.

Costa, W.J.E.M. and M.M. Cheffe. 2001. Three new annual fishes of the genus *Austrolebias* from the Laguna dos Patos system, southern Brazil, and a redescription of *A. adloffi* (AHL) (Cyprinodontiformes: Rivulidae). *Comun Museu de Ciências e Technologia PUCRS Sér Zoo, Porto Alegre* 14:179–200.

Coyne, J.A. and H.A. Orr. 2004. *Speciation*. Sunderland, MA: Sinauer, 545.

Denucé, J.M. 1989. Developmental changes in annual fish embryos kept in the proximity of adult fishes. *Acta Embryologiae et Morphologiae Experimentalis* 10:87–94.

Devlin, R.H. and Y. Nagahama. 2002. Sex determination and sex differentiation in fish: An overview of genetic, physiological and environmental influences. *Aquaculture* 208:191–364. doi: 10.1016/S0044-8486(02)00057-1

Dolfi, L., R. Ripa, and A. Cellerino. 2014. Transition to annual life history coincides with reduction in cell cycle speed during early cleavage in three independent clades of annual killifish. *EvoDevo* 5:32. doi: 10.1186/2041-9139-5-32

Fava, D. and M. Toledo-Piza. 2007. Egg surface structure in the annual fishes *Simpsonichthys* (subgenera *Opthalmolebia* and *Xenurolobias*) and *Nematolebias* (Teleostei: Cyprinodontiformes: Rivulidae): Variability and phylogenetic significance. *Journal of Fish Biology* 71:889–907. doi: 10.1111/j.1095-8649.2007.01572.x

Furness, A.I., D.N. Reznick, M.S. Springer, and R.W. Meredith. 2015. Convergent evolution of alternative developmental trajectories associated with diapauses in African and South American killifish. *Proceedings of the Royal Society of London B* 282:20142189. doi: 10.1098/rspb.2014.2189

García, D., M. Loureiro, and B. Tassino. 2008. Reproductive behavior in the annual fish *Austrolebias reicherti* Loureiro & García 2004 (Cyprinodontiformes, Rivulidae). *Neotropical Ichthyology* 6:243–248. doi: 10.1590/S1679-62252008000200012

García, G., M. Loureiro, N. Berois et al. 2009. Pattern of differentiation in the annual killifish genus *Austrolebias* (Cyprinodontiformes: Rivulidae) from a biosphere reserve site in South America: A multidisciplinary approach. *Biological Journal of the Linnean Society of London* 98:620–635. doi/10.1111/j.1095-8312.2009.01303

Genade, T., M. Benedetti, E. Terzibase et al. 2005. Annual fishes of the genus *Nothobranchius* as a model system for aging research. *Aging Cell* 4:223–233. doi: 10.1111/j.1474-9726.2005.00165

Grier, H.J. 1981. Cellular organization of the testis and spermatogenesis in fishes. *American Zoologist* 21:345–357. doi: 10.1093/icb/21.2.345

Hart, N.H. 1990. Fertilization in teleost fishes: Mechanisms of sperm-egg interactions. *International Review of Cytology* 121:1–66.

Iwamatzu, T. 2004. Stages of normal development in the medaka *Oryzias latipes*. *Zoological Science* 11:605–618. doi: 10.1016/j.mod.2004.03.012

Inglima, K., A. Perlmutter, and J. Markofsky. 1981. Reversible stage-specific embryonic inhibition mediated by the presence of adults in the annual fish *Nothobranchius guentheri*. *Journal of Experimental Zoology* 215:23–33. doi: 10.1002/jez.1402150104

Jamieson, B.G.M. 1991. *Fish Evolution and Systematics: Evidence from Spermatozoa*. Cambridge: Cambridge University Press.

Johnson, E.Z. and R.G. Werner. 1986. Scanning electron microscopy of the chorion of selected freshwater fishes. *Journal of Fish Biology* 29:257–265. doi: 10.1111/j.1095-8649.1986.tb04943

Kimmel, C.B., W.W. Ballard, S.R. Kimmel, B. Ullmann, and T.F. Schilling. 1995. Stages of embryonic development of the zebrafish. *Developmental Dynamics* 203:253–310. doi: 10.1002/aja.1002030302

Levels, P.J., R.E. Gubbels, and J.M. Denucé. 1986. Oxygen consumption during embryonic development of the annual fish *Nothobranchius korthausae* with special reference to diapause. *Comparative Biochemistry and Physiology. A Comparative Physiology* 84:767–770. doi: 10.1016/0300-9629(86)90403-2

Loureiro, M. and R.O. de Sá. 1996. External morphology of the chorion of the annual fishes *Cynolebias* (Cyprinodontiformes: Rivulidae). *Copeia* 4:1016–1022.

Markofsky, J. and J.R. Matias. 1977. The effects of temperature and season of collection on the onset and duration of diapause in embryos of the annual fish *Nothobranchius guentheri*. *Journal of Experimental Zoology* 202:49–56. doi: 10.1002/jez.1402020107

Meller, C.L., R. Meller, R.P. Simon, K.M. Culpepper, and J.E. Podrabsky. 2012. Cell cycle arrest associated with anoxia-induced quiescence, anoxic preconditioning, and embryonic diapause in embryos of the annual killifish *Austrofundulus limnaeus*. *Journal of Comparative Physiology B* 182:909–920. doi: 10.1007/s00360-012-0672-9

Myers, G.S. 1952. Annual fishes. *Aquarium Journal* 23:125–141.

Parenti, L.R. and J.H. Grier. 2004. Evolution and phylogeny of gonad morphology in bone fishes. *Integrative and Comparative Biology* 44:333–348. doi: 10.1093/icb/44.5.333.

Peters, N. 1963. Embryonale Anpassungen oviparer Zahnkarpfen aus periodisch austrocknenden Gewassern. *Internationale Revue des gesamten Hydrobiologie* 48:257–313. doi: 10.1002/iroh.19630480204

Podrabsky, J.E. 1999. Husbandry of the annual killifish *Austrofundulus limnaeus* with special emphasis on the collection and rearing of embryos. *Environmental Biology of Fishes* 54:421–431. doi: 10.1023/A:1007598320759

Podrabsky, J.E., J.F. Carpenter, and S. Hand. 2001. Survival of water stress in annual fish embryos: Dehydration avoidance and egg envelope amyloid fibers. *American Journal of Physiology Regulatory, Integrative and Comparative Physiology* 280:R123–131.

Podrabsky, J.E., I.D. Garrett, and Z.F. Kohl. 2010a. Alternative developmental pathways associated with diapause regulated by temperature and maternal influences in embryos of the annual killifish *Austrofundulus limnaeus*. *Journal of Experimental Biology* 213:3280–3288. doi: 10.1242/jeb.045906

Podrabsky, J.E. and S.C. Hand. 1999. The bioenergetics of embryonic diapauses in an annual killifish, *Austrofundulus limnaeus*. *Journal of Experimental Biology* 202:2567–2580.

Podrabsky, J.E., J.P. López, T.W. Fan, R. Higashi, and G.N. Somero. 2007. Extreme anoxia tolerance in embryos of the annual killifish *Austrofundulus limnaeus*: Insights from a metabolomics analysis. *Journal of Experimental Biology* 210:2253–2266. doi: 10.1242/jeb.005116

Podrabsky, J.E., A. Tingaud-Sequeira, and J. Cerdà. 2010b. Metabolic dormancy and responses to environmental desiccation in fish embryos. In *Topics in Current Genetics. Dormancy and Resistance in Harsh Environments*, eds. E. Lubzens, J. Cerdà, and M. Clark. The Netherlands: Springer, 203–226.

Polačik, M., R. Blažek, R. Rezucha, M. Vrtilek, E. Terzibasi Tozzini, and M. Reichard. 2014. Alternative intra-population life-history strategies and their trade-offs in an African annual fish. *Journal of Evolutionary Biology* 27:854–865. doi: 10.1111/jeb.12359

Pri-Tal, B.M., S. Blue, F.K. Pau, and J.E. Podrabsky. 2011. Hormonal components of altered developmental pathways in the annual killifish, *Austrofundulus limnaeus*. *General and Comparative Endocrinology* 174:166–174. doi: 10.1016/j.ygcen.2011.08.016

Spargo, S.C. and R.M. Hope. 2003. Evolution and nomenclature of the zona pellucida gene family. *Biology of Reproduction* 68:358–362. doi.org/10.1095/biolreprod.102.008086

Strüssmann, C.A. and N. Nakamura. 2002. Morphology, endocrinology, and environmental modulation of gonadal sex differentiation in teleost fishes. *Fish Physiology and Biochemistry* 26:13–29. doi: 10.1023/A:1023343023556

Taphorn, D.C. and J.E. Thomerson. 1978. A revision of the South American cyprinodont fishes of the genera *Rachovia* and *Austrofundulus*, with the description of a new genus. *Acta Biológica de Venezuela* 9:377–452.

Taylor, M.H. 1999. A suite of adaptations for intertidal spawning. *American Zoologist* 39:313–320.

Tingaud-Sequeira, A., J.J. Lozano, C. Zapater et al. 2013. A rapid transcriptome response is associated with desiccation resistance in aerially-exposed killifish embryos. *PLoS One* 8:e64410. doi: 10.1371/journal.pone.0064410

Tingaud-Sequeira, A., C. Zapater, F. Chauvigne, D. Otero, and J. Cerda. 2009. Adaptive plasticity of killifish (*Fundulus heteroclitus*) embryos: Dehydration-stimulated development and differential aquaporin-3 expression. *American Journal of Physiology. Regulatory, Integrative and Comparative Physiology* 296:1041–1052. doi: 10.1093/icb/39.2. 313

Trinkaus, J.P. 1951. A study of the mechanism of epiboly in the egg of *Fundulus heteroclitus*. *Journal of Experimental Zoology* 118:269–319. doi: 10.1002/jez.1401180204

Trinkaus, J.P. 1993. The yolk syncytial layer of *Fundulus*: Its origin and history and its significance for early embryogenesis. *Journal of Experimental Zoology* 265:258–284. doi: 10.1002/jez.1402650308

Valenzano, D.R., S. Sharp, and A. Brunet 2011. Transposon-mediated transgenesis in the short-lived African killifish *Nothobranchius furzeri*, a vertebrate model for aging. *G3 Genes Genomes Genetics* 1:531–538. doi: 10.1534/g3.111.001271

Vaz-Ferreira, R., B. Sierra, and S. Scaglia. 1964. Eco-etología de la reproducción en los peces del género *Cynolebias* Steindachner, 1876. *Apartados de los Archivos de la Sociedad de Biología de Montevideo* 26:44–49.

Volcan, M.V., L.A. Sampaio, D.C. Bongalhardo, and R.B. Robaldo. 2013. Reproduction of the annual fish *Austrolebias nigrofasciatus* (Rivulidae) maintained at different temperatures. *Journal of Applied Ichthyology* 29:648–652. doi: 10.1111/jai.12013

Wourms, J.P. 1964. Comparative observations on the early embryology of *Nothobranchius taeniopygus* and *Aplocheilicthys pumilis* with special reference to the problem of naturally occurring embryonic diapause in teleost fishes. *Annual Report East African Freshwater Fisheries Institute*, pp. 68–73.

Wourms, J.P. 1967. Annual fishes. In *Methods in Developmental Biology*, eds. F.H. Wilt and N. Wessels. New York: Thomas and Crowell Company, 123–137.

Wourms, J.P. 1972a. The developmental biology of annual fishes I. Stages in the normal development of *Austrofundulus myersi* Dahl. *Journal of Experimental Zoology* 182:143–168. doi: 10.1002/jez.1401820202

Wourms, J.P. 1972b. The developmental biology of annual fishes. II. Naturally occurring dispersion and reaggregation of blastomers during the development of annual fish eggs. *Journal of Experimental Zoology* 182:169–200. doi: 10.1002/jez.1401820203

Wourms, J.P. 1972c. The developmental biology of annual fishes. III. Pre-embryonic and embryonic diapause of variable duration in the eggs of annual fishes. *Journal of Experimental Zoology* 182:389–414. doi: 10.1002/jez.1401820310

Wourms, J.P. and H. Sheldon. 1976. Annual fish oogenesis. II. Formation of the secondary egg envelope. *Developmental Biology* 50:355–366.

Yamamoto, T. 1969. Sex differentiation. In *Fish Physiology Vol III*, eds. W.S. Hoar and D.J. Randall. New York: Academic Press, 117–175.

Yanagimachi, R., G. Cherr, T. Matsubara et al. 2013. Sperm attractant in the micropyle region of fish and insect eggs. *Biology of Reproduction* 88:47,1–11. doi: 10.1095/biolreprod.112.105072

Yokoya, S. 1966. Cell dissociation and reaggregation in early stage embryo of a teleost *Oryzias latipes*. *The Science Reports of the Tohoku University Ser* 32:229–236.

CHAPTER 3

Sperm and Egg Envelope Ultrastructure and Some Considerations on Its Evolutionary Meaning

Magela Rodao, Jimena Montagne, Graciela A. Clivio,
Nicolás G. Papa, and Gabriela Casanova Larrosa

CONTENTS

3.1 INTRODUCTION

Among vertebrates, fishes show the greatest biological, morphological, and ecological diversity, which is reflected in the variety of their reproductive strategies. Within this group, Cyprinodontiformes have developed unique reproductive strategies and, among Neotropical groups, the family Rivulidae stands out because it includes species with annual life cycles. This is the case of the genus *Austrolebias*, with 39 currently recognized species (Loureiro et al., 2011). This genus inhabits temporal ponds in Paraguay, Uruguay (Figure 3.1), southern Brazil, and reaching into north and northeast Argentina (Costa, 2006; Loureiro et al., 2011). Like other rivulids, *Austrolebias* has external fertilization, and the male gamete has remarkable morphological adaptations, including changes in size, shape, and ultrastructure that are probably related to fertilization efficiency.

In *Austrolebias*, as in other teleosts, the testes are paired structures located dorsally in the peritoneal cavity and surrounded by the tunica albuginea (Schulz et al., 2010). Each testis is composed of a germinal and an interstitial layer separated by a basal membrane. The germinal layer consists of clusters of germ cells, called cysts, which are surrounded by Sertoli cells. All the germ cells within a cyst are connected to each other through cytoplasmic bridges; consequently, they mature

Figure 3.1   Map of Uruguay showing collecting areas of *Austrolebias* specimens. Red star = males *A. luteo-flammulatus,* green circle = males and female of *A. charrua,* and violet triangle = *A. reicherti,* the three species were collected in the Department of Rocha; other individuals of *A. charrua* and *A. reicherti* were also found in ponds at the southeastern area of the Department of Treinta y Tres. Blue square = *A. affinis* collected in the central area of the Department of Durazno south of the Río Negro; orange square = *A. bellottii*; purple star = *A. nigripinnis* from the southwestern of Uruguay, between the Departments of Colonia and Soriano; yellow triangle = *A. melanoorus* collected in the Department of Tacuarembó; red circle = *A. vazferreirai* from northeast of Uruguay, area located in between the Departments of Tacuarembó, Rivera, and Cerro Largo, north of Río Negro.

synchronically within each cyst. The interstitial compartment consists of connective tissue, Leydig cells (steroidogenic), fibroblasts, collagen fibers, mast and macrophage cells, blood cells, blood vessels, and myelin fibers (Grier, 1981; Schulz et al., 2010).

Similar to previous reports for Atherinomorpha, the testes of *Austrolebias* belong to the "restricted lobular" type (Grier et al., 1980; Parenti and Grier, 2004). In this kind of testes, the spermatogonia are located at the distal end of the lobules near the tunica albuginea. As meiosis and spermiogenesis proceed, the cysts move closer to the sperm duct, and spermiation occurs when the cysts open, releasing the sperm cells into the sperm ducts. Therefore, in these species, the mature spermatozoon is free in the lumen of the sperm ducts (Grier et al., 1980).

3.2 SPERM CELLS OF ANNUAL FISH

The teleost fish sperm has high diversity, mainly among families, but there are two very different types depending on the type of fertilization the species has. The spermatozoa of species with internal fertilization usually have elongated nuclei and a larger midpiece region with many mitochondria. On the other hand, teleosts with external fertilization generally have sperm cells with smaller spherical nuclei, shorter midpiece, and few mitochondria (usually 1 to 6), and they can be mono- or biflagellated (Lahnsteiner and Patzner, 2008).

Annual fish sperm is the characteristic uniflagellate anacrosomal aqua sperm of teleosts with external fertilization (Jamieson, 1991; Berois et al., 2014). There are several reviews of the overall subcellular structure of fishes (Baccetti, 1991; Jamieson, 1991; Mattei, 1991); furthermore, a few studies have information on the spermatozoa of annual fishes (Thiaw et al., 1986; Arezo et al., 2007; García et al., 2009).

3.3 SPERM ULTRASTRUCTURE IN *AUSTROLEBIAS*

The ultrastructural morphology of the sperm is highly conserved, and it has been widely used in systematics and to assess phylogenetic relationships among fishes (Jamieson, 1991; Jamieson and Grier, 1993; Berois et al., 2011; Cassel et al., 2014). Here, the sperm of eight *Austrolebias* species were analyzed. The sperm of *Austrolebias* species have some shared characteristics as well as other species-specific traits. Sperm of all analyzed species have three distinct regions: small spherical head, short midpiece containing few mitochondria, and a single flagellum carrying lateral cell membrane expansions. In addition, some traits are species specific. The head of the sperm cell shows variation in several traits: the appearance of the nuclear chromatin, the size and distribution of electron-lucent areas, the shape and size of the nuclear fossa, the arrangement of the centrioles, and the presence of specialized structures associated with the centrioles. In the midpiece, the differences are mostly related to the size and appearance of the mitochondria. Finally, the flagellum may have one or more lateral membrane expansions (Table 3.1).

A similar ultrastructure was previously described for the sperm of other nonannual rivulids, for example, *Rivulus marmoratus* (Kweon et al., 1998) *Kryptolebias marmoratus* (Costa, 2004), and *Melanorivulus punctactus* (Cassel et al., 2014).

3.4 ORGANIZATION OF THE HEAD OF THE MALE GAMETE

The sperm of *Austrolebias* has a rounded head (2.5–3.5 µm diameter), lacks an acrosome, and is almost filled by a spherical nucleus. The chromatin is strongly compacted except in a few electron-lucent areas. The nucleus has a nuclear fossa on its caudal half, where the proximal and distal centrioles are found and obliquely positioned to each other.

Despite these similarities, the overall appearance of the chromatin of the sperm differs among species. The chromatin of *Austrolebias affinis* and *Austrolebias nigripinnis* appears homogenous, strongly compact (Figure 3.2c and d), and highly electron dense in comparison with that of other analyzed species. However, *A. nigripinnis* and *Austrolebias luteoflammulatus* differ from *A. affinis*, because their sperm nuclei possess electron-lucent areas of variable sizes (between 150 and 200 nm in diameter; Figure 3.2a), whereas in *A. affinis*, the interchromatin spaces do not contrast with the appearance of the compact chromatin, because they appear as gray areas (Figure 3.2c). In *Austrolebias vazfer-reirai* the chromatin is less compact and electron-lucent areas are usually smaller and more numerous than those observed in the aforementioned species (Figure 3.2b). *Austrolebias bellottii* has compact and homogenous chromatin, although it is less electron dense than other studied species (Figure 3.2f). Furthermore, it has few electron-dense areas that are of similar size to those of *A. vazferreirai*. Three of the eight species (*Austrolebias melanoorus*, *Austrolebias charrua*, and *Austrolebias reicherti*) have granular-like chromatin with different electron densities and many spaces that are particularly numerous and smaller in *A. charrua* and *A. reicherti* (Figure 3.2e, g, and h).

The presence of a nuclear fossa on the posterior region of the sperm head is found in all species. However, the fossa is significantly deeper and occupies approximately 25% of the nuclear diameter in *A. bellottii* and *A. nigripinnis* (Figure 3.2d and f). Furthermore, in *A. bellottii*, the fossa houses the centriolar complex, whereas in *A. nigripinnis*, only a proximal centriole is visible inside the fossa

Table 3.1 Ultrastructure Characteristics of the Spermatozoa in *Austrolebias*

	Head		Midpiece	Tail
	Nucleus/Chromatin	Nuclear Fossa/ Centriolar Complex	Mitochondria	Flagella/Lateral Fins
A. luteoflammulatus	Highly condensed chromatin with predominance of big electron-lucent interchromatin spaces	Moderately deep nuclear fossa partially housing the proximal centriole	A single layer of big mitochondria with crests inconspicuous	Axoneme conformation 9 + 2 with 2 and 3 lateral fins
A. vazferreirai	Chromatin with numerous electron-lucent and predominantly small spaces	Moderately deep nuclear fossa partially housing the proximal centriole	A single layer of small mitochondria with visible crests	Axoneme conformation 9 + 2 with 2 and 3 lateral fins
A. affinis	Highly condensed chromatin with predominance of small interchromatin spaces, few electron lucent	Very shallow nuclear fossa, with a centriolar complex outside it	A single layer of big very electron-dense mitochondria, with crests inconspicuous	Axoneme conformation 9 + 2 with 2 and 3 lateral fins
A. nigripinnis	Highly condensed chromatin with predominance of big electron-lucent interchromatin spaces	Very deep nuclear fossa housing the proximal centriole	A single layer of small few electron-dense mitochondria with crests clearly visible	Axoneme conformation 9 + 2 with 1, 2, and 3 lateral fins
A. melanoorus	Chromatin with granular appearance and numerous electron-lucent interchromatin spaces	Very shallow nuclear fossa, with a centriolar complex outside it	A single layer of small electron-dense mitochondria, with visible crests	Axoneme conformation 9 + 2 with 2 and 3 lateral fins
A. bellottii	Homogeneous, highly condensed chromatin with predominance of small electron-lucent interchromatin spaces, few numerous	Very deep nuclear fossa housing the centriolar complex with fibrillar material associated	A single layer of small few electron-dense mitochondria with crests clearly visible	Axoneme conformation 9 + 2 with 1, 2, and 3 lateral fins
A. charrua	Chromatin with granular appearance and numerous electron-lucent interchromatin spaces of different sizes	Very shallow nuclear fossa, with a centriolar complex outside it and with associated fibrillar material	A single layer of big mitochondria with visible crests	Axoneme conformation 9 + 2 with 1, 2, 3, and 4 lateral fins
A. reicherti	Chromatin with granular appearance and numerous electron-lucent interchromatin spaces of different sizes	Very shallow nuclear fossa, with a centriolar complex outside it and with associated fibrillar material	A single layer of big mitochondria with crests inconspicuous	Axoneme conformation 9 + 2 with 1, 2, and 3 lateral fins

(Figure 3.2d and f). *Austrolebias luteoflammulatus* and *A. vazferreirai* also have distinct nuclear fossa, although it is less deep and houses the proximal centriole (Figure 3.2a and b). In the other species studied, the nuclear fossa is quite shallow, and the centrioles are not in the fossa but in the cytoplasm facing the nuclear fossa. The centrioles of *A. bellottii*, *A. reicherti*, and *A. charrua* are surrounded by osmiophilic fibers that are more distinct and arranged radially in *A. bellottii* (Figure 3.2f–h).

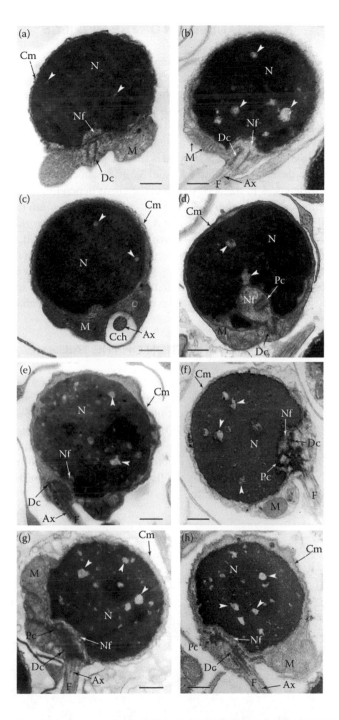

Figure 3.2 Transmission electron microscopy (TEM) micrographs of the head and midpiece of the sperm of eight *Austrolebias* species. All of them display a spherical head, a round nucleus with homogenously compacted chromatin interrupted by electron-lucent areas, (arrowheads) surrounded by a cytoplasmic membrane (Cm), a nuclear fossa (Nf) that in some species contains the proximal centriole (Pc); the distal centriole (Dc) serves as the base for the classical (9 + 2) axoneme (Ax); the midpiece is formed by a ring of mitochondria (M) separated from the flagellum (F) by a cytoplasmic channel (Cch). (a) *A. luteoflammulatus*; (b) *A. vazferreirai*; (c) *A. affinis*; (d) *A. nigripinnis*; (e) *A. melanoorus*; (f) *A. bellottii*; (g) *A. charrua*; (h) *A. reicherti*. Scale bars = 500 nm.

3.5 ORGANIZATION OF THE MIDPIECE

The sperm cells of *Austrolebias* have a short midpiece with mitochondria arranged in a single layer surrounding the base of the flagellum, characteristics common to fishes with external fertilization. These mitochondria are usually spherical and vary in number and diameter within the same individual. All the mitochondria of the midpiece appear to be disposed in a single plane, very close to one another and to the nuclear surface. Their distribution around the flagellum seems to be determined by the angle of insertion of the flagellum to the head. Transmission and scanning electron microscopy show that if the insertion is oblique, the number of mitochondria is smaller in the region where the flagellum forms an acute angle, relative to the nucleus (Figures 3.2 and 3.3). The subcellular analysis of the mitochondrial structure shows differences in the appearance of the crests depending on the species. The mitochondria of *A. luteoflammulatus*, *A. charrua*, *A. affinis*, and *A. reicherti* are usually larger and have barely visible crests (Figure 3.2a, c, g, and h), while *A. bellottii*, *A. vazferreirai*, *A. nigripinnis*, and *A. melanoorus* have small mitochondria with clearly visible crests (Figure 3.2b, d, e, and f).

While the mitochondrial subcellular morphology described here corresponds to sperm obtained from adult specimens collected during the breeding season, it is not yet known whether these differences in size and general appearance of the mitochondrial cristae has a species-specific value or reflects a particular metabolic state. More studies with the corresponding statistical analysis are necessary.

3.6 ORGANIZATION OF THE TAIL

The sperm tail of *Austrolebias* has a ribbon shape rather than the more common cylindrical shape. This is because the cell membrane side extensions have the form of two or more fins flanking the entire axonemal length. These fins are present in several other species of marine and freshwater fishes (Billard, 1978, 1983; Dreanno et al., 1999). The flagellum consists of an axoneme with the classical disposition of nine doublets of peripheral and a central pair of microtubules (9 + 2 pattern) surrounded by a cell membrane that strikingly increases its surface as it continues into lateral extensions, forming "fins" that vary in width along the flagellum. The number of side fins varies among species as well as within a single individual.

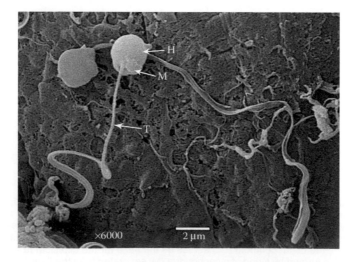

Figure 3.3   General view of *Austrolebias charrua* spermatozoa by scanning electron microscopy (SEM). Small spherical head (H), short midpiece (M), and tail (T) with a single flagellum.

The occurrence of two flagellar fins prevails among studied species. However, *A. luteoflammu-latus*, *A. affinis*, *A. vazferreirai*, and *A. melanoorus* have sperm cells with two and three flagellar fins (Figure 3.4a, b, c, and e), whereas *A. reicherti*, *A. bellottii*, and *A. nigripinnis* have tails with one, two, and three flagellar fins (Figure 3.4d, f, and h). Exceptionally, we observed up to four flagellar fins in the sperm of *A. charrua* (Figure 3.4g).

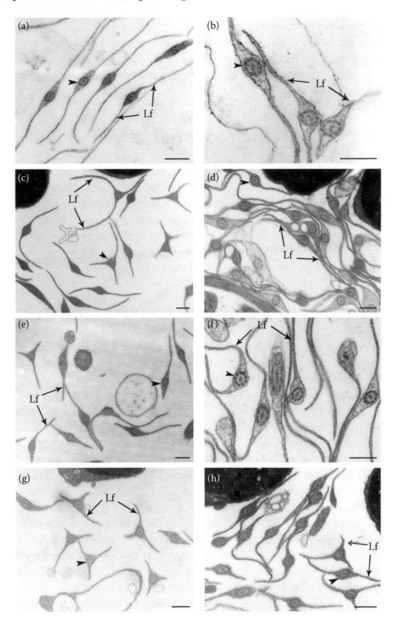

Figure 3.4 Cross sections of the tail of eight *Austrolebias* species (TEM). *Austrolebias affinis*, *A. luteo-flammulatus*, *A. melanoorus*, and *A. vazferreirai* display flagella with 2 or 3 lateral fins; whereas *A. reicherti*, *A. bellottii*, and *A. nigripinnis* show flagella with 1, 2, and 3 lateral fins; in *A. charrua* flagella have 1, 2, 3, and 4 lateral fins (data not shown). In all the cases the flagella with two fins are the most common. All the species present a classical (9 + 2) axoneme (arrowheads), with lateral fins (Lf). (a) *A. luteoflammulatus*; (b) *A. vazferreirai*; (c) *A. affinis*; (d) *A. nigripinnis*; (e) *A. melanoorus*; (f) *A. bellottii*; (g) *A. charrua*; (h) *A. reicherti*. Scale bars = 500 nm.

The role of the lateral fins of the sperm has been previously discussed by Cosson (2008) and Gillies et al. (2013), who argue that these structures may contribute to the efficient movement of sperm and also to the large increase in membrane surface (Cosson, 2008).

The characterization of the subcellular morphology of the sperm cells of the eight *Austrolebias* species studied allowed the identification of a set of typical features for each species. These features, together with karyotype and molecular approaches, provide complementary data for traditional morphological approaches to understanding phylogenetic relationships. Moreover, the development of complementary studies to understand the sperm activation physiology, coupled with the morphological characteristics described here, could contribute to a better understanding of the gamete interactions, encouraging further study about its relationship with the habitat and with the reproductive behavior of *Austrolebias*.

3.7 EGG ENVELOPE

The oocytes of annual fishes possess an egg envelope (also named chorion, eggshell, zona pellucida, or zona radiata) common to vertebrates and made of conserved glycoproteins of the ZP family (Litscher and Wassarman, 2007). The egg envelope is important in fertilization and embryonic development (Dumont and Brummet, 1980). Considering that annual fish embryos are subject to long dry periods buried in the mud, the protective function of this structure seems to be particularly important for survival. Herein, we will describe the ultrastructure of the egg envelope, in oocytes and embryos, in relation to its functions and evolutionary significance.

3.8 EGG ENVELOPE ORGANIZATION WITHIN THE OVARY

Histological analysis of *Austrolebias*' mature ovary shows follicles in every stage of development consistent with a daily spawning pattern. Germ cells are classified according to their size, nuclear and cytoplasmic morphology, and presence and structure of the egg envelope (Arezo et al., 2007). Oogonia are the smallest germ cells; they are grouped in clusters and free of follicular cells. Previtellogenic oocytes (pre-VO) are larger than oogonia, are associated with follicular cells, and have a basophilic cytoplasm and a nucleus with numerous nucleoli (Figure 3.5a). Vitellogenic oocytes (VO) undergo successive stages. In early VO, the cytoplasm has unstained droplets that correspond to yolk lipids (Figure 3.5a). Subsequently, a fine granular acidophilic material accumulates among these droplets and progressively fuses to form a fluid protein yolk that displaces the yolk lipids toward the cortex (Arezo et al., 2007; Figure 3.5a). This agrees with the unique fluid yolk characteristic previously reported for other atherinomorph fishes (Parenti and Grier, 2004).

An egg envelope is evident in VO. This structure, located between the cell membrane of the oocyte and the follicular cells, is acidophilic and eosin positive and appears radially striated (Figure 3.5a and b). Hair-like structures are seen on the surface of the chorion among the follicular cells (Figure 3.5b); these structures are refractive when observed in live eggs with bright-field light microscopy. Mature oocytes (fully grown) are the largest cells in the ovary, with the cytoplasm filled with acidophilic yolk fluid and yolk lipids and the nucleus displaced toward the periphery; they are surrounded by a thick egg envelope (Figure 3.5a). Finally, *A. charrua* atretic follicles show a disorganized granular yolk and fragmented egg envelope (unpublished data). This morphology has been previously described in other teleosts (Vizziano and Berois, 1990; Miranda et al., 1999; Santos et al., 2008).

Transmission electron microscope (TEM) analysis of the oocyte egg envelope shows several features related to the process of deposition. Egg envelope deposition begins in pre-VO in association with microvilli (Anderson, 1967; Wourms, 1976). Cell extensions across the egg envelope, presumably oocyte microvilli in contact with the overlying follicular cells, were reported in *Cynopoecilus*

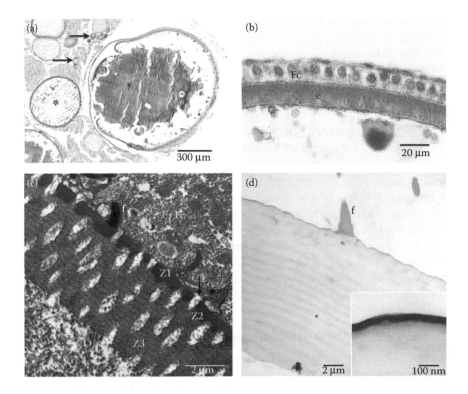

Figure 3.5   Egg envelope of *A. charrua*; (a) and (b) are light histological micrographs (hematoxylin–eosin technique) showing the organization of adult ovary of the species; (a) previtellogenic oocytes (arrows), vitellogenic oocyte containing droplets of lipid yolk (asterisk) and fully grown oocyte (right) with fluid yolk (y) and (b) egg envelope of fully grown oocyte (e), note the radial striation and the filaments among the follicular cells (Fc). (c) Transmission electron microscopy (TEM) micrograph of the egg envelope of a developing oocyte in ovary. Note the three concentric layers of different thickness and electron density (Z1, Z2, Z3). An electron-dense hair-like filament (f) appears on the outer surface. Microvilli (asterisk) cross the envelope and seem to contact with the surrounding follicular cells (arrow). (d) TEM micrograph of an advanced developing embryo's egg envelope; note the generally homogeneous appearance with concentric stripes of low electron density. Note scar-like structures (asterisk), a hair-like filament (f) and the outer thin electron-dense layer (inset).

(Wourms, 1976) and here in *Austrolebias* (Figure 3.5c). Both follicular and oocyte microvilli crossing the egg envelope and establishing cell–cell junctions were reported in Perciformes (Berois et al., 2007; Le Menn et al., 2007). These structures could explain the radial striation observed under light microscopy (Figure 3.5b). As mentioned above, the egg envelope in teleosts consists of ZP proteins. These proteins are synthesized in the liver or the ovary depending on the species (Modig et al., 2007). Preliminary evidence in *A. charrua* suggests that two cDNA, *achzpL* and *achzpH* (AN: KP083410, KP083411), found in liver with homology to *zp* genes are involved (N.G. Papa et al. unpublished data; see Chapter 2). In teleosts with a liver expression of the ZPs, the proteins are secreted and transported in the bloodstream to the ovary (Hamazaki et al., 1985; Modig et al., 2007).

The egg envelope of late developing oocytes shows three different layers: Z1, an outer, homogeneous and electron-dense layer; Z2, a medial and more electron-dense layer than the former; and Z3, the innermost, a heterogeneous and less electron-dense zone (Arezo et al., 2007; Figure 3.5c). Z3 is the thickest layer and the last to be deposited during choriogenesis. The entire envelope is crossed by channels occupied by microvilli (Figure 3.5c). Light and TEM micrographs of the egg envelope of developing oocytes show hair-like structures (electron dense in TEM) projecting toward and among the follicular cells (Figure 3.5b and c).

The three-layered structure disappears during the process of maturation and spawning, as previously described for teleosts (Anderson, 1967; Dumont and Brummet, 1980; Berois et al., 2007; Modig et al., 2007). In spawned oocytes, the envelope consists of two types of bands: wide and weak electron-dense bands that alternate with narrow and strongly electron-dense bands (Berois et al., 2007, 2011)

3.9 EGG ENVELOPE ORGANIZATION AND SURFACE ORNAMENTATIONS IN EMBRYOS

TEM micrographs show a stripped pattern in developing embryos (Figure 3.5d). This pattern consists of concentric stripes of low electron density. At higher magnifications, a thin, less than 100-nm-wide, more electron-dense outer layer can be distinguished (Figure 3.5d inset). Concentric stripes were reported in *Nothobranchius korthausae* (Schoots et al., 1982), where the thickness of the stripes decreases gradually toward the surface. These authors also report a very thin outer electron-dense layer similar to the one we are describing for *Austrolebias*. Scar-like structures are seen across the egg envelope, reminiscent of the channels formerly occupied by oocyte and follicular cell microvilli (Figure 3.5d). These observations of the *A. charrua* egg envelope were made on late developing embryos; it would be interesting to analyze the chorion of embryos at different developmental stages to track and understand these structural changes.

The ornamentation of the fish's egg envelope correlates with ecological features related to reproductive strategies (Rizzo et al., 2002). The chorion plays an important role in the interaction of embryos with the environment: providing mechanical resistance, allowing gas exchange, and helping to keep the eggs associated to the substrate (Wourms and Sheldon, 1976; Riehl and Patzner, 1998). Furthermore, fungicidal and bactericidal properties for the eggshell were reported for teleosts (Kudo and Inoue, 1989). The resistance of the egg envelope of annual fishes has been described previously (Wourms, 1967). Matias (1984) reported differential chorion resistance to enzymatic action during the three diapauses on the African genus *Nothobranchius*, suggesting changes in egg envelope structure during embryo development and participation of the egg envelope in diapause stages.

The association with the substrate is particularly important in annual fishes for survival of the embryos during the dry season. Previous studies analyzed the resistance of *Austrofundulus limnaeus* embryos at different stages of development exposed to different desiccating atmospheres (Podrabsky et al., 2001). Diapause II embryos are the most resistant to water loss, living more than 32 days in 50% relative humidity. Embryos at early stages of dispersion or reaggregation showed both less resistance than embryos in diapause II and more than embryos at diapause III. Diapause III embryos dehydrate and die in a few days depending on the relative air humidity. The perivitelline space disappears due to water loss during desiccation, but the embryo proper is not dehydrated (Podrabsky et al., 2001). Consequently, the latter evidence supports the authors' suggestion that the syncytial-enveloping layer of the embryo may function as a barrier to water loss (Podrabsky et al., 2001). In our work, *A. charrua* diapause III embryos showed very low resistance to desiccation, dying after 4 days exposed to ambient humidity. Thus, the egg envelope in itself would not prevent water loss, and probably contact of the chorion with the substrate in nature may provide the necessary humidity for survival. Taking that into account, to complete the life cycle of annual fishes in the laboratory, it is necessary to reserve embryos in a slightly humid substrate to mimic natural conditions (Chapter 7).

Species of annual fishes have distinct egg envelope ornamentations that are complex, variable, and detectable by light and electron microscopy (Wourms, 1976; Loureiro and de Sá, 1996, 2000; Fava and Toledo-Piza, 2007; Arezo et al., 2007; García et al., 2009; Berois et al., 2011, 2012). *Austrolebias* has various patterns of hair-like filaments; as an example, the egg envelope surfaces of two species of *Austrolebias* are shown that differ in length, diameter, and curvature

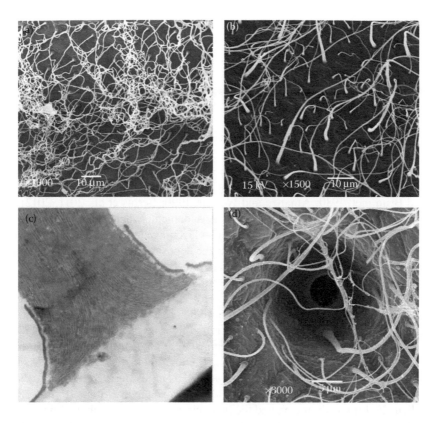

Figure 3.6 Egg envelope surface patterns of filaments of *A. affinis* (a) and *A. charrua* (b) obtained with scanning electron microscope (SEM). (c) Internal structure of one *A. charrua* chorion filament (transmission electron microscopy TEM): note the alternated stripes of different electron densities and the thin electron-dense layer covering the base of the filament. (d) Micropyle of *A. charrua* early fertilized egg: note the funnel shaped vestibule and the deeper circular opening (SEM). The images shown are from fertilized eggs' envelopes where best images are obtained.

(Figure 3.6a and b). The filaments of *A. affinis* are curly and have regular diameters at their bases; on the other hand, *A. charrua* filaments are straight and longer, with a variety of diameters and lengths (Figure 3.6a and b). Overall in *Austrolebias*, the distribution, thickness, and length of these filaments are diverse among species (Loureiro and de Sá, 1996, 2000; García et al., 2009). These filaments are homogeneously electron dense at low TEM magnifications (Figure 3.5c and d). At higher magnifications, the internal structure of these filaments appears to be organized into alternating stripes of different electron densities (Figure 3.6c). A similar structure was reported in the genus *Fundulus*, a nonannual cyprinodontiform (Anderson, 1966). The nature of the stripes is unknown, although electron-dense stripes may correspond to fibrillar proteins. A molecular approach could shed light on this matter.

In *Cynopoecilus*, *Leptolebias*, and *Simpsonichthys*, diverse "palm-like" structures (also called "mushroom-like" structures) and large, rounded protuberances with a regular arrangement have been described. These palm-like structures are conical or cylindrical projections that end with a crown of filaments (Wourms, 1976; Fava and Toledo-Piza, 2007; Costa and Leal, 2009). Interspecific variation in the length of the palm-like structures relative to egg diameter was reported in *Simpsonichthys* (Fava and Toledo-Piza, 2007). Large and rounded protuberances were described in *Nematolebias whitei* (Fava and Toledo-Piza, 2007; Loureiro and de Sá, 1996). The genus *Leptolebias* has surface ornamentations consisting of irregular pentagons or hexagons (Costa and Leal, 2009). Some *Nematolebias* species show both types of chorion ornamentation: surface and palm-like ornamentations (Costa and Leal, 2009). *A. myersi* lacks surface projections, while *Nothobranchius guentheri* and *N. korthausae* have them (Schoots et al., 1982; Wourms, 1967). The combination of ornamentation characteristics has been proposed as species specific and could be a complement when building and resolving phylogenies (Loureiro and de Sá, 1996; García et al., 2009; Fava and Toledo-Piza, 2007).

3.10 MICROPYLE

The teleostean micropyle is a unique oocyte structure among vertebrates. It is a small pore on the surface of the chorion, located at the animal pole, through which the sperm enters the oocyte (Dumont and Brummet, 1980; Amanze and Iyengar, 1990; Berois et al., 2011, 2012). It is formed concomitantly with the chorion during oocyte development, when the micropilar cell (or cells, depending on the species) derived from the follicular layer invades the envelope. The micropilar cell retains contact with the oocyte and thus prevents the deposition of the fibrillar material that forms the chorion (Wourms, 1976; Wallace and Selman, 1981; Le Menn et al., 2007).

Previous studies showed that the micropyle functions as a sperm-guiding system in teleosts; morphologically it is a funnel-shaped vestibule on the chorion surface (Dumont and Brummet, 1980) and there are grooves or ridges around it in some species (Amanze and Iyengar, 1990; Rizzo et al., 2002). In the micropyle or in the area surrounding it, glycoproteins attract the spermatozoa toward the orifice (Iwamatsu et al., 1997; Yanagimachi et al., 2013). The grooves may guide the spermatozoa as well (Amanze and Iyengar, 1990). The micropyle of *Austrolebias* is visible in SEM micrographs on the surface of spawned oocytes as well as in just fertilized eggs. In early fertilized eggs of *A. charrua*, the micropyle is a tunnel with a funnel-shaped vestibule (Figure 3.6d); the outer opening is about 20.4 µm in diameter, and the inner circular opening is about 4.6 µm. In addition to the probable presence of species-specific glycoproteins, the size of the micropyle may be a barrier to hybridization among species inhabiting the same ephemeral ponds. *Austrolebias charrua* is a species that shows morphological and molecular variable features (García et al., 2009). Consequently, a more extensive study of micropyle size within and among populations of *A. charrua* may be interesting to test its correlation to the reported morphological and molecular variation in the species. The micropyle of *Nematolebias* and *Simpsonichthys* is surrounded by concentric rows of spiny projections that are absent in *Austrolebias* (Fava and Toledo-Piza, 2007). The egg envelope and the micropyle undergo changes during fertilization, when the micropyle is plugged (Coward et al., 2003). Morphological changes and fate of the micropyle of annual fishes are unknown.

3.11 CONCLUSION

Considering that *Austrolebias* species have external fertilization, it is reasonable that the sperm tail plays a key role in reaching and fertilizing the egg. Correlating sperm, oocyte, and micropyle morphologies may identify species specific and phylogenetic informative biological traits. The size of the sperm head may be directly correlated with the size of the inner opening of the micropyle system. Species-specific molecular interactions among sperm and micropyles of teleosts have been reported (Yanagimachi et al., 2013), but this remains to be studied in annual fishes. Egg fertilization in annual fishes occurs underground in a very "muddy" environment; for this reason, it would be interesting to evaluate morphological and molecular gamete adaptations to this unique context. Hybridization, at least under laboratory conditions, has been reported among *Austrolebias* species (see Chapter 17), suggesting that molecular and behavioral barriers among species may be weak under certain conditions. Gamete morphology and gamete–oocyte interactions may contribute to understanding of speciation and reproductive isolation barriers in annual fishes.

ACKNOWLEDGMENTS

We thank the book editors for inviting us to present our work on Neotropical annual fishes; a fascinating group for evolutionary biology. We thank Martín Astesiano for his assistance with Figures 3.5 and 3.6. We also thank Lic. Camila Davison, Msc. Marcie Jiménez, and Professor Damian Pisoni for their collaboration on language revision. Finally, we thank Dr. Nibia Berois for her research guidance and support over the years.

Note: MR and GCL are responsible for the work on sperm ultrastructure, JM, GAC, and NGP are responsible for the work on egg ultrastructure. All authors contributed equally to the work.

REFERENCES

Amanze, D., and A. Iyengar. 1990. The micropyle: A sperm guidance system in teleost fertilization. *Development* 109:495–500. PMID:2401208

Anderson, E. 1966. A study of the fibrillar appendages associated with the surface of eggs of the killifish, *Fundulus heteroclitus*. *The Anatomical Record, 154,* 308–309. In: Wourms, J. P., and Sheldon, H. 1976. *Annual Fish Oogenesis II. Formation of the Secondary Egg Envelope. Developmental Biology* 50:355–356. doi: 10.1016/0012-1606(76)90157-3

Anderson, E. 1967. The formation of the primary envelope during oocyte differentiation in teleosts. *Journal of Cell Biology* 35:193–212. doi: 10.1083/jcb.35.1.193

Arezo, M.J., S. D'Alessandro, N.G. Papa, R.O. de Sá, and N. Berois. 2007. Sex differentiation pattern in the annual fish *Austrolebias charrua* (Cyprinodontiformes: Rivulidae). *Tissue and Cell* 39:89–98. doi: 10.1016/j.tice.2007.01.004

Baccetti, B. 1991. *Comparative Spermatology 20 years After*. Raven Press, New York.

Berois, N., M.J. Arezo, C. Chalar, M.M. Brauer, and C. Barros. 2007. Ultrastructure and protein composition of the oocyte envelope in the whitemouth croaker (*Micropogoniasfurnieri*, Desmarest, 1823, Sciaenidae, Perciformes). *Journal of Applied Ichthyology* 23:34–39. doi: 10.1111/j.1439-0426.2006.00762.x

Berois, N., M.J. Arezo, and R.O. de Sá. 2014. The neotropical genus *Austrolebias*: An emerging model of annual killifishes. *Cell & Developmental Biology* 3:1–9. doi: 10.4172/2168-9296.1000136

Berois, N., M.J. Arezo, and N.G. Papa. 2011. Gamete interactions in teleost fish: The egg envelope. Basic studies and perspectives as environmental biomonitor. *Biological Research* 44:119–124. doi: 10.4067/s0716-97602011000200002

Berois, N., M.J. Arezo, N.G. Papa, and G.A. Clivio. 2012. Annual fish: Developmental adaptations for an extreme environment. *Wiley Interdisciplinary Reviews. Developmental Biology* 1:595–602. doi: 10.1002/wdev.39

Billard, R. 1978. Changes in structure and fertilizing ability of marine and fresh water fish spermatozoa diluted in media of various salinities. *Aquaculture* 14:187–198. doi: 10.1016/0044-8486(78)90094-7

Billard, R. 1983. Ultrastructure of trout spermatozoa: Changes after dilution and deep-freezing. *Cell and Tissue Research* 228:205–218. doi: 10.1007/BF00204873

Cassel, M., A. Ferreira, and M. Mehanna. 2014. Ultrastructural features of spermatogenesis in *Melanorivulus punctatus* (Cyprinodontiformes: Rivulidae). *Micron* 62:1–6. doi: 10.1016/j.micron.2014.02.012

Cosson, J.J. 2008. The motility apparatus of fish spermatozoa. In *Fish Spermatology*, ed. S.M.H. Alavi et al., pp. 281–316. Oxford, UK: Alpha Science Ltd.

Costa, W.J.E.M. 2004. A new killifish genus and species from the coastal plains of north-eastern Brazil (Teleostei: Cyprinodontiformes: Rivulidae). *Zootaxa* 642:1–10.

Costa, W.J.E.M. 2006. The South American annual killifish genus *Austrolebias* (Teleostei: Cyprinodontiformes: Rivulidae): Phylogenetic relationships, descriptive morphology and taxonomic revision. *Zootaxa* 1213:1–162. ISSN 1175–5334.

Costa, W.J.E.M., and F. Leal. 2009. Egg surface morphology in the Neotropical seasonal killifish genus *Leptolebias* (Teleostei: Aplocheiloidei: Rivulidae). *Vertebrate Zoology* 59:25–29.

Coward, K., N.R. Bromage, O. Hibbitt, and J. Parrington. 2003. Gamete physiology, fertilization and egg activation in teleost fish. *Reviews in Fish Biology and Fisheries* 12:33–58. doi: 10.1023/A:1022613404123

Dreanno, C., J. Cosson, M. Suquet, F. Seguin, G. Dorange, and R. Billard. 1999. Nucleotide content, oxidative phosphorylation, morphology, and fertilizing capacity of turbot (*Psetta maxima*) spermatozoa during the motility period. *Molecular Reproduction and Development* 53:230–243. doi: 10.1002/(SICI)1098-2795(199906)53:2<230::AID-MRD12>3.0.CO;2-H

Dumont, J.N., and A.R. Brummet. 1980. The vitelline envelope, chorion, and micropyle of *Fundulus heteroclitus* eggs. *Gamete Research* 3:25–44. doi: 10.1002/mrd.1120030105

Fava, D., and M. Toledo-Piza. 2007. Egg surface structure in the annual fishes *Simpsonichthys* (subgenera *Ophthalmolebias* and *Xenurolebias*) and *Nematolebias* (Teleostei: Cyprinodontiformes: Rivulidae): Variability and phylogenetic significance. *Journal of Fish Biology* 71:889–907. doi: 10.1111/j.1095-8649.2007.01572.x

García, G., M. Loureiro, N. Berois, M.J. Arezo, G. Casanova, G. Clivio, and Olivera. 2009. Pattern of differentiation in the annual killifish genus *Austrolebias* (Cyprinodontiformes: Rivulidae) from a biosphere reserve site in South America: A multidisciplinary approach. *Biological Journal of the Linnean Society* 98:620–635. doi: 10.1111/j.1095-8312.2009.01303.x

Gillies, E.A., V. Bondarenko, J. Cosson, and A.A. Pacey. 2013. Fins improve the swimming performance of fish sperm: A hydrodynamic analysis of the Siberian sturgeon *Acipenser baerii*. *Cytoskeleton* 70:85–100. doi: 10.1002/cm.21093

Grier, H.J. 1981. Cellular organization of the testis and spermatogenesis in fishes. *American Zoologist* 21:345–357. doi: 10.1093/icb/21.2.345

Grier, H.J., L.R. Linton, J.F. Leatherland, and V.L. De Vlaming. 1980. Structural evidence for two different testicular types in teleost fishes. *American Journal of Anatomy* 159:331–345. doi: 10.1002/aja.1001590307

Hamazaki, T., I. Iuchi, and K. Yamagami. 1985. A spawning female-specific substance reactive to anti-chorion (egg envelope) glycoprotein antibody in the teleost Oryziaslatipes. *Journal of Experimental Zoology* 235:269–279. doi: 10.1002/jez.1402350214

Iwamatsu, T., N. Yoshizaki, and Y. Shibata. 1997. Changes in the chorion and sperm entry into the micropyle during fertilization in the teleostean fish, *Oryziaslatipes*. *Development Growth Differentiation* 39:33–41. doi: 10.1046/j.1440-169x.1997.00005.x

Jamieson, B.G.M. 1991. *Fish Evolution and Systematics: Evidence from Spermatozoa. With a Survey of Lophophorate, Echinoderm and Protochordale Sperm and an Account of Gamete Cryopreservation.* (Chapters 5 and 17). Cambridge University Press, Cambridge.

Jamieson, B.G.M., and H.J. Grier. 1993. Influences of phylogenetic position and fertilization biology on spermatozoal ultrastructure exemplified by exocoetoid and poeciliid fish. *Hydrobiologia* 271:11–25. doi: 10.1007/bf00005691

Kudo, S., and M. Inoue. 1989. Bacterial action of fertilization envelope extract from eggs of the fish *Cyprinus carpio* and *Plecoglossus altivelis*. *Journal of Experimental Zoology* 250:219–228. doi: 10.1002/jez.1402500214

Kweon, H.S., E.H. Park, and N. Peters. 1998. Spermatozoon ultrastructure in the internally self-fertilizing hermaphroditic teleost, *Rivulus marmoratus* (Cyprinodontiformes, Rivulidae). *Copeia* 4:1101–1106. doi: 10.2307/1447367

Lahnsteiner, F., and R.A. Patzner. 2008. Sperm morphology and ultrastructure in fish. In *Fish Spermatology*, eds. S.M.H. Alavi et al. 1–61. Oxford, UK: Alpha Science Ltd.

Le Menn, F., J. Cerdà, and P.J. Babin. 2007. Ultrastructural aspects of the ontogeny and differentiation of ray-finned fish ovarian follicles. In *The Fish Oocyte: From Basic Studies to Biotechnological Applications*, eds. J. Cerdà, P.J. Babin, and E. Lubzens, 1–38. Dorecht, the Netherlands: Springer. doi: 10.1007/978-1-4020-6235-3_1

Litscher, E.S., and P.M. Wassarman. 2007. Egg extracellular coat proteins: From fish to mammals. *Histology and Histopathology* 22:337–347.

Loureiro, M., and R.O. de Sá. 1996. External morphology of the chorion of the annual fishes *Cynolebias* (Cyprinodontiformes: Rivulidae). *Copeia* 4:1016–1022. doi: 10.2307/1447669

Loureiro, M., and R.O. de Sá. 2000. Ornamentación del corion en los huevos de especies de *Austrolebias* y *Sympsonichthys* (Rivulidae: Cynolebiatinae). *FACENA* 16:85–91.

Loureiro, M., A. Duarte, and M. Zarucki. 2011. A new species of *Austrolebias*, Costa (Cyprinodontiformes: Rivulidae) from northeastern Uruguay, with comments on distribution patterns. *Neotropical Ichthyology* 9:335–342. doi: 10.1590/S1679-62252011000200010

Matias, J.R. 1984. The stage-dependent resistance of the chorion to external chemical damage and its relationship to embryonic diapause in the annual fish, *Nothobranchiusguentheri*. *Experientia* 40:753–754. doi: 10.1007/bf01949761

Mattei, X. 1991. Spermatozoon ultrastructure and its systematic implications in fishes. *Canadian Journal of Zoology* 69:3038–3055. doi: 10.1139/z91-428

Miranda, A., N. Bazzoli, and E. Rizzo. 1999. Ovarian follicular atresia in two teleost species: A histological and ultrastructural study. *Tissue and Cell* 31:480–488. doi: 10.1054/tice.1999.0045

Modig, C., L. Westerlung, and P.E. Olsson. 2007. Oocyte zonapellucida proteins. In *The Fish Oocyte: From Basic Studies to Biotechnological Applications*, eds. J. Cerdà, P.J. Babin, and E. Lubzens, 113–139. Dordrecht, the Netherlands: Springer. doi: 10.1007/978-1-4020-6235-3_5

Parenti, L.R., and H.J. Grier. 2004. Evolution and phylogeny of gonad morphology in bony fishes. *Integrative and Comparative Biology* 44:333–348. doi: 10.1093/icb/44.5.333

Podrabsky, J.E., J.F. Carpenter, and S.C. Hand. 2001. Survival of water stress in annual fish embryos: Dehydration avoidance and egg envelope amyloid fibers. *The American Journal of Physiology. Regulatory, Integrative and Comparative Physiology* 280:R123–R131. http://ajpregu.physiology.org/content/280/1/R123.

Riehl, R., and R.A. Patzner. 1998. Minireview: The modes of egg attachment in teleost fishes. *Italian Journal of Zoology* 65(Suppl.1):415–420. doi: 10.1080/11250009809386857

Rizzo, E., Y. Sato, B.P. Barreto, and H.P. Godinho. 2002. Adhesiveness and surface patterns of eggs in neotropical freshwater teleosts. *Journal of Fish Biology* 61:615–632. doi: 10.1016/s0022-1112(02)92085-4

Santos, H.B., R.G. Thomé, F.P. Arantes, Y. Sato, N. Bazzoli, and E. Rizzo. 2008. Ovarian follicular atresia is mediated by heterophagy, autophagy, and apoptosis in *Prochilodus argenteus* and *Leporinus taeniatus* (Teleostei: Characiformes). *Theriogenology* 70:1449–1460. doi: 10.1016/j.theriogenology.2008.06.091

Schoots, A.F.M., J.J.M. Stikkelbroeck, J.F. Bekhuis, and Denucé. 1982. Hatching in teleostean fishes: Fine structural changes in the egg envelope during enzymatic breakdown *in vivo* and *in vitro*. *Journal of Ultrastructure Research* 80:185–196. doi:10.1016/S0022-5320(82)90017-X

Schulz, R.W., L.R. de França, J.J. Lareyre et al. 2010. Spermatogenesis in fish. *General and Comparative Endocrinology* 165:390–411. doi: 10.1016/j.ygcen.2009.02.013

Thiaw, O.T., X. Mattei, R. Romand, and B. Marchand. 1986. Reinvestigation of spermatic flagella structure: The teleostean Cyprinodontidae. *Journal of Ultrastructure and Molecular Structure Research* 97:109–118. doi: 10.1016/s0889-1605(86)80011-8

Vizziano, D., and N. Berois. 1990. Histología del ovario de *Macrodonancylodon* (Bloch y Schneider, 1801) Teleostei: Sciaenidae. Ovogénesis. Folículos post-ovulatorios. Atresia. *Revista Brasileira de Biología* 50:523–536.

Wallace, R.A., and K. Selman. 1981. Cellular and dynamic aspects of oocyte growth in teleosts. *American Zoologist* 21:325–343. doi: 10.1093/icb/21.2.325

Wourms, J.P. 1967. Annual fishes. In *Methods in Developmental Biology*, eds. F.H. Wilt and N. Wessells, 123–137. New York: Thomas and Crowell Company. doi:10.1086/406095

Wourms, J.P. 1976. Annual fish oogenesis I. differentiation of the mature oocyte and formation of the primary envelope. *Developmental Biology* 50:338–354. doi:10.1016/0012-1606(76)90156-1

Wourms, J.P., and H. Sheldon. 1976. Annual fish oogenesis II. Formation of the secondary egg envelope. *Developmental Biology* 50:355–356. doi:10.1016/0012-1606(76)90157-3

Yanagimachi, R., G. Cherr, T. Matsubara et al. 2013. Sperm attractant in the micropyle region of fish and insect eggs. *Biology of Reproduction* 88:1–11. doi: 10.1095/biolreprod.112.105072

CHAPTER 4

Alternative Developmental Pathways

Jason E. Podrabsky, Amie L. Romney, and Kristin M. Culpepper

CONTENTS

4.1 INTRODUCTION

Annual killifishes exploit ephemeral and marginal habitats that experience a great deal of spatial and temporal variation. However, to describe their habitat as unpredictable would be a gross oversimplification. In reality, there are patterns of rather predictable changes in their environment that are mixed with stochastic events that are essentially unpredictable. Both the predictable and unpredictable aspects of their environment have likely played a critical role in the evolution of the unique suite of characters that define annual killifishes. Since most, if not all, of the adaptations used to survive dehydrating conditions have been invested in embryonic stages, the production of drought-tolerant diapausing embryos is of key importance to survival of these species and likely has been under intense selection pressures.

4.1.1 Predictability in the Annual Killifish Environment

In many climates where annual killifishes are successful, rainfall patterns and the availability of inundated ponds are at least somewhat unpredictable. This may be due to a number of factors but is likely highly influenced by the microhabitat requirements of the fishes, long durations of drought punctuated by large rain events, and a complex interplay of rain events that hydrate the soils to the point that they can hold water. Some of the unpredictable nature comes from the properties of the vertisol clay and alluvial soils that appear to be a key requirement in the survival of embryos through the dry season. Depending on the length and intensity of the dry season, it may take several rain events to hydrate the soil to levels that will lead to standing water. Although the probability of a rain event occurring may be quite predictable on an annual basis, intervals between events and local intensity of rain events add a great deal of uncertainty to the overall seasonal predictability of pond inundation. Therefore, annual killifishes must respond to cues that provide general information about the probability that rain events will occur with an adequate intensity and frequency to leave standing water, but they must also be able to endure a great deal of temporal and spatial heterogeneity in their environment. Thus, it is not surprising that these species have evolved a diverse suite of life history traits that appear to maximize the probability of populations surviving in a variety of environmental conditions.

4.1.2 Facultative versus Obligate Diapause

Much attention has been paid in the literature in defining cessation of development in diapause as either facultative or obligate. In fact, some species of the African *Aphyosemion* group have been defined as "facultative annuals." As with many concepts in biology, these definitions represent the black-and-white extremes of a very gray phenotype landscape. It is highly probable that a true obligate diapause stage does not exist in annual killifishes. Instead, we would suggest that diapause can be obligate under certain environmental or ecological circumstances and thus is always facultative. This is almost certainly the case for entrance into diapause I, which appears to be induced by low temperatures, presence of adult fish in the water, and perhaps anoxia (Peters, 1965; Wourms, 1972; Levels et al., 1986b). Diapause III may be the closest thing to an obligate diapause stage that can be found in annual killifishes, as most embryos from most species will enter dormancy at diapause III for a variable, at least a short, length of time under a variety of environmental conditions (Wourms, 1972; Murphy and Collier, 1997). However, hatching can easily be induced in many species of annual killifishes by exposing the embryos to increased temperatures or hypoxia, and it appears to be easier to induce hatching during early diapause III compared with later (J. E. Podrabsky, personal observation; Levels and Denuce, 1988). Thus, even for diapause III, entrance into dormancy is only obligate for a defined range of environmental variables. Entrance into diapause II is dependent on both environmental and maternal factors and will be the focus of this chapter.

4.2 ALTERNATIVE DEVELOPMENTAL PATHWAYS

Embryos of annual killifishes can develop along two alternative developmental trajectories that Wourms referred to as the diapause and "escape" trajectories because the direct developing embryos escape dormancy in diapause II (Figure 4.1; Wourms, 1972; Podrabsky et al., 2010). These two trajectories differ in the time required to complete development, but more importantly they also differ in the relative timing of morphological and physiological development (Podrabsky et al., 2010; Furness et al., 2015), a process defined as heterokairy (Spicer and Rundle, 2007). Morphologically, the relative development of anterior and posterior structures is different during the segmentation period of development, the time when the somites (balls of cells that will contribute to the muscle and bone of the body) are being formed (Figure 4.2a). Physiologically, heart development and metabolic rate are both incredibly accelerated in escape embryos compared with those that will enter diapause II (Figure 4.2b; Podrabsky et al., 2010; Furness et al., 2015). This difference in developmental timing has been confirmed in species of annual killifishes from several groups in both Africa and South America that appear to have evolved diapause through convergence (Furness et al., 2015). This consistency across multiple evolutionary events suggests something fundamentally important or necessary to this unique shift in timing. The consistency of the timing and patterning of somite development (somitogenesis) in vertebrates is typically highly conserved within a species and thus is often used as an indicator of developmental stage. However, in annual killifish embryos developing

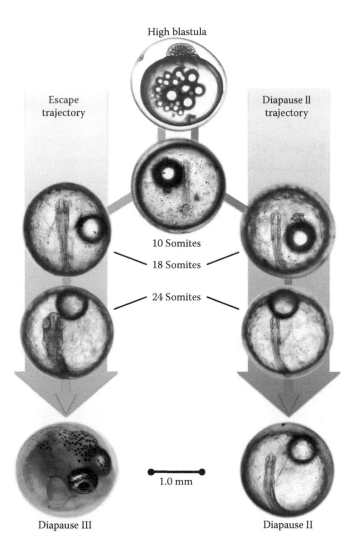

High blastula

Escape
trajectory

Diapause ll
trajectory

10 Somites

18 Somites

24 Somites

1.0 mm

Diapause III

Diapause II

Figure 4.1 Two alternative developmental pathways possible in embryos of annual killifishes. One pathway includes developmental and metabolic dormancy in diapause II, while on the other pathway embryos "escape" from diapause and develop directly to diapause III. Importantly, these two pathways appear to be indistinguishable prior to about the 10-somite stage of development. (Adapted from Podrabsky, J. E., I. D. F. Garrett and Z. F. Kohl. 2010. *Journal of Experimental Biology* 213:3280–3288.)

along the diapause and escape trajectories, the timing is so different that the two groups of embryos could easily be mistaken for different species.

4.3 PHYSIOLOGY OF THE DIAPAUSE AND ESCAPE TRAJECTORIES

Diapause is a state of profound metabolic depression in embryos of annual killifishes (Levels et al., 1986a; Podrabsky and Hand, 1999). During entrance into diapause II in *Austrofundulus limnaeus* the ontogenetic increase in DNA content normally associated with development is arrested, and for both diapause II and diapause III more than 90% of the cells arrest in the G_1 phase of the cell

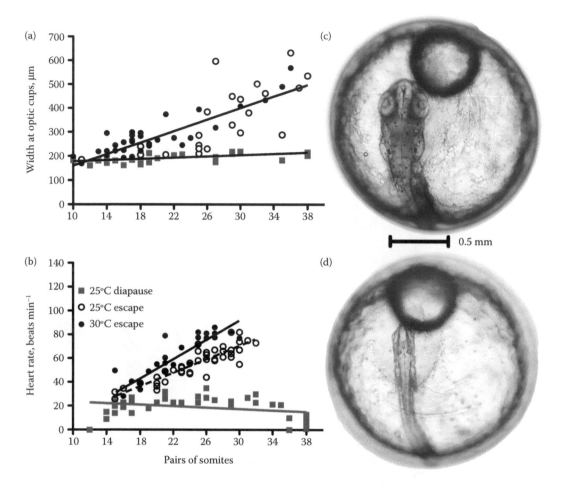

Figure 4.2 Morphological and physiological differences in *Austrofundulus limnaeus* embryos developing along the escape and diapause II trajectories. (a) When compared at the same number of somite pairs, development of anterior structures such as the head are advanced in escape embryos compared with embryos bound to enter diapause II. (b) Physiological development is advanced in escape embryos, as evidenced by increased heart rate in embryos with the same number of somite pairs. Heart rates were measured at a common temperature of 25°C, and thus it is also apparent that incubation at 30°C has an additional effect on increasing heart rate. Embryos enter diapause II with 38 pairs of somites in *A. limnaeus*, but escape at the 24-somite stage (c), and diapause II-bound (d) embryos are easily distinguished by their morphological and physiological differences. (Adapted from Podrabsky, J. E., I. D. F. Garrett and Z. F. Kohl. 2010. *Journal of Experimental Biology* 213:3280–3288.)

cycle (Podrabsky and Hand, 1999; Culpepper and Podrabsky, 2012; Meller et al., 2012). Heart rate is significantly reduced and sometimes completely absent during diapause II and is greatly reduced during diapause III (Podrabsky and Hand, 1999; Anderson and Podrabsky, 2014). In *A. limnaeus*, diapause II embryos have rates of oxygen consumption that are reduced by greater than 94% compared with the peak of prediapause II development during the dispersion and reaggregation stages and greater than 84% during prolonged sojourns in diapause III (Podrabsky and Hand, 1999). During diapause II and III, reduced metabolic rates are likely supported by reduced activity of mitochondrial respiratory complexes, especially the ATP synthase (Duerr and Podrabsky, 2010), and reduced (diapause III) or almost absent (diapause II) capacity for activity of ion-motive ATPases such as the Na$^+$-K$^+$-ATPase (Machado and Podrabsky, 2007). During diapause II, it is also known that a greater than 90% decrease in the rate of protein synthesis contributes to greater than 30% of the metabolic depression (Podrabsky and Hand, 2000).

The metabolic pathways active during diapause are currently not fully understood, but there is clearly a large contribution from anaerobic pathways even when embryos are held under aerobic conditions (Podrabsky and Hand, 1999). Embryos that enter into diapause II have a strong anaerobic poise to their metabolism with an estimated 50 times greater capacity for anaerobic metabolic pathways (lactate production) compared with aerobic metabolic pathways (citric acid cycle) (Chennault and Podrabsky, 2010). Given that fish embryos are known to rely heavily on the metabolism of yolk proteins to fuel development, it is not surprising that free amino acids accumulate during diapause (Podrabsky and Hand, 2000; Podrabsky et al., 2007), which may indicate a complex pattern of the use of some amino acids coupled to the production of others to fuel the small amount of metabolic activity required during diapause II (Podrabsky and Hand, 2015).

The physiology of escape embryos has received very little attention. From a metabolic rate perspective, the patterns of oxygen consumption appear to mimic those of other continuously developing fish embryos (Furness et al., 2015). The metabolic poise of escape embryos shifts from values around 50 to only about 25 times more capacity for anaerobic lactate production when compared with embryos entering diapause II (Chennault and Podrabsky, 2010). Thus, it appears that escape embryos are likely more dependent on aerobic metabolism earlier in development compared with embryos that will enter diapause II. The long-term effects of this shift in metabolic rate and metabolic poise at a relatively early stage in development may contribute to the observed differences in larval and adult physiology that are associated with escape embryos (see Section 4.4).

It is important to note that all of the metabolic data presented above were collected on embryos incubated under aquatic conditions, and it is highly likely that exposure to aerial conditions will further reduce metabolic rate and alter embryonic physiology in significant ways. The interplay between the endogenous regulation of metabolism during diapause and the shifts that may be imposed by aerial incubation or heat stress conditions experienced in their native habitat have yet to be explored in any detail and should be a rich area for future explorations.

4.4 MATERNAL EFFECTS ON DEVELOPMENTAL TRAJECTORY

Development along the escape or diapause trajectory can be regulated by maternal effects (Podrabsky et al., 2010). Female age (Podrabsky et al., 2010) and photoperiod (Markofsky and Matias, 1977; Markofsky et al., 1979) have both been illustrated to alter the proportion of embryos that enter or escape diapause II. In contrast, adult spawning temperature appears to have very little effect on developmental trajectory (Markofsky and Matias, 1977). Thus, it is clear that the environmental conditions experienced by adults can affect developmental trajectory. The mechanism by which this information experienced by the adults is integrated and transferred into the embryo has yet to be identified. It is clear that maternal provisioning or programming of some kind is involved, which leads to the need for a better understanding of the epigenetic influences that adults have on the production of gametes. We believe it is likely that the production of an RNA, protein, or hormone/small molecule packed into the embryo is essential for the production of diapausing embryos and that the absence of this signal leads to escape embryo production. This prediction, although complete conjecture at this point, is supported by the conclusion that the escape embryo developmental trajectory is likely the ancestral state (Furness et al., 2015). This promises to be an exciting area for future studies.

In laboratory stocks of A. limnaeus held under near-constant conditions (26–28°C, 14L:10D), there is a clear pattern for a greater production of escape embryos in young females and a nearly exclusive production of diapausing embryos in older females (Podrabsky et al., 2010). However, the pattern for any given female on any given day is nearly impossible to predict, and often a single female will produce mixed clutches of embryos developing along both developmental trajectories (Figure 4.3). This

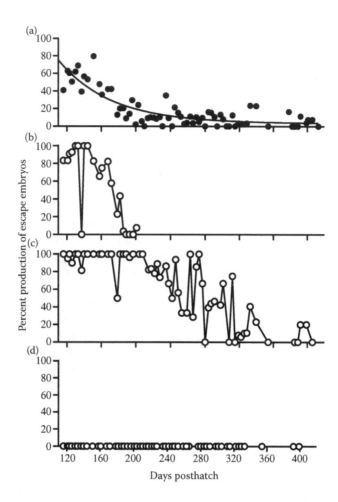

Figure 4.3 Maternal influences on developmental trajectory in *Austrofundulus limnaeus*. (a) Young females produce a high percentage of embryos that develop along the escape trajectory, while old females produce mostly embryos that enter diapause II. (b–d) There is a great deal of intra- and inter-individual variation in the pattern of escape embryo production, such that it is hard to predict the mix of phenotypes that may be produced by an individual on any given spawning event. (Adapted from Podrabsky, J. E., I. D. F. Garrett and Z. F. Kohl. 2010. *Journal of Experimental Biology* 213:3280–3288.)

effect appears to be independent of the male, as an individual female's pattern of escape embryo production was not correlated with her male partner (Podrabsky et al., 2010). Given the constant laboratory conditions used in this study it appears likely that physiological changes in the female reproductive system associated with aging might explain the age-specific pattern of escape embryo production.

Plasma and ovarian levels of steroid hormones have been shown to decrease during the lifetime of female *A. limnaeus*, and thus it was hypothesized that high steroid hormone levels in young females might lead to the production of escape embryos (Pri-Tal et al., 2011). However, no relationship was found between the absolute levels of any steroid hormones in ovarian tissue or eggs and developmental trajectory. In contrast, the ratio of 17β-estradiol to testosterone was positively correlated with the production of escape embryos (Pri-Tal et al., 2011). This evidence suggests that differences in aromatase activity in an individual female could alter ovarian gene expression and maternal packaging in a way that leads to development along the escape trajectory. This interesting hypothesis has yet to be tested.

4.5 EMBRYONIC ENVIRONMENT AND DEVELOPMENTAL TRAJECTORY

While maternal influences might "program" the embryo for one trajectory, the environment experienced by the embryo can alter that initial program. The variables that appear to have the strongest effects are temperature and photoperiod. In addition, the hormonal status of the embryo may also influence developmental trajectory.

Incubation temperature has received the most attention in terms of studies investigating developmental trajectory in annual killifish embryos. In general, lower temperatures favor entry of embryos into diapause II in both African and South American lineages of annual killifishes, while even short exposures to higher temperatures can lead to the production of escape embryos that do not enter diapause II (Levels and Denuce, 1988; Podrabsky et al., 2010). The exact temperatures required to alter developmental trajectory are species specific. In *A. limnaeus*, an embryonic incubation temperature of 20°C yields 100% embryos that enter diapause II, while 30°C leads to 100% escape embryos (Figure 4.4; Podrabsky et al., 2010). In *Nothobranchius korthausae,* temperatures from 18°C to 25°C favor diapause II embryos, while even short exposures to 30°C favor escape embryos (Levels and Denuce, 1988). In *N. guentheri*, incubation temperatures of 19.4°C yielded a high proportion of embryos entering diapause II, while temperatures of 22.7°C and 26.5°C yielded a high proportion of escape embryos (Markofsky and Matias, 1977). The environmental relevance of this temperature effect is difficult to interpret at this point without detailed information about the thermal environment experienced by embryos under natural conditions.

As discussed above, it is likely that the hormonal environment of the female may influence the production of embryos that are programmed to develop along a specific trajectory. It also appears that the hormonal environment of the embryo can affect developmental trajectory, perhaps through different mechanisms than those important for maternal programming. Incubation of *N. korthausae* embryos in exogenous deoxycorticosterone, cortisol, testosterone, dehydroepiandrosterone, pregnenolone, progesterone, and β-estradiol during the dispersion/reaggregation phases of development results in embryos following the escape trajectory (Levels, 1988). However, most of these hormones were only effective at extremely high doses that also resulted in high levels of mortality and abnormal development. The one exception was β-estradiol, which was effective at low dosages (Levels, 1988). Similarly in *A. limnaeus*, incubation in high doses of exogenous 17β-estradiol and cortisol resulted

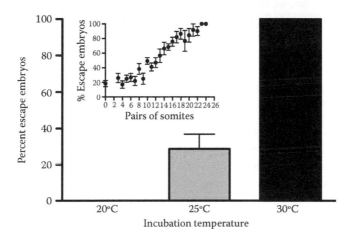

Figure 4.4 Effect of incubation temperature on developmental trajectory in embryos of *Austrofundulus limnaeus*. Increasing incubation temperature favors development along the escape trajectory. Inset: When embryos are incubated at 30°C and transferred to 25°C at different developmental stages, it is clear that the effect of temperature on developmental trajectory occurs in embryos containing 10–20 pairs of somites. (Adapted from Podrabsky, J. E., I. D. F. Garrett and Z. F. Kohl. 2010. *Journal of Experimental Biology* 213:3280–3288.)

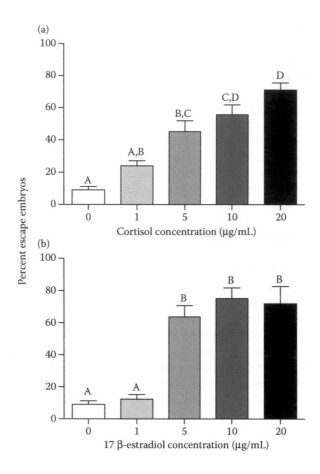

Figure 4.5 Effect of exogenous steroid hormones on developmental trajectory in *Austrofundulus limnaeus*. Exposure to elevated levels of exogenous (a) cortisol and (b) 17 β-estradiol both cause embryos to develop along the escape trajectory. Cortisol data are from B. M. Pri-Tal and J. E. Podrabsky (Personal observation), while the 17β-estradiol data are from Pri-Tal and Podrabsky (2011). Bars are means ± SEM ($n = 11$ for cortisol and $n = 12$ for 17β-estradiol).

in a dramatic increase in embryos that develop along the escape trajectory (Figure 4.5; Pri-Tal et al., 2011). In seeming conflict with these data is the fact that embryos of *A. limnaeus* incubated at 30°C, and thus developing along the escape trajectory, metabolize maternally packaged 17β-estradiol at a faster rate during early development compared with those that will enter diapause II. This leads to low levels of 17β-estradiol at the 5–10 somite stage in escape embryos, the stage at which the decision to follow either trajectory appears to first be made (Pri-Tal et al., 2011). Thus, the physiology of steroid hormone metabolism and signaling and its role in regulating developmental trajectory in annual killifishes is complex and requires a great deal of further study. However, we feel that at a minimum, the evidence available points to a likely role for steroid hormone signaling in the coordination and regulation of the physiological changes associated with escape embryo physiology, even if these hormones are not found to be responsible for the ultimate decision of which trajectory to follow.

4.6 CONSEQUENCES OF DEVELOPING ALONG ALTERNATIVE TRAJECTORIES

Differential timing of developmental progression, or heterokairy (Spicer and Rundle, 2007), may have a number of important consequences that are relevant to the development and survival of annual killifishes in their natural environment. First, differences in developmental timing may alter when embryos are most vulnerable to influences from environmental stress or variation. This could in theory lead to differential investment in particular structures or organ systems. While this topic has received very little attention, it is clear that in embryos of *A. limnaeus*, escape embryos complete development with a lower metabolic capacity in both aerobic and anaerobic metabolic pathways (Chennault and Podrabsky, 2010). This could result in a competitive disadvantage immediately after hatching if performance is also hindered by this apparently reduced capacity. The significance of this difference from a performance perspective has not been addressed in *A. limnaeus*. However, in *Nothobranchius furzeri*, it is clear that escape embryos reach sexual maturity more quickly, obtain smaller adult sizes, and have significantly shorter lifespans (Polačik et al., 2014; Furness et al., 2015).

It was noted very early in the study of development in annual killifishes that a great deal of variation could be observed in the overall developmental rate of embryos from a single clutch of eggs produced by a single female (Wourms, 1972). This high degree of variation can be attributed to two different mechanisms: (1) the facultative entrance into three different diapause stages and (2) variation in the length of time that any individual embryo may stay in diapause. Wourms referred to this as the "multiplier effect" (Wourms, 1972), which likely plays a significant role in ensuring that developmental progression is absolutely *not* synchronous in a clutch of embryos, and thus the probability that all of the embryos sense and respond to any given environmental cue in an identical manner is greatly reduced. This *reduces* the probability of population extinction due to a single hatching event in response to any given pond inundation. In fact, the age structure for populations of *Nothobranchius* in Mozambique supports a lack of hatching synchrony in many populations (Polačik et al., 2011). In addition, lack of developmental synchrony *increases* the probability that any single extreme event, for example, high/low temperature, oxygen, dehydration, etc., will be sensed by embryos at a variety of developmental stages. Thus if there are sensitive "windows" of development when these events can alter developmental outcomes, the whole population does not respond in kind and risk is spread across the entire "population" of embryos encased in the mud.

One interesting observation about the effects of environmental variables and incubation of embryos in steroid hormones on developmental trajectory is a general lack of "binary" reactions by a group of embryos to any particular stimulus. For instance, there is a great deal of interindividual variation in response to temperature transfer from 30°C to 25°C, which creates a rather broad "window" of development when incubation temperature can affect developmental trajectory (Figure 4.4 inset). This could again be viewed as a mechanism that ensures that not all embryos respond to similar environmental cues in the same manner. While the mechanism for this pattern is unknown, we hypothesize that degradation of a maternally packaged diapause-specific "factor" (protein or mRNA, for example) would logically lead to this type of pattern. In this scenario, small differences in maternal packaging of this "factor" and alterations in the rate of degradation of this factor would work in combination to direct the developmental trajectory of the embryo. This type of epigenetic mechanism would almost certainly ensure a high degree of variation in developmental timing and response to any given environmental situation. It could also be a way to integrate maternal nutritional and physiological status into the type of offspring produced.

4.7 CONCLUSIONS

The study of these two alternative developmental pathways has only just begun, and there is still much to explore concerning their physiological, ecological, and evolutionary significance. It is clear that the developmental program of annual killifishes is poised to generate diversity in developmental rates, and perhaps in embryonic physiology. This pattern and strategy for development is in stark contrast to most experimental models of vertebrate development, including the zebrafish. Like most other model organisms in biology, zebrafish develop quickly and express rather low variation in developmental traits. Studies of development in annual killifishes offer a unique opportunity to explore maternal and environmental effects on development in an ecologically relevant framework. Future studies on the developmental mechanisms that support alternative developmental trajectories, and the ecological and evolutionary consequences these trajectories enable, will yield fundamental insights on how vertebrate embryos develop and how maternal and environmental information interact with the genome during development.

ACKNOWLEDGMENTS

This work was supported by NIH grant HL095454 and NSF grant IOS 1354549 to JEP.

REFERENCES

Anderson, S. N. and J. E. Podrabsky. 2014. The effects of hypoxia and temperature on metabolic aspects of embryonic development in the annual killifish. *Austrofundulus limnaeus. Journal of Comparative Physiology B* 184:355–370. doi: 10.1007/s00360-014-0803-6

Chennault, T. and J. E. Podrabsky. 2010. Aerobic and anaerobic capacities differ in embryos of the annual killifish *Austrofundulus limnaeus* that develop on alternate developmental trajectories. *Journal of Experimental Zoology Part A: Ecological Genetics and Physiology* 313A:587–596. doi: 10.1002/jez.632

Culpepper, K. M. and J. E. Podrabsky. 2012. Cell cycle regulation during development and dormancy in embryos of the annual killifish *Austrofundulus limnaeus. Cell Cycle* 11:1697–1704. doi: 10.4161/cc.19881

Duerr, J. M. and J. E. Podrabsky. 2010. Mitochondrial physiology of diapausing and developing embryos of the annual killifish *Austrofundulus limnaeus*: Implications for extreme anoxia tolerance. *Journal of Comparative Physiology B: Biochemical, Systemic, and Environmental Physiology* 180:991–1003. doi: 10.1007/s00360-010-0478-6

Furness, A. I., D. N. Reznick, M. S. Springer and R. W. Meredith. 2015. Convergent evolution of alternative developmental trajectories associated with diapause in African and South American killifish. *Proceedings of the Royal Society B: Biological Sciences* 282:20142189. doi: 10.1098/rspb.2014.2189

Levels, P. J. 1988. *An Experimental Study of Diapause in Annual Fishes*. Nijmegen: Katholieke Universiteit te Nijmegen.

Levels, P. J. and J. M. Denuce. 1988. Intrinsic variability in the frequency of embryonic diapauses of the annual fish *Nothobranchius korthausae*, regulated by light: Dark cycle and temperature. *Environmental Biology of Fishes* 22:211–223. doi: 10.1007/BF00005382

Levels, P. J., R. E. M. B. Gubbels and J. M. Denuce. 1986a. Oxygen consumption during embryonic development of the annual fish *Nothobranchius korthausae* with special reference to diapause. *Comparative Biochemistry and Physiology—Part A: Molecular & Integrative Physiology* 84A:767–770. doi: 10.1016/0300-9629(86)90403-2

Levels, P. J., L. J. H. Van Tits and J. M. Denuce. 1986b. The effect of the presence of adult fishes, gonad homogenates, and embryo homogenates on the dispersion-reaggregation phase during early embryonic development of the annual fish *Nothobranchius korthausae. Journal of Experimental Zoology* 240:259–264. doi: 10.1002/jez.1402400212

Machado, B. E. and J. E. Podrabsky. 2007. Salinity tolerance in diapausing embryos of the annual killifish *Austrofundulus limnaeus* is supported by exceptionally low water and ion permeability. *Journal of Comparative Physiology B: Biochemical, Systemic, and Environmental Physiology* 177:809–820. doi: 10.1007/s00360-007-0177-0

Markofsky, J. and J. R. Matias. 1977. The effects of temperature and season of collection on the onset and duration of diapause in embryos of the annual fish *Nothobranchius guentheri. Journal of Experimental Zoology* 202:49–56. doi: 10.1002/jez.1402020107

Markofsky, J., J. R. Matias, K. Inglima, J. H. Vogelman and N. Orentreich. 1979. The variable effects of ambient and artificial light: Dark cycles on embryonic diapause in a laboratory population of the annual fish *Nothobranchius guentheri. Journal of Experimental Biology* 83:203–215.

Meller, C. L., R. Meller, R. P. Simon, K. M. Culpepper and J. E. Podrabsky. 2012. Cell cycle arrest associated with anoxia-induced quiescence, anoxic preconditioning, and embryonic diapause in embryos of the annual killifish *Austrofundulus limnaeus*. *Journal of Comparative Physiology B* 182:909–920. doi: 10.1007/s00360-012-0672-9

Murphy, W. J. and G. E. Collier. 1997. A molecular phylogeny for Aplocheiloid fishes (Atherinomorpha, Cyprinodontiformes): The role of vicariance and the origins of annualism. *Molecular Biology and Evolution* 14:790–799. doi: 10.1093/oxfordjournals.molbev.a025819

Peters, N. 1965. Diapause und embryonale missbildung bei eierlegenden zahnkarpfen. *Roux' Archiv für Entwicklungsmechanik* 156:75–87. doi: 10.1007/BF00576719

Podrabsky, J. E. and S. C. Hand. 1999. The bioenergetics of embryonic diapause in an annual killifish, *Austrofundulus limnaeus*. *Journal of Experimental Biology* 202:2567–2580.

Podrabsky, J. E. and S. C. Hand. 2000. Depression of protein synthesis during diapause in embryos of the annual killifish, *Austrofundulus limnaeus*. *Physiological and Biochemical Zoology* 73:799–808. doi: 10.1086/318106

Podrabsky, J. E., I. D. F. Garrett and Z. F. Kohl. 2010. Alternative developmental pathways associated with diapause regulated by temperature and maternal influences in embryos of the annual killifish, *Austrofundulus limnaeus*. *Journal of Experimental Biology* 213:3280–3288. doi: 10.1242/jeb.045906

Podrabsky, J. E. and S. C. Hand. 2015. Physiological strategies during animal diapause: Lessons from brine shrimp and annual killifish. *Journal of Experimental Biology* 218:1897–1906.

Podrabsky, J. E., J. P. Lopez, T. W. M. Fan, R. Higashi and G. N. Somero. 2007. Extreme anoxia tolerance in embryos of the annual killifish, *Austrofundulus limnaeus*: Insights from a metabolomics analysis. *Journal of Experimental Biology* 210:2253–2266. doi: 10.1242/jeb.005116

Polačik, M., R. Blažek, R. Řežucha, M. Vrtílek, F. Tozzini and M. Reichard. 2014. Alternative intrapopulation life-history strategies and their trade-offs in an African annual fish. *Journal of Evolutionary Biology* 27:854–865. doi: 10.1111/jeb.12359

Polačik, M., M. Donner and M. Reichard. 2011. Age structure of annual *Nothobranchius* fishes in Mozambique: Is there a hatching synchrony? *Journal of Fish Biology* 78:796–809. doi: 10.1111/j.1095-8649.2010.02893.x

Pri-Tal, B. M., S. Blue, F. K. Y. Pau and J. E. Podrabsky. 2011. Hormonal components of altered developmental pathways in the annual killifish, *Austrofundulus limnaeus*. *General and Comparative Endocrinology* 174:166–174. doi: 10.1016/jgcen.2011.08.016

Spicer, J. I. and S. D. Rundle. 2007. Plasticity in the timing of physiological development: Physiological heterokairy—What is it, how frequent is it, and does it matter? *Comparative Biochemistry and Physiology—Part A: Molecular and Integrative Physiology* 148:712–719. doi: 10.1016/j.cbpa.2007.05.027

Wourms, J. P. 1972. The developmental biology of annual fishes III. Pre-embryonic and embryonic diapause of variable duration in the eggs of annual fishes. *Journal of Experimental Zoology* 182:389–414. doi: 10.1002/jez.1401820310

CHAPTER 5

Sex Determination and Differentiation in Annual Fishes

María José Arezo, Nicolás G. Papa, Nibia Berois, Graciela García, Verónica Gutiérrez, and Marcelo Antonelli

CONTENTS

5.1 INTRODUCTION

Sex determination, the process involved in establishing the gender of an organism, is a simple binary fate decision: male or female (Penman and Piferrer, 2008). Sex determination mechanisms are responsible for producing the sex ratio, a critical demographic parameter for population viability (Opslna-Alvarez and Piferrer, 2008). The two major types of sex-determining mechanisms are: genotypic sex determination (GSD), where the sex is determined at fertilization, and environmental sex determination (ESD), where sex is determined by environmental cues after fertilization. These mechanisms commit the morphologically undifferentiated bipotential gonad primordium, not yet determined, to a gonadal rudiment (ovary or testis fate). Sex differentiation is the process that transforms those determined and undifferentiated rudiments into ovaries or testes. Embryonic gonads represent the unique organ primordium that could develop into two different adult tissues. Sexual fate is accomplished by activating the ovarian or testicular pathway and at the same time repressing the alternative differentiation cascade (Gilbert, 2013). In this context, activation or suppression of a variety of genes in a strict spatiotemporal dimension is needed; epigenetic mechanisms for the regulation of gene expression (e.g., DNA methylation, posttranscriptional histone modifications, and regulatory noncoding RNAs) become of critical importance (Piferrer, 2013).

Fishes represent the most basal and diverse vertebrate group, encompassing nearly 28,000 species (Nelson, 2006). This diversity is also revealed in different sex determination mechanisms (i.e., environmental and genetic) and sex differentiation strategies (e.g., unisexuality; hermaphroditism: synchronic or sequential; and gonochorism: undifferentiated or differentiated). Environmental factors related to sex determination are water pH, oxygen concentration, growth rate, density, social state, and temperature. Genetic sex determination includes monofactorial systems involving a single master sex gene or polyfactorial systems with several genes on multiple autosomal chromosomes and/or the presence of sex chromosomes. Sex chromosomes could be cytogenetically distinguishable (heteromorphic) or identical (homomorphic). Nevertheless, in both cases one sex produces two types of gametes (heterogametic) and the other only one type (homogametic). Moreover, some species display more complex systems involving more than one chromosome pair (Devlin and Nagahama, 2002). Sex chromosomes in fish are generally poorly differentiated. For this reason, they are considered evolutionarily young and might have emerged independently in different fish lineages. Therefore, studies on different fish species are essential to understand the evolutionary dynamics of sex determination in this group (Tomaszkiewicz et al., 2014).

Sex determination systems also differ in closely related fish species and even in populations of the same species (Conover and Kynard, 1981; Devlin and Nagahama, 2002). Populations that inhabit ecosystems characterized by variable environmental temperature exhibit genetic sex determination mechanisms, while in populations residing in more stable climatic conditions, sex is determined by environmental clues. This pattern was originally observed in populations of the teleost fish *Menidia menidia* (Conover and Kynard, 1981; Conover and Heins, 1987). The evolutionary basis of this variability could be explained by ancient genomic duplication events that caused extra gene copies capable of acquiring new functions and probably added plasticity to sex determination gene networks. The ability to modify sex determination control could be selected in response to environmental disturbances affecting sex ratios (Volff, 2005; Mank et al., 2006). Therefore, fishes are an attractive group for studying the evolution of sex determination/differentiation mechanisms (Devlin and Nagahama, 2002; Heule et al., 2014).

Opsina-Alvarez and Piferrer (2008) reevaluated the prevalence of temperature as a sex determination factor in fishes. These authors applied two independent criteria to discriminate between true cases of temperature-dependent sex determination (TSD) and genetic sex determination influenced by temperature (thermal effects; GSD + TE). First criterion implied that the presence of sex chromosomes is strong evidence for a genetic sex determination mechanism (Valenzuela et al., 2003). The second criterion determines that to assign TSD for a given species, sex ratio shifts must occur in response to exposed temperatures during development in nature. This criterion is extremely important because the thermosensitive period occurs during early ontogeny, that is, embryo development or larval stages depending on the species (Strüssmann et al., 1997; Koumoundouros et al., 2002). Opsina-Alvarez and Piferrer (2008) suggest that TSD has evolved independently many times in fishes. It is found in only four nonclosely related fish orders, and the phylogenetic distribution suggests that TSD is a derived mechanism rather than an ancestral mechanism. Nevertheless, the available data on sex determination is still scarce, and our understanding of the distribution of sex determination mechanisms may vary as additional species are examined (Opsina-Alvarez and Piferrer, 2008). The putative link between ESD and GSD are steroid hormones, or the regulation of their production, aromatase *cyp19a1* being a key candidate (Valenzuela et al., 2013).

Among vertebrates, different master genes that participate in sex determination have been identified. In most mammals, *Sry* (testis-determining gene), located in the Y chromosome, is the transcription factor that triggers the testis-determining cascade. The first nonmammalian master gene was discovered in fishes (i.e., *Oryzias latipes, dmy*, described below). Among birds, *Dmrt1* (double-sex-mab-3 related transcription factor 1), located in the Z chromosome, plays this role, while in the frog *Xenopus laevis DM-W*, a truncated copy of *Dmrt1*, found in the W chromosome, determines formation of the ovary (reviewed by Graves and Peichel, 2010).

Recently, four master sex-determining candidate genes were identified in fishes: *amhy* in *Odontesthes hatcheri* (Hattori et al., 2012), *gsdf* in *Oryzias luzonensis* (Kamiya et al., 2012), *amhr2* in *Takifugu rubripes* (Myosho et al., 2012), and *sdY* in *Oncorhynchus mykiss* (Yano et al., 2012). Three of these genes (*amhy*, *gsdf*, and *amhr2*) code for growth factors and one of their receptors, demonstrating that different genes, other than transcription factors, can be recruited at the top of the sex determination cascade (reviewed by Kikuchi and Hamaguchi, 2013). In spite of this diversity, it has been demonstrated that the core of gene regulatory cascades in determination/differentiation pathways are conserved in vertebrates and invertebrates (Smith et al., 1999; Graves and Peichel, 2010). However, recent data from various fish species suggest that their role and position in these cascades might vary in different organisms; for example, this is evidenced by different and species-specific expression patterns during development (Herpin et al., 2013; Heule et al., 2014).

Among these genes, a DM gene family, *Dmrt* (doublesex-mab-3 related transcription factor), is expressed in association with the development of sex-specific organs in all studied animals (Kopp, 2012). The first identified member of this family: *Dmrt1*, exhibits an expression pattern mainly involved in postnatal (or posthatching) male gonad development (Hodgkin, 2002; Yamaguchi et al., 2006). In some fishes, its expression pattern is restricted to testes (e.g., *O. latipes*; Kobayashi et al., 2004), while in others it is expressed in both gonads (e.g., *Odontesthes bonariensis*; Fernandino et al., 2008). In hermaphrodite fishes, *Dmrt1* expression is related to male differentiation phases (Herpin and Schartl, 2011). The interest in this gene was greater after the discovery that *Dmrt1* paralogues (genes related by duplication within a genome) have moved up in the regulatory hierarchy from downstream position in gonad differentiation to the top of the sex determination cascade in at least three distantly related organisms, *O. latipes*, *Xenopus laevis*, and chicken (Kopp, 2012). In *O. latipes*, the male sex-specific copy of *Dmrt1*, *Dmrt1b(Y)* (Nanda et al., 2002), or *dmy* (Matsuda et al., 2002) is located in the Y chromosome and has an equivalent function to mammalian *Sry* (male master sex determining gene). The *Dmrt1b(Y)* gene is expressed in male embryos before gonadal differentiation in the neurula stage (Nanda et al., 2002). It is involved in the specification and maintenance of Sertoli cells and inhibits male germ cell division at the beginning of gonadal differentiation (Herpin et al., 2007). The autosomal paralog *Dmrt1a* begins expression between 20 and 30 days post hatching during testicular differentiation (Kobayashi et al., 2004). In adult testes, both paralogues are expressed, but *Dmrt1a* predominates (Hornung et al., 2007). In addition, an insertion of a transposable element is responsible for the regulation of *Dmrt1bY* expression and also contributed to the establishment of this new regulatory hierarchy (Herpin et al., 2010). Recently, another transposable element was suggested to be involved in mobilize and spread Y-specific fertility genes in the Cyprinodontiform *Xiphophorus maculatus* (Tomaszkiewicz et al., 2014).

5.2 SEX DETERMINATION IN *AUSTROLEBIAS CHARRUA*

Chromosomal studies and DNA analyses in *A. charrua* demonstrated high genetic variation (García, 2006) and the existence of ancestral polymorphisms in certain populations. These data suggest the need to incorporate this species in biodiversity conservation programs (García et al., 2009). Moreover, fishes are becoming important indicators of environmental contamination, since pollutants (e.g., xenoestrogens and agrochemicals) produced by human activities may affect gene regulatory networks that impact reproduction in fishes. In this context, knowledge about reproductive strategies, including sex determination mechanisms, is essential at two levels: to contribute to the understanding of the evolution of these mechanisms and to assess possible contamination effects (Devlin and Nagahama, 2002). Furthermore, these data are also important to evaluate species liability associated with climate change, since the viability of some fish populations with TSD or GSD + TE can be compromised through alterations in their sex ratios as a response to predicted temperature fluctuations (Opsina-Alvarez and Piferrer, 2008). These data are particularly relevant to design conservation strategies and management plans (Volcan et al., 2013).

5.2.1 Thermosensibility Evaluation in Sex Determination

The effect of constant temperatures (19°C and 25°C) on sex ratio was analyzed. This approach does not consider the environmental temperature variability experienced by the embryos in nature. However, these studies are still useful for identifying or discarding TSD in a given species (Warner and Shine, 2011). Temperature selection was based on three considerations. First, *A. charrua* embryos cultured under 17°C enter into diapause I (dispersed phase: annual fishes late blastula stage). The embryos may remain arrested for at least 8 months (De la Piedra, personal communication, September 2006). Second, high embryo mortality was observed at 28°C and 29°C (M.J. Arezo et al., unpublished data). Third, that temperatures correspond to those experienced by embryos of *A. charrua* during development in natural ponds (T. Van Dooren, personal communication, May, 2009).

Statistical analysis (Poisson regression) of 252 sexed individuals demonstrated that the interaction between temperature and sex ($p = 0.108$) was not statistically significant (p-value > 0.05). This result indicates that temperatures of 19°C and 25°C did not shift sex ratio from 1:1 in the analyzed embryo population. In nature, five ponds were sampled at the end of the rainy season. In three of them sex ratio was 1:1 ($p = 0.150$; $p = 0.330$; $p = 0.670$, p-values >0.05), while in the other two ponds, there was a higher proportion of females than males and these differences were statistically significant ($p = 0.028$; $p = 0.007$; p-values <0.05) (Arezo, 2012, Appendix). *A. charrua* males are probably more exposed to predators because of their body colorations and elaborate courtship behaviors (Belote and Costa, 2004; D'Anatro and Loureiro, 2005; García et al., 2008). It is also possible that the high energy cost exerted by courtship behavior could shorten the male lifespan, as described for *A. reicherti* (Passos et al., 2014). Furthermore, males of polygamist species, including annual fishes, have shorter lifespan than females (Clinton et al., 1993). Declines in the number of males late in the breeding season were reported in the African annual fish genus *Nothobranchius* (Reichard et al., 2009). All these facts could explain female-biased sex ratios found in two ponds at the end of the life cycle.

5.2.2 Is Sex Determination of *A. charrua* Genetic or Temperature Dependent?

In agreement with the criterion proposed by Valenzuela et al. (2003), which emphasizes the presence of sex chromosomes as strong evidence for genetic sex determination, sex chromosomes were not cytogenetically identified in *A. charrua* (García, 2006). This observation is in agreement with data obtained from most-studied teleosts (reviewed by Devlin and Nagahama, 2002). Differentiated sex chromosomes are sporadic in different fish taxa, suggesting a recent and polyphyletic origin of sex chromosomes in this vertebrate group (Mank et al., 2006). Fish sex chromosomes are difficult to evidence by classic cytogenetic techniques because of their small size and their putative recent origin (homomorphic chromosomes). Ultrastructural analyses of synaptonemic complexes formed during meiosis revealed tiny unpaired regions, allowing sex chromosome pair recognition in the heterogametic sex (Devlin and Nagahama, 2002). Consequently, we could not discard the presence of homomorphic sex chromosomes in *A. charrua* using the aforementioned analysis. Moreover, even in the absence of sex chromosomes, GSD could be demonstrated by the construction of a microsatellite-based linkage map to define genome regions involved in sex determination. This approach allowed the identification of the male sex-determining nonrecombinant region in *Nothobranchius furzeri* (Valenzano et al., 2009), similar to those identified in other phylogenetically related fishes (e.g., *O. latipes*: Matsuda et al., 2002; Nanda et al., 2002; *Gasterosteus aculeatus*: Peichel et al., 2004; *Xiphophorus maculatus*: Schultheis et al., 2009; and *Poecilia reticulata*: Tripathi et al., 2009).

According to Opsina-Avarez and Piferrer's (2008) second criterion, it is necessary to take into account that TSD could be proposed as the predominant strategy when a sex ratio shift occurs in response to temperatures to which the embryo was exposed in nature during the thermolabile period. Thermal treatments were performed applying temperatures experienced in nature by embryos between the dispersed phase stage and 1 month postfertilization to cover the whole putative thermolabile period (Arezo et al., 2007). It was found that the treatments did not affect the 1:1 sex ratio in the assessed embryo population.

Taken together, these data suggest that *A. charrua* undergoes a genetic sex determination mechanism. This is consistent with the heterogeneous environmental habitat of *A. charrua* and with the assumption that TSD in fishes is an exception (Opsina-Alvarez and Piferrer, 2008). Nevertheless, and since the 252 embryos analyzed were obtained from three *A. charrua* couples belonging to populations that inhabit a restricted geographic area, similar studies are needed to include individuals from the entire distribution of the species to validate the results.

5.2.3 Searching for a Candidate Sex-Specific Gene

Genetic sex markers have not been identified in *Austrolebias* and to date there are no complete genome sequences available. Therefore, the presence of a candidate master gene involved in *A. charrua*'s sex determination and its expression during ontogenesis was initially explored (Arezo et al., 2014). This analysis employed *dmrt1bY*-specific oligonucleotides designed to isolate this male-specific sex-determining gene in *O. latipes* (Beloniformes; Nanda et al., 2002). Cyprinodontiformes, Beloniformes, and Atheriniformes belong to the monophyletic Atherinomorpha series, with Atherinimorpha being the sister group to a clade consisting of Beloniformes and Cyprinodontiformes (Parenti, 2005; Nelson, 2006; Setiamarga et al., 2008).

PCR amplifications of *A. charrua* detected two fragments of 1000 and 900 base pairs (bp). This pattern was similar to that found in the genome of *O. latipes* males, where a fragment of 1289 nucleotides corresponds to the *Dmrt1* gene and another fragment of 965 nucleotides corresponds to the male-specific duplicated gene version, *Dmrt1bY* (Nanda et al., 2002). The same pattern was also found in the genome of *A. charrua* females, but nucleotide sequence comparison of the larger fragments showed only 42% similarity, while the smaller fragments are 99% similar. Thus, the smaller fragment is present in both sexes in *A. charrua* (Arezo et al., 2014).

In fishes, five classes of *Dmrt* genes have been described: *Dmrt* 1, 2, 3, 4, and 5 (Huang et al., 2002). Thus, to resolve putative relationships among the *A. charrua* isolated genomic fragments described above and the *Dmrt* gene family sequences, a phylogenetic analysis using two model-based approaches: maximum-likelihood (ML) and Bayesian inference (BI) phylogenetic tree reconstruction were performed (Figure 5.1). They showed a topology with three main supported clades, and a moderate similarity was detected among the *A. charrua* genomic fragments and the *Dmrt1* sequences. These results suggest an inconclusive relationship between the isolated fragments and the *Dmrt1* from different fish groups (Arezo et al., 2014)

The RT-PCR amplifications with the same oligonucleotides (Nanda et al., 2002) revealed an expressed 205-bp cDNA fragment in embryo stages from dispersed phase (late blastula) to pre-hatching embryos. In juveniles and adults, the fragment was only detected in males. Phylogenetic analyses (MP, ML, and BI) grouped the *A. charrua* fragment with *Dmrt1* alternative-splicing versions *c* and *d* from *O. latipes* (Beloniformes) instead of with *Dmrt1* sequences from more closely related species of Cyprinodontiformes (*K. marmoratus*, *X. maculatus*, and *P. reticulata*; Figure 5.2). Potential contaminations were checked and discarded. Consequently, this discordance may be explained by maintenance of ancestral polymorphism, since in the resulting tree, the fragment from *A. charrua* grouped with those of *O. latipes Dmrt1 c* and *d* alternative-splicing versions as well as *Drosophila dsx*. Furthermore, this clade is basal to *Odontesthes* (Atheriniformes) *Dmrt1* sequences. Discordance between *Dmrt1* sequence relationships was also evidenced in phylogenetic

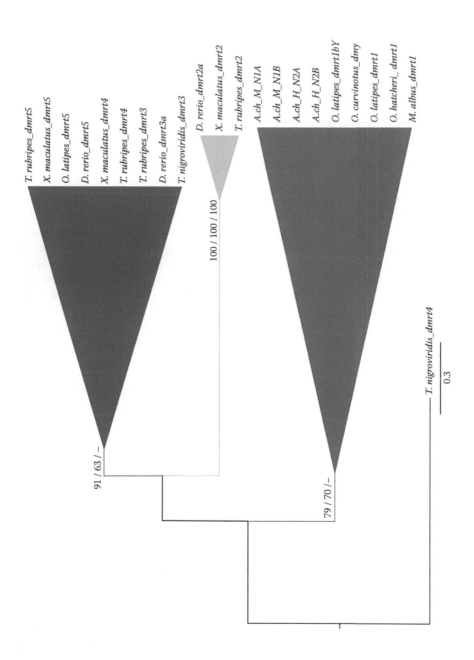

Figure 5.1 Maximum likelihood analyses [under general time-reversible model with gamma correction (GTR + G)], based on partial genome sequences isolated from *Austrolebias charrua* and *Dmrt* gene family sequences retrieved from GenBank. Numbers above the branches (left to right) correspond to the Bayesian posterior probability for clades obtained using BEAST 1.5.4v, followed by ML and MP bootstrap support (>60%), respectively, recovered in 1000 replicates. *A.ch_M_ N1A*, *A.ch_M_ N1B*, *A.ch_H_ N2A*, and *A.ch_H_ N2B*: *Austrolebias charrua* genomic fragments. (Sequence data from Arezo, M.J. et al. 2014. *Genetics and Molecular Biology* 37:364–374.)

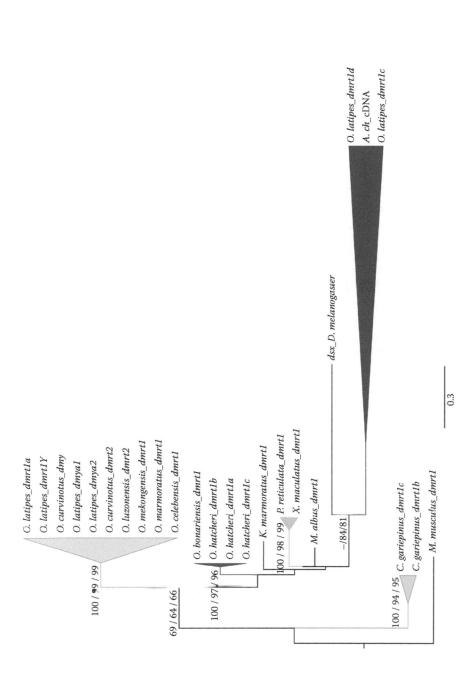

Figure 5.2 Maximum likelihood analyses (under K81 + G model with gamma correction), based on the expressed partial sex-specific cDNA sequence isolated from *Austrolebias charrua* (*A.ch*_cDNA), *Dmrt1/Dmrt1bY* mRNA-like sequences and *dsx* from *Drosophila* retrieved from GenBank. Numbers above the branches (left to right) correspond to the Bayesian posterior probability for clades obtained using BEAST 1.5.4v, followed by ML and MP bootstrap support (>60%), respectively, recovered in 1000 replicates. (Sequence data from Arezo, M.J. et al. 2014. *Genetics and Molecular Biology* 37:364–374.)

analyses performed with *Dmrt1* of *O. bonariensis* (Atheriniformes) and other fishes (Fernandino et al., 2006). All phylogenetic methods revealed that the isolated partial cDNA sequence is more closely related to the small fragments amplified from male (M N1B) and female (H N2B) genomes (HPP > 96) and supported the relationships of these sequences with the *dsx D. melanogaster* gene (HPP > 97) (Figure 5.3). In this context, it is interesting to consider that the *Dmrt1* gene is closest in terms of structure and function to *Drosophila dsx* and *Caenorhabditis elegans mab-3* (Herpin and Schartl, 2011). Furthermore, all phylogenetic analyses straightforwardly grouped *Drosophila dsx* with *A. charrua*-expressed cDNA partial fragments and amplified sequences from male (M N1B) and female (H N2B) genomes, showing high posterior probability of occurrence for such a clade (Arezo et al., 2014).

The expression pattern of cDNA partial sequence during the ontogeny of the studied species showed that this sequence is detected starting at the dispersed phase (late blastula). This suggests that its expression depends on the zygotic genome activation (Arezo et al., 2014). In *O. latipes*, the *Dmrt1bY* expression begins during the neurula stage (1 day postfertilization), while *Dmrt1a* starts its expression during testis differentiation (about 20 days posthatching) (Nanda et al., 2002; Kobayashi et al., 2004; Iwamatsu, 2004). In *A. charrua*, cDNA partial fragment expression was detected in an even earlier developmental stage than *Dmrt1bY*. This observation contrasts with those reported for other vertebrate species (reviewed by Arezo et al., 2014).

Consequently, the finding that this cDNA partial sequence is expressed very early during *A. charrua* development and is closely related to the *dsx* gene of *D. melanogaster* may indicate that it could play an essential role during the sex determination process and might be located near the top of the sex determination cascade in this species. Taken together, these results suggest that both isolated *A. charrua* genomic sequences and the expressed cDNA partial sequence are more closely related to *dsx* from *D. melanogaster* than the two alternative-splicing *Dmrt1* sequences from *O. latipes*. The observed male-specific expression pattern supports the hypothesis that this sequence

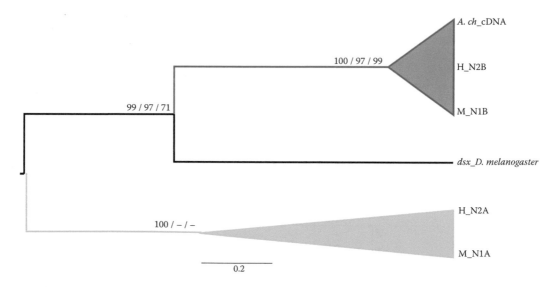

Figure 5.3 Tree topology generated using the HKY85 model of molecular evolution based on male (MN1A and MN1B) and female (H N2A and H N2B) genomes, on the expressed partial sex-specific cDNA sequence isolated from *Austrolebias charrua* (*A. ch* cDNA) and the *dsx_ D. melanogaster* sequence. Bayesian phylogeographic inference framework implemented in BEAST 1.5.4v. Numbers above the branches (left to right) correspond to the Bayesian posterior probability for clades obtained using BEAST 1.5.4v, followed by ML and MP bootstrap support (>60%), respectively, recovered in 1000 replicates. (Sequence data from Arezo, M.J. et al. 2014. *Genetics and Molecular Biology* 37:364–374.)

probably belongs to an alternative-splicing *Dmrt1* gene version in *A. charrua*. This gene is present in both sexes and represents the most-conserved downstream gene member implicated in male sex development during vertebrate evolution. Nevertheless, it is also possible that the isolated cDNA sequence belongs to a noncoding RNA with potential regulatory function (Arezo et al., 2014).

5.3 SEX DIFFERENTIATION

Studies examining the origin and development of cells involved in the formation of the gonad are critical to understanding sex determination/differentiation processes (Devlin and Nagahama, 2002). In this context, the type and timing of gonadal differentiation, sexual strategy, and gametogenesis in *A. charrua* were previously studied, starting with the identification of presumptive primordial germ cell (pPGC) populations (Arezo et al., 2007).

5.3.1 Primordial Germ Cells

Primordial germ cells (PGCs) are among the most-studied cell lineages (Extravour, 2007). They are a highly specialized cell population that differentiates into gametes and transmits genetic and epigenetic information between generations, ensuring the survival of the species (Magnúsdóttir and Surani, 2014). Primordial germ cells are set aside from the somatic cells and kept transcriptionally quiescent. This segregation is an effective way to protect genome integrity from potential mutations generated during differentiation and cell division. Since germ cells are responsible for reproduction and genetic continuity of the species, their mode of specification remains a central problem in developmental and evolutionary biology. The most crucial aspect of germ cell development for understanding the evolution of this cell population relates to the first specification event that separates the germinal and somatic lines (Extravour, 2007). At least two different modes of germ cell specification are found among organisms. In some species, germ cells are identified very early in embryogenesis by localized maternally inherited determinants (preformation), whereas in other species, germ cells are observed in later developmental stages as a result of inductive signals (epigenesis). The germline can also be formed even in adulthood from multipotent stem cell precursors, as in hydra (Extravour and Akam, 2003). The key difference between epigenesis and preformation is the time of gene expression and localization of germ cell-specific gene products (Extravour, 2007). Among vertebrates, an epigenetic mode of PGC specification is the ancestral pattern, whereas preformation evolved independently by convergence across lineages (Johnson et al., 2003). Preformation is associated with greater speciation, suggesting that this specification mechanism confers a selective advantage (Evans et al., 2014). However, in the diverse and specious group of bony fishes (Nelson, 2006), it is unclear whether germ cell segregation by preformation is common to all teleosts. Therefore, studies on PGC specification modes in a variety of fish species may provide experimental evidence for how widespread preformation is in teleost fishes and whether it is the ancestral mechanism of germ cell determination in this group (Extravour and Akam, 2003; Extravour, 2007).

The PGC populations of different organisms share morphological, behavioral, and molecular features. A group of conserved molecular determinants expressed by germ cells include the Vasa, Nanos, Tudor, and Piwi protein families. These determinants are localized in a specific cytoplasmic region of the egg called the germplasm or nuage. The product of *vasa* gene is a DEAD-box helicase involved in RNA binding. *Drosophila* vasa homologs have been identified as germline-specific molecular markers throughout the animal kingdom, indicating the highly conserved nature of this gene (Raz, 2000; Lin et al., 2012).

Based on morphological characteristics, the PGCs of teleost fishes were recognized at early somitogenesis stages (Timmermans and Taverne, 1989; Gevers et al., 1992). The identification of the first molecular marker, *Drosophila vasa* homologue gene, for the germline in zebrafish (Yoon et al., 1997) and medaka (Shinomiya et al., 2000) has allowed the analysis of PGCs from earlier stages of embryonic development. This protein is the most-documented molecular marker of teleosts germ cells (Lin et al., 2012). Primordial germ cells migrate autonomously across the embryo through somatic cells to reach somatic gonadal precursors. This cell population displays a peculiar migration pattern from the position where they are specified toward the future gonad depending on the species. Studies using model organisms have revealed that several features of PGC migration are conserved in different organisms (Richardson and Lehmann, 2010).

We performed an immunocytochemistry study of *A. charrua* using an antibody generated against *Danio rerio* Vasa protein to identify PGC population, (M.J. Arezo et al., unpublished data; see Appendix). Zebrafish Vasa protein is distributed uniformly in all embryonic cells at early stages up to late blastula (Braat et al., 2000). The expression pattern observed during early stages of *A. charrua* development was similar to that described in *D. rerio* embryos despite the fact that the antibodies used in our study recognized a different protein region. In early blastula stages, all *A. charrua* cells were uniformly Vasa-positive (Figure 5.4a). The first differential expression pattern was detected during epiboly (late blastula stage; Figure 5.4b). In the reaggregation phase (gastrula stage), the label was restricted to fewer cells (Figure 5.4c) located outside the aggregation zone. During somitogenesis and early organogenesis (2 weeks postfertilization), Vasa-positive cells remained outside the embryo (Figure 5.4d). Histological analyses showed that PGCs arrive and incorporate into the gonadal ridge at 3 weeks postfertilization (Figure 5.4e). At the same developmental stage, PGCs were recognized based on morphological features (large cell size, large nucleus, darkly stained nucleoli, definite nuclear and cellular membranes, and hyaline cytoplasm; Arezo et al., 2007). In 4 weeks post fertilization embryos, a Vasa-positive zone was also evidenced in the developing gonadal region defined previously by histology (Figure 5.4f).

Microinjection of *Danio rerio* 3′UTR mRNA vasa-GFP construct (Knaut et al., 2002) was also performed, (M.J. Arezo et al., unpublished data), but the embryos survived only to reaggregation phase. The expression pattern was similar to whole-mount immunodetections using Vasa antibody at the same developmental stages. Primordial germ cells are visualized *in vivo* using RNA constructs that contain 3′UTR *vasa* sequences derived from highly divergent taxonomic groups (Yoshizaki et al., 2005). These results suggest that both experimental approaches, the antibody generated against *D. rerio* Vasa protein and the injection of 3′UTR mRNA vasa-GFP construct, allow the identification of a cell population with similar characteristics in *A. charrua*. The loss of the label in most embryonic cells, documented in both assays, could be due to gradual degradation process of maternal Vasa protein in somatic cell lineage and maintenance in PGCs population as a result of a degradation-protection process as previously described in other fishes (Wolke et al., 2002).

The pattern observed in whole-mount immunodetections identifying the putative endogenous Vasa protein agrees with the *in vivo* microinjection assays (data not shown). These results suggest that these positive cells constitute the presumptive PGCs of *A. charrua* and that preformation is the specification mechanism the species.

5.3.2 Sex Differentiation Strategy

Studies in teleost species showed that the presence of gonadal structures (i.e., ovarian cavity and efferent duct formation), differences in germ cell number, and female precocious meiosis are useful to determine gonadal sex at an early phase in development. Nevertheless, the most reliable criterion is variable depending on the studied species. In most teleosts, germ cells in putative ovaries outnumbered those in putative testes (reviewed by Nakamura et al., 1998; Strüssman and Nakamura, 2002). The first evidence of histological sex differentiation in *A. charrua* embryos, using the difference in

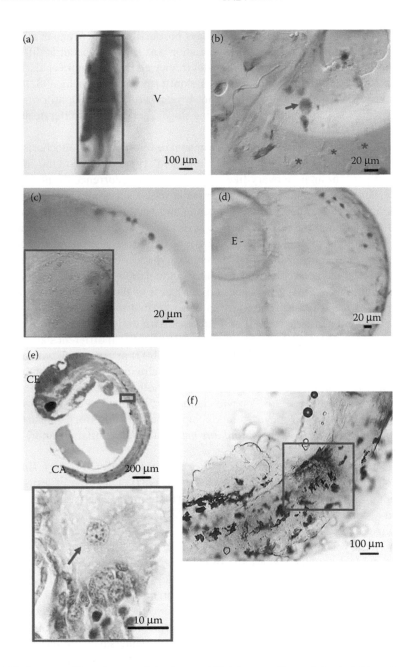

Figure 5.4 Whole-mount Vasa immunodetection and histology sections of *Austrolebias charrua* embryos. (a) Early blastula stage. V: yolk. Square: Vasa-positive blastomeres. (b) Epiboly—late blastula stage. Arrow: Vasa-positive cell. Asterisks: Vasa-negative cells. (c) Reaggregation phase (gastrula stage). Square: reaggregate. Red Vasa-positive cells are seen in the egg periphery. (d) Somitic embryo. Red Vasa-positive cells are seen outside the embryo. E: somitic embryo. (e) Upper image: histology of 3-weeks postfertilization embryo (longitudinal section, hematoxylin and eosin). CE: cephalic region; CA: caudal region. Square: gonadal ridge. Lower image: arrow: PGC arriving at gonadal ridge (transverse section, hematoxylin and eosin). (f) Four weeks postfertilization embryo. Square: Vasa-positive zone coincident with gonadal ridge region identified by histology.

germ cell number criterion, was observed at 30 days postfertilization (Arezo et al., 2007). In most gonochoric teleosts examined to date sex differentiation was shown to occur at different posthatching times beginning earlier in females than in males (reviewed by Nakamura et al., 1998; Strüssman and Nakamura, 2002). Conversely, in *O. latipes* (Beloniformes, Hamaguchi, 1982), *Gambusia affinis* (Cyprinodontiformes, Koya et al., 2003), and *A. charrua* (Cyprynodontiformes, Arezo et al., 2007) sex differentiation occurs during prehatching stages (germ cells proliferate at a higher rate in presumptive ovaries).

Yamamoto (1969) identified two patterns of gonadal development in gonochorists. Early gonads directly develop into a testis or ovary in the "differentiated" type, whereas in the "undifferentiated" type, gonads first pass through an ovary-like stage. Transient occurrence of intersex gonads is not present in the differentiated gonochorists and sporadic intersexes would appear only in "undifferentiated" gonochorists (Yamamoto, 1969; Strüssman and Nakamura, 2002). Histological analyses of gonadal differentiation in the 341 individuals examined of *A. charrua* from 3 weeks postfertilization embryos to adults showed neither 100% ovary-like stage nor intersex gonads. Therefore, *A. charrua* sexual strategy corresponds to a "differentiated gonochoric" pattern (Arezo et al., 2007).

Gonad morphology has been employed to identify annual reproductive cycles, onset of reproductive maturity, spawning rhythms, and various other aspects of the reproductive biology. Moreover, histological analyses are extensively used as an accurate method to determine reproductive patterns in teleosts (Parenti and Grier, 2004). *A. charrua* spawn daily from the time they reach sexual maturity until senescence. Oocytes and spermatogenetic cysts in all different stages of development were observed during the breeding months (Arezo et al., 2007) in agreement with an asynchronous spawning pattern for females (reviewed by Wallace and Selman, 1981) and a continuous spawning for males. *A. charrua* ovarian duct network opens into a hollow structure, "oocyte chamber" as in other *Austrolebias* species. Since this "oocyte chamber" contained ovulated oocytes, it was suggested that it could function as an oocyte storage compartment (L. Pereiro, personal communication, 2002). In the nonannual cyprinodontid *Fundulus heteroclitus*, a similar structure called "ovisac" was found (Brummett et al., 1982).

The organization of *A. charrua* testes corresponds to the restricted pattern since spermatogonia are confined to the distal end of the lobules (Grier, 1981). This type of testis organization is uniquely present in fishes belonging to the series Atherinomorpha, whereas the unrestricted pattern (spermatogonia occurring along the testicular lobules) is present in most teleosts (Parenti and Grier, 2004). The main cellular aspects of gametogenesis in *A. charrua* can be summarized as: First, during vitellogenesis, the yolk appears as small globules that continuously fuse. This process of deposition results in a fluid mass that occupies almost the entire mature oocyte (Arezo et al., 2007). The fluid yolk character has been identified only in Atherinomorpha (Parenti and Grier, 2004). These authors attribute phylogenetic value to this condition that, along with the restricted spermatogonial pattern, appears to be uniquely derived characteristics for Atherinomorpha. Second, the eggs of *A. charrua* are laid in the bottom mud, and the ultrastructure of the outer surface of the vitelline envelope contains adhesive hair-like filaments according to their demersal condition (Arezo et al., 2007).

Early sexual maturation observed in *A. charrua* is one of the typical features shared by organisms inhabiting temporary and unpredictable habitats (Stearns, 1992; Winemiller and Rose, 1992). Early maturation was also documented in two African annual fish species showing the earliest known maturation schedule: 18 days in *Nothobranchius furzeri* and 17 days in *Nothobranchius kadleci*, the shortest generation time among vertebrates (Blazek et al., 2013).

5.4 CONCLUSIONS AND PERSPECTIVES

Information generated concerning sex determination and differentiation is a useful contribution to the knowledge of reproductive biology of *A. charrua*. This knowledge is essential for planning conservation and management strategies. Nevertheless, many important questions, particularly related to sex determination, remain to be answered. For example, studies are needed to determine presence or absence of sex chromosomes in *A. charrua* karyotype; to evaluate thermosensibility, including that of individuals from the whole distribution area of this species; to elucidate whether the male sex-specific expressed sequence belongs to an alternative splicing *Dmrt1* gene version or to a noncoding RNA with potential epigenetic regulatory function and to clarify its hierarchical level in the sex determination cascade of this sequence. It is essential to obtain more sequence information (e.g., stage-specific transcriptomes) to reveal this sequence identity. Currently, there is increasing evidence that inherited variation in ecologically relevant traits can be generated through epigenetic mechanisms. Thus, epigenetics can significantly improve our understanding of the environmental versus genetic influences on sex determination of sensitive species and the responses of organisms under a global climate change scenario. The growing interest in this research area will undoubtedly generate many new insights on the role of epigenetic modifications on sex determination (Piferrer, 2013).

The identification of PGCs in living embryos by the injection of biomarkers as fusion protein Vasa-GFP has been widely utilized for studying the PGC migration and for cryopreservation of this cell population in various fish species. This nontransgenic method is especially useful for the conservation of genetic resources in endangered natural fish populations, since it is not possible to cryopreserve teleost oocytes and embryos. Cryopreservation of PGCs is an alternative approach, since PGCs are capable of differentiation in either sperm or oocytes (Yoshizaki et al., 2005). This method is the best alternative for the conservation of the diploid genome in fish until embryo cryopreservation is achieved (Riesco et al., 2012).

ACKNOWLEDGMENTS

The authors wish to thank Cora Chalar for her key assistance in technical approaches (RNA isolation and RT-PCR); Nuria Lahuerta for her important help in setting up the molecular techniques; MSc. Gustavo Saona for his invaluable assistance with the statistical analyses; Dr. Miguel Concha and coworkers for technical support from Facultad de Medicina (Universidad de Chile) in microinjections of one-celled embryos; and AMSUD-PASTEUR Regional Research training fellowships program/young researchers, International Union of Biochemistry and Molecular Biology (IUBMB), Wood-Whelan Research Fellowships, Agencia Nacional de Investigación e Innovación (ANII FCE2007), Programa de Desarrollo de las Ciencias Básicas (PEDECIBA), and Facultad de Ciencias—Universidad de la República, Montevideo, Uruguay for financial support. The original findings presented in this chapter were part of a PEDECIBA (Programa de Desarrollo Ciencias Básicas) PhD thesis by MJA (Facultad de Ciencias—Universidad de la República).

APPENDIX: THERMOSENSIBILITY EVALUATION IN *AUSTROLEBIAS CHARRUA* SEX DETERMINATION

Adult individuals of *Austrolebias charrua* (three males and three females) were collected in temporal ponds from Rocha Department, Uruguay (GPS: 33°54′09″S/53°40′38″W). Reproduction was carried out using one male and one female per aquarium containing peat moss during 1 month as previously described (Arezo et al., 2005, 2007). For each couple, the embryos (arrested in diapause *I* = late blastula stage) buried in the peat moss were collected and divided into two experimental groups, one cultured at 19°C and the other at 25°C. The cultures were carried out in Yamamoto solution until the first histological sign of gonadal sex differentiation (Arezo et al., 2007), then the embryos were placed in sterilized peat moss. In order to simulate the natural dry season, peat moss was dried gradually and kept in darkness for 4 months. After this period, the peat moss was placed in dechlorinated water where the embryos hatched in 12–24 h. The fries were cultured as described (Arezo et al., 2007), and sexual maturity was reached 3 months later. The adults were sexed macroscopically, and dead juveniles were fixed and processed in order to sex them by histology (Arezo et al., 2007). The sex ratio was calculated, and statistical analyses were carried out using STATA12 statistical software (STATACorp. LP, MSc. Gustavo Saona license). Statistical analysis employed in the present work was Poisson regression, a type of generalized linear model (GLM). The GLMs are the most suitable statistical tests because they were developed to analyze proportion data and also to allow study of whether a variable (e.g., sex) depends or not on another variable (e.g., temperature) (Wilson and Hardy, 2002).

To analyze the ratio of males to females in nature, five ponds from the same distribution area were sampled at the end of the rainy season (GPS: 33°54′09″S/53°40′38″W). A chi-square (goodness of fit) test was applied using R 2.7.0 (R Development Core Team).

A.1 Whole-Mount Immunostaining

Antibodies were raised in a rabbit against the MITTY-2 region of the zebrafish Vasa sequence (Yoon et al., 1997). For Vasa protein staining, *A. charrua* embryos were fixed in PFA and 4% PBS overnight at room temperature. After dechorionation and blocking of endogenous peroxidase activity (4 parts methanol, 1 part H_2O_2 3%), embryos were exposed to blocking solution (20% fetal bovine serum, 1% DMSO, 0.1% Tween in PBS). Incubation with the primary polyclonal antibody (anti-Vasa, kindly donated by Dr. Nancy Hopkins) was done at a 1:1000 dilution in blocking solution at 4°C overnight, and washed with PBST and 1% DMSO five times for 20 min each. Positive and negative controls were included. Incubation with a second antibody (goat anti-rabbit IgG coupled with peroxidase, SIGMA) was performed at 1:250 dilution during 4 h at room temperature. Vasa-positive cells were detected by peroxidase substratum AEC (3-amino-9-ethylcarbazole, SIGMA).

REFERENCES

Arezo, M.J. 2012. Análisis de los mecanismos de determinación y diferenciación del sexo en *Austreolebias charrua* (Cyprinodontiformes: Rivulidae). (Unpublished doctoral dissertation thesis). Universidad de La República. Montevideo, Uruguay.

Arezo, M.J., S. D'Alessandro, N. Papa, R. de Sà, and N. Berois. 2007. Sex differentiation pattern in the annual fish *Austrolebias charrua* (Cyprinodontiformes: Rivulidae). *Tissue and Cell* 39:89–98. doi: 10.1016/j.tice.2007.01.004

Arezo, M.J., N. Papa, V. Gutiérrez, G. García, and N. Berois. 2014. Sex determination in annual fishes: Searching for the master sex-determining gene in *Austrolebias charrua* (Cyprinodontiformes, Rivulidae). *Genetics and Molecular Biology* 37:364–374. doi: 10.1590/s1415-47572014005000009

Arezo, M.J., L. Pereiro, and N. Berois. 2005. Early development in the annual fish *Cynolebias viarius* (Cyprinodontiformes: Rivulidae). *Journal of Fish Biology* 66:1357–1370. doi: 10.1111/j.1095-8649.2005.00688

Braat, A.K., S. van de Water, H. Goos, J. Bogerd, and D. Zivkovic. 2000. Vasa protein expression and localization in the zebrafish. *Mechanisms of Development* 95:271–274. doi: 10.1016/S0925-4773(00)00344-0

Belote, D.F., and W.J.E.M. Costa. 2004. Reproductive behavior patterns in three species of the South American annual fish genus *Austrolebias* Costa, 16 1998 (Cyprinodontiformes, Rivulidae). *Boletim do Museu Nacional*, Nova 17 Serie, Rio de Janeiro 514:1–7.

Blazek, R., M.J. Polacik, and M. Reichard. 2013. Rapid growth, early maturation and short generation time in African annual fishes. *EvoDevo* 4:24. doi: 10.1186/2041-9139-4-24.

Brummett, A.R., J.N. Dumont, and J.R. Larkin. 1982 The ovary of *Fundulus heteroclitus*. *Journal of Morphology* 173:1–16. doi: 10.1002/jmor.1051730102

Clinton, W.L., J. Burney, and J. Le Boeuf. 1993. Sexual selection effects on male life history and the pattern of male mortality. *Ecology* 74:1884–1892. doi.org/10.2307/1939945

Conover, D.O., and S.W. Heins. 1987. Adaptative variation in environmental and genetic sex determination in a fish. *Nature* 326:496–498. doi: 10.1038/326496a0

Conover, D.O., and B.E. Kynard. 1981. Environmental sex determination: Interaction of temperature and genotype in a fish. *Science* 213:577–579. doi: 10.1126/science.213.4507.577

D'Anatro, A., and M. Loureiro. 2005. Geographic variation in the annual killifish *Austrolebias luteoflamulatus* Vaz Ferreira, Sierra, Scaglia (Cyprinodontiformes, Rivulidae). *Journal of Fish Biology* 67:849–865. doi: 10.1111/j.0022-1112.2005. 00791

Devlin, R.H., and Y. Nagahama. 2002. Sex determination and sex differentiation in fish: An overview of genetic, physiological and environmental influences. *Aquaculture* 208:191–364. doi: 10.1016/s0044-8486(02)00057

Evans, T., C.M. Wade, F.A. Chapman, A.D. Johnson, and M. Loose. 2014. Acquisition of germ plasm accelerates vertebrate evolution. *Science* 344:200–2003. doi: 10.1126/science.1249325

Extravour, C.G., and M. Akam. 2003. Mechanisms of germ cell specification across the metazoans: Epigenesist and preformation. *Development* 130:5869–5884. doi: 10.1242/dev.00804

Extravour, C.G.M. 2007. Evolution of the bilaterian germ line: Lineage origin and modulation of specification mechanisms. *Integrative and Comparative Biology* 47:770–785. doi: 10.1093/icb/icm027

Fernandino, J.I., L.G. Guilgur, and G.M. Somoza. 2006. *Dmrt1* expression analysis during spermatogenesis in pejerrey, *Odontesthes bonariensis*. *Fish Physiology and Biochemistry* 32:231–240. doi: 10.1007/s10695-006-9005-9

Fernandino, J.I., R.S. Hattori, T. Shinoda et al. 2008. Dimorphic expression of *dmrt1* and *cyp19a1* (ovarian aromatase) during early gonadal development in pejerrey, *Odontesthes bonariensis*. *Sex Development* 2:316–324. doi: 10.1159/000195681

García, G. 2006. Multiple simultaneous speciation in killifishes of the *Cynolebias adloffi* species complex (Cyprinodontiformes, Rivulidae) from phylogeography and chromosome data. *Journal of Zoological Systematics and Evolutionary Research* 44:75–87. doi: 10.1111/j.1439-0469.2005.00346

García, G., M. Loureiro, N. Berois et al. 2009. Pattern of differentiation in the annual killifish genus *Austrolebias* (Cyprinodontiformes; Rivulidae) from a biosphere reserve site in South America: A multidisciplinary approach. *Biological Journal of the Linnean Society* 98:620–635. doi: 10.1111/j.1095-8312.2009.01303

García, D., M. Loureiro, and B. Tassino. 2008. Reproductive behavior in the annual fish *Austrolebias reicherti* (Cyprinodontiformes, Rivulidae). *Neotropical Ichthyology* 6:243–248. doi.org/10.1590/S1679-622520 08000200012

Gevers, P., J. Dulos, H. Schipper, and L. Timmermans. 1992. Origin of primordial germ cells, as characterized by the presence of nuage, in embryos of the teleost fish *Barbus conchonius*. *European Journal of Morphology* 30:195–204.

Gilbert, S.F. 2013. *Developmental Biology*. 10th edition. Sunderland, MA: Sinauer Associates, Inc.

Graves, J.A.M., and C. Peichel. 2010. Are homologies in vertebrate sex determination due to shared ancestry or to limited options? *Genome Biology* 11:205. doi: 10.1186/gb-2010-11-4-205

Grier, H.J. 1981. Cellular organization of the testis and spermatogenesis in fishes. *American Zoologist* 21:345–357. doi: 10.1093/icb/21.2.345

Hamaguchi, S. 1982. A light- and electron-microscopic study on the migration of primordial germ cells in the teleost, *Oryzias latipes*. *Cell and Tissue Research* 227:139–151. doi: 10.1007/BF00206337

Hattori, R.S., Y. Murai, M. Oura et al. 2012. A Y-linked anti-Mullerian hormone duplication takes over a critical role in sex determination. *Proceedings of the National Academy of Sciences of the United States of America* 109:2955–2959. doi: 10.1073/pnas.1018392109

Herpin, A., M.C. Adolfi, B. Nicol et al. 2013. Divergent expression regulation of gonad development genes in medaka shows incomplete conservation of the downstream regulatory network of vertebrate sex determination. *Molecular Biology and Evolution* 30:2328–2346. doi: 10.1093/molbev/mst130

Herpin, A., I. Braasch, M. Kraeussling et al. 2010. Transcriptional rewiring of the sex determining *dmrt1* gene duplicate by transposable elements. *PloS Genetics* 6:e100844. doi:10.1371/journal.pgen.1000844.http://www.plosgenetics.org/article/fetchObject.action?uri=info:doi/10.1371/journal.pgen.1000844&representation

Herpin, A., S. Rohr, D. Riedel, N. Kluever, E. Raz, and M. Schartl. 2007. Specification of primordial germ cells in medaka (*Oryzias latipes*). *BMC Development Biology* 7:3. doi: 10.1186/1471-213X-7-3

Herpin, A., and M. Schartl. 2011. *Dmrt1* genes at the crossroads: A widespread and central class of sexual development factors in fish. *Federation of European Biochemical Societies Journal* 278:1010–1019. doi: 10.1111/j.1742-4658.2011.08030

Heule, C, W. Salzburger, and A. Böhne. 2014. Genetics of sexual development: An evolutionary playground for fish. *Genetics* 196:579–591. doi: 10.1534/genetics.114.161158

Hodgkin, J. 2002. The remarkable ubiquity of DM domain factors as regulators of sexual phenotype: Ancestry or aptitude? *Genes Development* 16:2322–2326. doi:10.1101/gad.1025502

Hornung, U., A. Herpin, and M. Schartl. 2007. Expression of the male determining gene *dmrt1bY* and its autosomal coorthologue *dmrt1a* in medaka. *Sex Development* 1:197–206. doi: 10.1159/000102108

Huang, X., H. Cheng, Y. Guo et al. 2002. A conserved family of doublesex-related genes from fish. *Journal of Experimental Zoology* 294:63–67. doi: 10.1002/jez.10044

Iwamatsu, T. 2004. Stages of normal development in the medaka *Oryzias latipes*. *Zoological Sciences* 11:825–839. doi: 10.1016/j.mod.2004.03.012

Johnson A. D., M. Drum, R.F. Bachvarova et al. 2003. Evolution of predetermined germ cells in vertebrate embryos: Implications for macroevolution. *Evolution & Development* 5:414–431.

Kamiya, T., W. Kai, S. Tasumi et al. 2012. A trans-species missense SNP in *Amhr2* is associated with sex determination in the tiger pufferfish, *Takifugu rubripes* (Fugu). *PLoS Genetics* 8:e1002798. doi: 10.1371/journal.pgen.1002798.

Kikuchi, K., and S. Hamaguchi. 2013. Novel sex-determining genes in fish and sex chromosome evolution. *Developmental Dynamics* 242:339–353. doi: 10.1002/dvdy.23927

Knaut, H., H. Steinbeisser, H. Schwarz, and C. Nüsslein-Volhard. 2002. An evolutionarily conserved region in the vasa 3′UTR targets RNA translation to the germ cells in the zebrafish. *Current Biology* 12:454–466.

Kobayashi, T., M. Matsuda, H. Kajiura-Kobayashi et al. 2004. Two DM domain genes, *dmy* and *dmrt1*, involved in testicular differentiation and development in the medaka, *Oryzias latipes*. *Developmental Dynamics* 231:518–526. doi: 10.1002/dvdy.20158

Kopp, A. 2012. *Dmrt* genes in the development and evolution of sexual dimorphism. *Trends in Genetics* 28:175–184. doi: 10.1016/j.tig.2012.02.002

Koya, Y., A. Fujita, F. Niki, E. Ishihara, and H. Piyama. 2003. Sex differentiation and puberty development of gonads in the viviparous mosquitofish *Gambusia affinis*. *Zoological Sciences* 20:1231–1242. doi: 10.2108/zsj.20.1231

Koumoundouros, G., M. Pavlidis, L. Anezaki et al. 2002. Temperature sex determination in the European sea bass, *Dicentrarchus labrax* (L.,1758) (Teleostei, Perciformes, Moronidae): Critical sensitive ontogenetic phase. *Journal of Experimental Zoology* 292:573–579. doi: 10.1002/jez.10095

Lin, F., S. Xu, D. Ma et al. 2012. Germ line specific expression of a *vasa* homologue gene in turbot (*Scophthalmus maximus*): Evidence for *vasa* localization at cleavage furrows in euteleostei. *Molecular Reproduction and Development* 79:803–813. doi: 10.1002/mrd.22120

Magnúsdóttir, E., and M.A. Surani. 2014. How to make a primordial germ cell. *Development* 141:245–252. doi: 10.1242/dev.098269

Mank, J.E., D.E.L. Promislow, and J.C. Avise. 2006. Evolution of alternative sex-determining mechanisms in teleost fishes. *Biological Journal of the Linnean Society* 87:83–93. doi: 10.1111/j.1095-8312.2006.00558

Matsuda, M.Y., A. Nagahama, T. Shinomiya, and C. Sato. 2002. *Dmy* is a Y-specific DM-domain gene required for male development in the medaka fish. *Nature* 417:559–563. doi :10.1038/nature751

Myosho, T., H. Otake, H. Masuyama et al. 2012. Tracing the emergence of a novel sex determining gene in medaka, *Oryzias luzonensis*. *Genetics* 191:163–170. doi: 10.1534/genetics.111.137497

Nanda, I., M. Kondo, U. Hornung et al. 2002. A duplicated copy of *dmrt1* in the sex determining region of the Y chromosome of the medaka, *Oryzias latipes*. *Proceedings of the National Academy of Sciences of the United States of America* 99:11778–11783. doi: 10.1073/pnas.182314699

Nakamura, M., T. Kobayashi, X. Chang, and Y. Nagahama. 1998. Gonadal sex differentiation in teleost fish. *Journal of Experimental Zoology* 28:362–372. doi: 10.1002/(SICI)1097-010X(19980801)281:5<362::AID-JEZ3>3.0.CO;2-M

Nelson, J.S. 2006. *Fishes of the World*. 4th edition. Hoboken, New Jersey: John Wiley and Sons, Inc.

Opsina-Alvarez, N., and F. Piferrer. 2008. Temperature-dependent sex determination in fish revisited: Prevalence, a single sex ratio response pattern, and possible effects of climate change. *PlosOne* 3:e2837. doi:10.1371/journal.pone.0002837

Parenti, L.R. 2005. The phylogeny of atherinomorphs: Evolution of a novel fish reproductive system. In *Viviparous Fishes*, eds. M.C. Uribe and H.J. Grier. 13–30. Homestead, Florida: New Life Publications.

Parenti, L.R., and H.J. Grier. 2004. Evolution and phylogeny of gonad morphology in bone fishes. *Integrative and Comparative Biology* 44:333–348. doi: 10.1093/icb/44.5.333

Passos, C., B. Tassino, F. Reyes, and G.G. Rosenthal. 2014. Seasonal variation in female mate choice and operational sex ratio in wild populations of an annual fish, *Austrolebias reicherti*. *PLoS ONE* 9: e101649. doi:10.1371/journal.pone.0101649

Peichel, C.L., J.A Ross, C.K. Matson et al. 2004. DM: The master sex-determination locus in threespine sticklebacks is on a nascent Y chromosome. *Current Biology* 14:1416–1424. doi: 10.1016/j.cub.2004.08.030

Penman, D.J., and F. Piferrer. 2008. Fish gonadogenesis. Part I: Genetic and environmental mechanisms of sex determination. *Reviews in Fisheries Science* 16:14–32. doi: 10.1080/10641260802324610

Piferrer, F. 2013. Epigenetics of sex determination and gonadogenesis. *Development Dynamics* 242:360–370. doi: 10.1002/dvdy.23924

Raz, E. 2000. The function and regulation of *vasa*-like genes in germ-cell development. *Genome Biology* 1:1017.1–1017.6. doi: 10.1186/gb-2000-1-3-reviews1017

Reichard, M., M. Polacik, and O. Sedlácek. 2009. Distribution, colour polymorphism and habitat use of the African killifish *Nothobranchius furzeri*, the vertebrate with the shortest life span. *Journal of Fish Biology* 74:198–212. doi: 10.1111/j.1095-8649.2008.0212

Richardson, B.E. and R. Lehmann. 2010. Mechanisms guiding primordial germ cell migration: Strategies from different organisms. *Nature Reviews Molecular Cell Biology* 11:37–49. doi: 10.1038/nrm2815

Riesco, M.F., F. Martinez-Pastor, O. Chereguini, and V. Robles. 2012. Evaluation of zebrafish (*Danio rerio*) PGCs viability and DNA damage using different cryopreservation protocols. *Theriogenology* 77:122–130. doi: 10.1016/j.theriogenology.2011.07.024

Schultheis, C., A. Böhne, M. Schartl, J.N. Volff, and D. Galiana-Arnoux. 2009. Sex determination diversity and sex chromosome evolution in poeciliid fish. *Sex Development* 3:68–77. doi: 10.1159/000223072

Setiamarga, D.H.E, M. Miya, Y. Yamanoue et al. 2008. Interrelationships of Atherinomorpha (medakas, flyingfishes, killifishes, silversides, and their relatives): The first evidence based on whole mitogenome sequences. *Molecular Phylogenetics and Evolution* 49:598–605. doi: 10.1016/j.ympev.2008.08.008

Shinomiya, A., M. Tanaka, T. Kobayashi, Y. Nagahama, and S. Hamaguchi. 2000. The vasa-like gene, olvas, identifies the migration path of primordial germ cells during embryonic body formation stage in the medaka, *Oryzias latipes*. *Development Growth and Differentiation* 42:317–326. doi: 10.1046/j.1440-169x.2000.00521

Smith, C.A., P.J. McClive, P.S. Western, K.J. Reed, and A.H. Sinclair. 1999. Evolution: Conservation of a sex-determining gene. *Nature* 402:601–602. doi:10.1038/45130

Stearns, S.C. 1992. *The Evolution of Life Histories*. Oxford, UK: Oxford University Press.

Strüssman, C.A., and M. Nakamura. 2002. Morphology, endocrinology, and environmental modulation of gonadal sex differentiation in teleost fishes. *Fish Physiology* 26:13–29. doi: 10.1023/A:1023343023556

Strüssmann, C.A., T. Saito, M. Usui, H. Yammada, and F. Takashima. 1997. Thermal thresholds and critical period of thermolabile sex determination in two atherinid fishes, *Odontesthes bonariensis* and *Patagonia hatcheri*. *Journal of Experimental Zoology* 278:167–177.

Timmermans, L., and N. Taverne. 1989. Segregation of primordial germ cells: Their numbers and fate during early development of *Barbus conchonius* (Cyprinidae, Teleostei) as indicated by 3H-thymidine incorporation. *Journal of Morphology* 202:225–237. doi: 10.1002/jmor.1052020209

Tomaszkiewicz, M., D. Chalopin, M. Schartl, D. Galiana, and J.N. Volff. 2014. A multicopy Y-chromosomal SGNH hydrolase gene expressed in the testis of the platyfish has been captured and mobilized by a Helitron transposon. *BMC Genetics* 15:44. doi: 10.1186/1471-2156-15-44

Tripathi, N., M. Hoffmann, D. Weigel, and C. Dreyer. 2009. Linkage analysis reveals the independent origin of Poeciliid sex chromosomes and a case of atypical sex inheritance in the guppy (*Poecilia reticulata*). *Genetics* 182:365–374. doi: 10.1534/genetics.108.098541

Valenzano, D.R., J. Kirschner, R. Kamber et al. 2009. Mapping loci associated with tail color and sex determination in the short-lived fish *Nothobranchius furzeri*. *Genetics* 183:1385–1395. doi: 10.1534/genetics.109.108670

Valenzuela, N., D.C. Adams, and F.J. Janzen. 2003. Pattern does not equal process: Exactly when is sex environmentally determined? *The American Naturalist* 161:676–683. doi: 10.1086/368292

Valenzuela, N., J. Neuwald, and R. Literman. 2013. Transcriptional evolution underlying vertebrate sexual development. *Development Dynamics* 242:307–319. doi: 10.1002/dvdy.23897

Volcan, M.V., L.A. Sampaio, D.C. Bongalhardo, and R.B. Robaldo. 2013. Reproduction of the annual fish *Austrolebias nigrofasciatus* (Rivulidae) maintained at different temperatures. *Journal of Applied Icthyology* 29:648–652. doi: 10.1111/jai.12013

Volff, J.N. 2005. Genome evolution and biodiversity in teleost fish. *Heredity* 94:280–294. doi:10.1038/sj.hdy.6800635

Wallace, R.A., and K. Selman. 1981. Cellular and dynamic aspects of oocyte growth in teleosts. *American Society of Zoologists* 21:325–343. doi:10.1093/icb/21.2.325

Warner, D.A., and R. Shine. 2011. Interactions among thermal parameters determine offspring sex under temperature-dependent sex determination. *Proceedings of the Royal Society of the Biological Sciences* 278:256–265. doi: 10.1098/rspb.2010.1040

Wilson, K., and I.C.W. Hardy. 2002. Statistical analyses of sex ratios: An introduction. In *Ratios: Concepts and Research Methods*, ed. I.C.W. Hardy. 48–92. Cambridge: Cambridge University Press.

Winemiller, K.O., and K.A. Rose. 1992. Patterns of life history diversification in North American fishes: Implications for population growth. *Canadian Journal of Fisheries and Aquatic Sciences* 49:2196–2218. doi: 10.1139/f92-242

Wolke, U., G. Weidinger, M. Köprunner, and E. Raz. 2002. Multiple levels of posttranscriptional control lead to germ line-specific gene expression in the zebrafish. *Current Biology* 12:289–294. doi:10.1016/S0960-9822(02)00679-6

Yamaguchi, A., K.H. Lee, H. Fujimoto, K. Kadomura, S. Yasumoto, and M. Matsuyama. 2006. Expression of the *dmrt* gene and its roles in early gonadal development of the Japanese pufferfish *Takifugu rubripes*. *Comparative Biochemistry and Physiology—Part D: Genomics and Proteomics* 1:59–68. doi: 10.1016/j.cbd.2005.08.003

Yamamoto, T. 1969. Sex differentiation. In *Fish Physiology*, eds. W.S. Hoar and D. Randall. Vol. 3, 117–175. New York: Academic Press.

Yano, A., R. Guyomard, B. Nicol et al. 2012. An immune-related gene evolved into the master sex-determining gene in rainbow trout, *Oncorhynchus mykiss*. *Current Biology* 22:1423–1428. doi: 10.1016/j.cub.2012.05.045

Yoon, C., K. Kawakami, and N. Hopkins. 1997. Zebrafish vasa homologue RNA is localized to the cleavage planes of 2- and 4-cell-stage embryos and is expressed in the primordial germ cells. *Development* 124:3157–3166.

Yoshizaki, G., Y. Tago, Y. Takeuchi, E. Sawatari, T. Kobayashi, and T. Takeuchi. 2005. Green fluorescent protein labelling of primordial germ cells using a nontransgenic method and its application for germ cell transplantation in Salmonidae. *Biology of Reproduction* 73:88–93. doi: 10.1095/biolreprod.104.034249

CHAPTER 6

Annual Fishes of the Genus *Nothobranchius* as an Experimental Model

Alessandro Cellerino

CONTENTS

6.1 INTRODUCTION

Annual fishes of the genus *Nothobranchius* (especially *N. furzeri*) are emerging as a model system in various biological disciplines. They are short-lived, their husbandry is relatively easy, their transgenic methods are established, their genome was recently sequenced and assembled, and they are much cheaper to house than laboratory mice. *Nothobranchius* fishes are, therefore, ideal organisms to study the effects of genetic and nongenetic interventions on aging and aging-related pathologies.

In addition, large differences exist in life expectancy between different species, and this offers a unique paradigm to investigate the genetic architecture of longevity in natural conditions.

6.2 *NOTHOBRANCHIUS*: A HISTORICAL PERSPECTIVE

Nothobranchius annual fishes experienced an initial moment of interest in the 1970s and 1980s. *N. guentheri* was first studied in the context of sexual selection by Haas (2006) and were then used in some exploratory aging studies but never became established, possibly due to the relatively long lifespan of this species, which is about 18 months (Bailey, 1972; Markofsky and Perlmutter, 1973; Markofsky, 1976; Markofsky and Milstoc, 1979; Balmer, 1982; Cooper et al., 1983). Several species of *Nothobranchius* were studied during the same period as a model system for embryological studies, especially in the context of early cleavage, morphogenesis, and diapause (Lesseps et al., 1975, 1979; Van Haarlem, 1979; Van Haarlem et al., 1981, 1983a,b; Inglima et al., 1981; Matias, 1984; Levels et al., 1986; Levels and Denucé, 1988). Interest in *Nothobranchius* vanished in the 1990s and was revived only after 2003 by the description of the extremely short lifespan of *N. furzeri* (Valdesalici and Cellerino, 2003). Ever since 2006, there has been a continuous increase in the number of publications on *Nothobranchius*, and during 2014, on average there was one paper published every 15 days.

N. furzeri is currently the best-studied species, and its genome was recently sequenced, assembled, annotated, deposited, and made available to the scientific community (Harel et al., 2015). A second release of the genome from the Fritz Lipmann Institute for Age Research in Jena (Reichwald et al., 2009) was expected during 2015. For this reason, I will mainly concentrate on *N. furzeri* in this chapter, even if other species of *Nothobranchius* have been used in recent years.

6.3 *NOTHOBRANCHIUS* AS A MODEL IN DEVELOPMENTAL BIOLOGY

The vast majority of studies on embryonic development of annual fishes used Neotropical forms, for example, *Austrofundulus* and *Austrolebias* species. However, the recent development of microinjection techniques for *N. furzeri* initiated the use of this species for developmental biology.

Early cleavage of teleost embryos is characterized by a very fast cell cycle (15–30 min) and lack of G_1 and G_2 phases (Newport and Kirschner, 1982; Kimmel, 1995). On the other hand, annual killifishes are characterized by a slow cell cycle (60–90 min). To investigate whether killifish embryos lack G_1 and G_2 phases, *N. furzeri* embryos were microinjected with mRNAs coding for the fluorescent ubiquitination-based cell cycle indicator (FUCCI) system for imaging cell cycle progression (Sugiyama et al., 2009). In this system, a red and a green reporter are fused to protein motives that drive degradation in the G_1 and the S phases, respectively. From these experiments, it is clear that *N. furzeri* embryos do not have a G_1 phase during cleavage and, further, that the first five divisions are synchronous, with asynchrony starting from the sixth division (Dolfi et al., 2014).

This very slow cell cycle makes *N. furzeri* a unique model system for evo-devo of early embryonic stages.

6.4 *NOTHOBRANCHIUS FURZERI* AS AN AGING MODEL

Aging of *N. furzeri* is characterized by degenerative processes that are evident at the macroscopic level, such as emaciation and spinal curvature. At the behavioral level, there is reduced locomotor activity and impairment in learning paradigms (Valenzano et al., 2006). Histopathological examinations have revealed accumulation of lipofuscin (Terzibasi et al., 2008) and age-dependent lesions in heart, liver, and kidney that are similar to age-dependent lesions described in other small teleosts (Di Cicco et al., 2011). A peculiar characteristic of *N. furzeri* aging is a high incidence of spontaneous neoplasias that makes it a unique model for spontaneous tumorigenesis (Di Cicco et al., 2011).

At the molecular level, aging of *N. furzeri* is also associated with increased apoptosis (Di Cicco et al., 2011); telomere erosion (despite expression of somatic telomerase activity; Hartmann et al., 2009); reduced mitochondrial function (Hartmann et al., 2011); and expression of genes that act as negative regulators of the cell cycle, such as cyclin-dependent kinase inhibitor 1A (*CDKN1A*) and growth arrest DNA damage 45 (*GADD45*), indicating an increase in cellular senescence *in vivo* (Graf et al., 2013) that is corroborated by increased expression of the senescence marker senescence-associated beta-galactosidase (SA-βGal; Genade et al., 2005). In addition, typical markers for adult neuronal stem cells are expressed in *N. furzeri* brain, and aging is associated with reduced function of adult neuronal stem cells and enhanced expression of glial fibrillary acidic protein (*GFAP*; Tozzini et al., 2012).

6.5 GENOME-WIDE GENE EXPRESSION STUDIES ON AGING

The introduction of new sequencing technologies has allowed genome-wide quantification of gene expression in *N. furzeri* both for coding and noncoding RNAs (Baumgart et al., 2012, 2014). MicroRNAs (miRNAs) are short noncoding RNAs that regulate gene expression by binding to target mRNAs. It was found that tumor suppressor miRNAs with known negative interactions with MYC and/or positive interactions with TP53 are often upregulated miRNAs (e.g., miR-23a, miR-26a/b, miR-29a/b, miR-101a). Conversely, oncogenic miRNAs that are MYC targets are often downregulated miRNAs (miR-7a, members of miR cluster 17–92). These latter were previously shown to be downregulated during cellular senescence. In addition, three regulated miRNAs (miR-181c, miR-29a, and miR-338) are known to be age-regulated and globally contribute to regulation of their targets in the human brain. Therefore, there appears to be a degree of evolutionary conservation in age-dependent miRNA expression between humans and *N. furzeri*.

Comparison of genome-wide regulation of protein-coding transcripts with human data sets revealed conserved upregulation of ribosome, lysosome, and complement activation and conserved downregulation of synapse, mitochondrion, proteasome, and spliceosome. Downregulated genes differ in their temporal profiles: neurogenesis and extracellular matrix genes showed rapid decay, synaptic and axonal genes a progressive decay. A substantial proportion of differentially expressed genes (~40%) showed inversion of their temporal profiles in the last time point: spliceosome and proteasome showed initial downregulation and stress response genes initial upregulation. Extensive regulation was detected for chromatin remodelers of the DNA methyltransferase (DNMT) and chromobox homolog (CBX) families as well as members of the polycomb repressive complex and was mirrored by an upregulation of trimethylation of histone H3 at lysine 27 (H3K27me3).

6.6 *NOTHOBRANCHIUS* AS A MODEL FOR QUANTITATIVE GENETICS

Annual fishes of the genus *Nothobranchius* show large variations in lifespan and expression of age-related phenotypes between closely related populations/species that broadly correlate with duration of their wild habitats (Tozzini et al., 2013). This makes them a unique model to investigate the genetic architecture of naturally evolved differences in aging (Valenzano et al., 2009).

In *N. furzeri*, a particularly short-lived strain (GRZ) was isolated (Terzibasi et al., 2008). By crossing it with a longer lived strain, a genetic map comprising 355 markers was built, and this map was then used to identify quantitative trait loci (QTL) for lifespan. This study identified one significant and three suggestive loci showing that (1) lifespan determination in *N. furzeri* is polygenic and (2) candidate protein-coding genes can be identified by cross-species analysis including fishes and humans (Kirschner et al., 2012).

A second study involved crossing of *N. furzeri* with *N. kadleci* (Ng'oma et al., 2014). Lipofuscin and apoptotic cells were then quantified in F1 and F2 hybrids, and a genome-wide scan for QTL was performed. Surprisingly, lipofuscin and apoptosis appeared to be markers of different age-dependent biological processes controlled by different genetic mechanisms. Indeed, inheritance of lipofuscin accumulation showed hyperdominance and inheritance of apoptosis rate showed additivity, and these traits were uncorrelated in F2 hybrids and mapped to different loci (Ng'oma et al., 2014).

6.7 *NOTHOBRANCHIUS* AS AN EXPERIMENTAL MODEL: NONGENETIC INTERVENTIONS

The short lifespan and possibility of raising them in large numbers make *N. furzeri* a particularly useful model to investigate the effects of experimental manipulations on lifespan and aging-related phenotypes. It was shown that aging in *N. furzeri* can be influenced by temperature, drugs such as resveratrol, and dietary restriction (Valenzano et al., 2006; Terzibasi et al., 2009). These manipulations not only prolong lifespan but also influence both behavioral and cellular aging-related phenotypes.

6.8 *NOTHOBRANCHIUS* AS AN EXPERIMENTAL MODEL: GENETIC INTERVENTIONS

A breakthrough in the use of *N. furzeri* as an experimental model came from the establishment of gene manipulation technologies. The use of the Tol2 transposon system allows the insertion of an expression cassette in a random position of the genome of *N. furzeri* (Valenzano et al., 2011; Hartmann and Englert, 2012; Allard et al., 2013). Of seminal importance was the recent development of CRISPR/CAS9 techniques that allow targeted genome editing. This technique allows a facile generation of knockout lines and even the insertion of human mutations into the *N. furzeri* endogenous loci, inducing phenotypes that closely resemble the human pathology (Harel et al., 2015).

6.9 *NOTHOBRANCHIUS*: FUTURE PERSPECTIVES

The recent sequencing of the *N. furzeri* genome and the development of genome-editing techniques make *N. furzeri* a unique platform for investigations into the genetic control of aging. A number of interesting candidates have already emerged from whole-genome expression studies. In the future, genomic comparative studies will allow the identification of loci associated with naturally evolved differences in lifespan that will provide further candidates for functional studies.

In addition, genome-editing techniques will undoubtedly become popular for modeling human diseases in *N. furzeri* and then testing the effects of pharmacological interventions on disease-associated phenotypes.

REFERENCES

Allard, J.B., H. Kamei, and C. Duan. 2013. Inducible transgenic expression in the short-lived fish *Nothobranchius furzeri*. *Journal of Fish Biology* 82: 1733–1738. doi: 10.1111/jfb.12099

Bailey, R.G. 1972. Observations on the biology of *Nothobranchius guentheri* (Pfeffer) (Cyprinodontidae), an annual fish from the coastal region of east Africa. *The African Journal of Tropical Hydrobiology and Fisheries* 2: 33–43.

Balmer, R. 1982. The effect of age on body energy content of the annual cyprinodont fish, *Nothobranchius guentheri*. *Experimental Gerontology* 17: 139–143. doi: 0531-5565(82)90048-1 [pii]

Baumgart, M., M. Groth, S. Priebe et al. 2012. Age-dependent regulation of tumor-related microRNAs in the brain of the annual fish *Nothobranchius furzeri*. *Mechanisms of Ageing and Development* 133: 226–233. doi: 10.1016/j.mad.2012.03.015

Baumgart, M., M. Groth, S. Priebe et al. 2014. RNA-seq of the aging brain in the short-lived fish *N. furzeri*—Conserved pathways and novel genes associated with neurogenesis. *Aging Cell* 13: 965–974. doi: 10.1111/acel.12257

Cooper, E.L., A. Zapata, M. Garcia Barrutia, and J.A. Ramirez. 1983. Aging changes in lymphopoietic and myelopoietic organs of the annual cyprinodont fish, *Nothobranchius guentheri*. *Experimental Gerontology* 18: 29–38. doi: 0531-5565(83)90048-7 [pii]

Di Cicco, E., E.T. Tozzini, G. Rossi, and A. Cellerino. 2011. The short-lived annual fish *Nothobranchius furzeri* shows a typical teleost aging process reinforced by high incidence of age-dependent neoplasias. *Experimental Gerontology* 46: 249–256. doi: S0531-5565(10)00386-4 [pii]

Dolfi, L., R. Ripa, and A. Cellerino. 2014. Transition to annual life history coincides with reduction in cell cycle speed during early cleavage in three independent clades of annual killifish. *EvoDevo* 5: 32. doi: 10.1186/2041-9139-5-32

Genade, T., M. Benedetti, E. Terzibasi et al. 2005. Annual fishes of the genus *Nothobranchius* as a model system for aging research. *Aging Cell* 4: 223–233. doi: ACE165 [pii]10.1111/j.1474-9726.2005.00165.x.

Graf, M., N. Hartmann, K. Reichwald, and C. Englert. 2013. Absence of replicative senescence in cultured cells from the short-lived killifish *Nothobranchius furzeri*. *Experimental Gerontology* 48: 17–28. doi: S0531-5565(12)00058-7 [pii]

Haas, R. 1976. Sexual selection in *Nothobranchius guentheri* (Pisces: cyprinodontidae). *Evolution* 30(3):614–622.

Harel, I., B.A. Benayoun, B. Machado et al. 2015. A platform for rapid exploration of aging and diseases in a naturally short-lived vertebrate. *Cell* 160: 1013–1026. http://dx.doi.org/10.1016/j.cell.2015.01.038

Hartmann, N., and C. Englert. 2012. A microinjection protocol for the generation of transgenic killifish (species: *Nothobranchius furzeri*). *Developmental Dynamics* 241: 1133–1141. doi: 10.1002/dvdy.23789

Hartmann, N., K. Reichwald, A. Lechel et al. 2009. Telomeres shorten while Tert expression increases during ageing of the short-lived fish *Nothobranchius furzeri*. *Mechanisms of Ageing and Development* 130: 290–296. doi: S0047-6374(09)00008-6 [pii]10.1016/j.mad.2009.01.003

Hartmann, N., K. Reichwald, I. Wittig et al. 2011. Mitochondrial DNA copy number and function decrease with age in the short-lived fish *Nothobranchius furzeri*. *Aging Cell* 10: 824–831. doi:10.1111/j.1474-9726.2011.00723.x.

Inglima, K., A. Perlmutter, and J. Markofsky. 1981. Reversible stage-specific embryonic inhibition mediated by the presence of adults in the annual fish *Nothobranchius guentheri*. *Journal of Experimental Zoology* 215: 23–33. doi: 10.1002/jez.1402150104

Kimmel, C.B., W.W. Ballard, S.R. Kimmel, B. Ullmann, and T.F. Schilling. 1995. Stages of embryonic development of the zebrafish. *Developmental Dynamics* 203: 253–310.

Kirschner, J., D. Weber, C. Neuschl et al. 2012. Mapping of quantitative trait loci controlling lifespan in the short-lived fish *Nothobranchius furzeri*—A new vertebrate model for age research. *Aging Cell* 11: 252–261. doi: 10.1111/j.1474-9726.2011.00780.x.

Lesseps, R.J., M. Hall, and M.B. Murnane. 1979. Contact inhibition of cell movement in living embryos of an annual fish, *Nothobranchius korthausae*: Its role in the switch from persistent to random cell movement. *Journal of Experimental Zoology* 207: 459–470.

Lesseps, R.J., A.H. van Kessel, and J.M. Denuce. 1975. Cell patterns and cell movements during early development of an annual fish, *Nothobranchius neumanni*. *Journal of Experimental Zoology* 193: 137–146. doi: 10.1002/jez.1401930203.

Levels, P.J., and J.M. Denucé. 1988. Intrinsic variability in the frequency of embryonic diapauses of the annual fish *Nothobranchius korthausae*, regulated by light: Dark cycle and temperature. *Environmental Biology of Fishes* 22: 211–224.

Levels, P.J., R.E. Gubbels, and J.M. Denuce. 1986. Oxygen consumption during embryonic development of the annual fish *Nothobranchius korthausae* with special reference to diapause. *Comparative Biochemistry and Physiology—Part A: Comparative Physiology* 84: 767–770.

Markofsky, J. 1976. Longitudinal and cross-sectional observations of growth and body composition with age in laboratory populations of the male annual cyprinodont fish, *Nothobranchius guentheri*. *Experimental Gerontology* 11: 171–177. doi: 0531-5565(76)90030-9 [pii]

Markofsky, J., and M. Milstoc. 1979. Aging changes in the liver of the male annual cyprinodont fish, *Nothobranchius guentheri*. *Experimental Gerontology* 14: 11–20.

Markofsky, J., and A. Perlmutter. 1973. Growth differences in subgroups of varying longevities in a laboratory population of the male annual cyprinodont fish, *Nothobranchius guentheri* (Peters). *Experimental Gerontology* 8: 65–73.

Matias, J.R. 1984. The stage-dependent resistance of the chorion to external chemical damage and its relationship to embryonic diapause in the annual fish, *Nothobranchius guentheri*. *Experientia* 40: 753–754.

Newport, J., and M. Kirschner. 1982. A major developmental transition in early *Xenopus* embryos. *Cell* 30: 675–686.

Ng'oma, E., K. Reichwald, A. Dorn et al. 2014. The age related markers lipofuscin and apoptosis show different genetic architecture by QTL mapping in short-lived *Nothobranchius* fish. *Aging (Albany NY)* 6: 468–480.

Reichwald, K., C. Lauber, I. Nanda et al. 2009. High tandem repeat content in the genome of the short-lived annual fish *Nothobranchius furzeri*: A new vertebrate model for aging research. *Genome Biology* 10: R16. doi: gb-2009-10-2-r16 [pii] 10.1186/gb-2009-10-2-r16

Sugiyama, M., A. Sakaue-Sawano, T. Limura et al. 2009. Illuminating cell-cycle progression in the developing zebrafish embryo. *Proceedings of the National Academy of Sciences of the United States of America* 106: 20812–20817. doi: 10.1073/pnas.0906464106

Terzibasi, E., C. Lefrançois, P. Domenic et al. 2009. Effects of dietary restriction on mortality and age-related phenotypes in the short-lived fish *Nothobranchius furzeri*. *Aging Cell* 8: 88–99. doi: ACE455 [pii]10.1111/j.1474-9726.2009.00455.x

Terzibasi, E., D.R. Valenzano, M. Benedetti et al. 2008. Large differences in aging phenotype between strains of the short-lived annual fish *Nothobranchius furzeri*. *PLoS ONE* 3: e3866. doi: 10.1371/journal.pone.0003866

Tozzini, E.T., M. Baumgart, G. Battistoni, and A. Cellerino. 2012. Adult neurogenesis in the short-lived teleost *Nothobranchius furzeri*: Localization of neurogenic niches, molecular characterization and effects of aging. *Aging Cell* 11: 241–251. doi: 10.1111/j.1474-9726.2011.00781.x

Tozzini, E.T., A. Dorn, E. Ng'oma et al. 2013. Parallel evolution of senescence in annual fishes in response to extrinsic mortality. *BMC Evolutionary Biology* 13: 77. doi: 1471-2148-13-77 [pii]10.1186/1471-2148-13-77

Valdesalici, S., and A. Cellerino. 2003. Extremely short lifespan in the annual fish *Nothobranchius furzeri*. *Proceedings of the Royal Society of London. Series B. Biological Sciences* 270: S189–191.

Valenzano, D.R., J. Kirschner, R.A. Kamber et al. 2009. Mapping loci associated with tail color and sex determination in the short-lived fish *Nothobranchius furzeri*. *Genetics* 183: 1385–1395. doi: genetics.109.108670 [pii]10.1534/genetics.109.108670

Valenzano, D.R., S. Sharp, and A. Brunet. 2011. Transposon-mediated transgenesis in the short-lived African killifish *Nothobranchius furzeri*, a vertebrate model for aging. *G3 (Bethesda)* 1: 531–538. doi: 10.1534/g3.111.001271GGG_001271 [pii]

Valenzano, D.R., E. Terzibasi, A. Cattaneo, L. Domenici, and A. Cellerino. 2006. Temperature affects longevity and age-related locomotor and cognitive decay in the short-lived fish *Nothobranchius furzeri*. *Aging Cell* 5: 275–278. doi: ACE212 [pii] 10.1111/j.1474-9726.2006.00212.x

Valenzano, D.R., E. Terzibasi, T. Genade et al. 2006. Resveratrol prolongs lifespan and retards the onset of age-related markers in a short-lived vertebrate. *Current Biology* 16: 296–300. doi: S0960-9822(06)01020-7 [pii]10.1016/j.cub.2005.12.038

van Haarlem, R. 1979. Contact inhibition of overlapping: One of the factors involved in deep cell epiboly of *Nothobranchius korthausae*. *Developmental Biology* 70: 171–179.

Van Haarlem, R., J.G. Konings, and R. Van Wijk. 1983a. Analysis of the relationship between the variation in intercleavage times and cell diversification during the cleavage stages of the teleost fish *Nothobranchius guentheri*. *Cell Tissue Kinetics* 16: 167–176.

Van Haarlem, R., R. Van Wijk, and A.H. Fikkert. 1981. Analysis of the variability in cleavage times and demonstration of a mitotic gradient during the cleavage stages of *Nothobranchius guentheri*. *Cell Tissue Kinetics* 14: 285–300.

Van Haarlem, R., R. Van Wijk, and J.G. Konings. 1983b. Analysis of the variability and of the lengthening of intercleavage times during the cleavage stages of *Nothobranchius guentheri*. *Cell Tissue Kinetics* 16: 177–187.

CHAPTER 7

Husbandry, Reproduction, and Embryo Culture of Annual Fishes under Laboratory Conditions

Nicolás G. Papa, Graciela A. Clivio, and Jimena Montagne

CONTENTS

7.1 INTRODUCTION

Annual fishes have displayed ecological and developmental adaptations that must be considered when kept and bred under laboratory conditions. Herein, we summarize our experience in handling and breeding annual fishes at the Biología Celular de la Reproducción y del Desarrollo en Peces laboratory. Particularly, we summarize data and give advice on maintenance, reproduction, and embryo culture of annual fishes of the Neotropical genus *Austrolebias*. At the same time, we will review and compare husbandry of *Austrolebias* with the available information for other genera of annual fishes.

7.2 COLLECTION AND SETUP

Annual fishes of the genus *Austrolebias* are collected from temporal ponds during the rainy season (May to October in Uruguay). Ponds are shallow, usually being less than 0.6-m deep, with soft-muddy substrates and areas with different accumulations of macrophytes. Occasionally macrophytes form small islands separated by areas of cattle-made holes. Pond structure varies with geographical conditions, principally with locality and annual rain cycle. Fish use plants as shelter or they burrow in the mud to escape predation. In such environments, collection is difficult when using a regular fishing net, so a hand-fishing net specially designed for this purpose is used instead

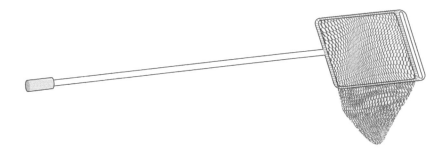

Figure 7.1 Hand-fishing net designed for collecting annual fish in muddy ponds.

(Figure 7.1). The handle of the net is made of 1-cm diameter aluminum or a galvanized iron tube. The net frame consists of a 30 × 20-cm rectangle made of 6-mm diameter iron bars. An anterior bar, anterior to the net's frame, is used to repeatedly knock the muddy bottom, causing the fish to leave the substrate in which they are hiding. Furthermore, the bar protects the net from damage. After collection, to minimize stress and allow oxygenation, fish are transported to the laboratory in containers half-filled with water from their natural habitat (e.g., ten 3-cm-long fish per 3-L container). The density of fish in these pots is subject to fish size and hours of travel to the lab. Collecting trips usually last an entire day; for longer trips we carry a large can filled with potable and dechlorinated water, and we change one-third of the water daily. In the laboratory, fish are gradually adapted to aquarium water by placing them in plastic bottles half-filled with water from their natural habitat. The bottles are floated inside the aquarium to allow the gradual exchange of water through a small hole in the bottom of the bottle. Aquarium water is supplemented with table salt (at a ratio of 5–10 g/L) during a quarantine week to prevent opportunistic diseases.

Nevertheless, white spots sometimes appear on the fish's epidermis, a common ciliated protozoan of the genus *Ichthyophthirius* usually causes these lesions. In such cases, the first measure is isolation of the infected fish. Daily changing of the salt-supplemented aquarium water is recommended, as is monitoring of the rest of the fish. The aquarium net and any other equipment used should also be disinfected by immersion in a salt-saturated water solution. We prefer not to introduce any other chemical to the system to avoid spurius molecular experimental results. We emphasize that the best course of action for prevention of diseases is accomplished by preserving a clean aquarium at all times.

7.3 GENERAL CONDITIONS OF MAINTENANCE

Adult fish are maintained in glass aquaria, separated by sex to maximize egg production when needed. The density of fish in each aquarium is determined by body size, sex, and level of cannibalism of the species. Overall, fish up to 3 cm in body length require 3 L per fish for females and 5 L per fish for males. This difference in space requirement is due to the territorial behavior exhibited by males. Males of both the African *Nothobranchius* and the Neotropical *Austrolebias* genera display aggressive behaviors that include attacking other males and even females unwilling to spawn (Genade et al., 2005; Passos et al., 2013). Three or more males can be placed in each aquarium if the water-to-fish ratio is kept as previously described; this is possible because male aggressiveness diminishes after hierarchies are established (Chapter 12). Apart from periods when eggs are being obtained, for instance, when setting experimental groups, males and females can be kept together. We use 20-L aquaria with central 5-cm glass divisions at the bottom; other researchers use smaller aquaria, for example, 3.8-L aquaria for *Austrofundulus* pairs (Pri-Tal et al., 2011). On one side of the division, we place an undergravel filter consisting of plastic grills covered with small pebbles

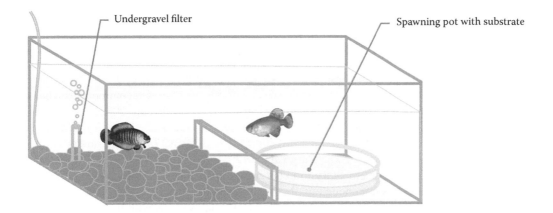

Figure 7.2 Breeding aquarium model. Note the glass division that separates the undergravel filter from the spawning and feeding area.

and an uplift tube connected to an air pump (Figure 7.2). Excessive turbulence should be avoided to minimize stress and prevent exhaustion. Similar recommendations were provided for researchers working with species of *Nothobranchius* (Genade et al., 2005), although other authors use a gentle recirculating filtration system (Pri-Tal et al., 2011). Water is changed partially, one-third of the volume, at least twice a week subject to the feeding regimen (frequency and type of food offered). The other side of the aquarium is used for routine feeding and to place spawning pots when needed. A few plants are added for environmental enrichment, shelter, stress relief, and to decrease the accumulation of dissolved nitrogenated compounds in the water (Nijman and Heuts, 2000; Williams et al., 2009). The most widely and successfully used plants are *Elodea* sp. and Java moss (*Hypnaceae*).

Aquarium water is reverse-osmosis water-added with synthetic marine salt (Tetra Marine Salt Pro, Tetra) and subsequently stored in a 150-L tank. The resultant synthetic freshwater should have a 300 ± 50 μS conductivity and 6.5–7.5 pH. Aquaria are located in a room maintained at 19–20°C with a natural photoperiod. This temperature range is suitable for *A. charrua* and other species of *Austrolebias* with which we have worked (i.e., *A. viarius*, *A. affinis*, *A. duraznensis*, *A. cheradophilus*, *A. reicherti*, etc.), and the same range was established for *A. nigrofasciatus* (Volcan et al., 2013) and *A. wolterstorffi* (Peres da Fonseca et al., 2013). The ideal temperature for *Austrofundulus limnaeus* is 27°C (Podrabsky, 1999), whereas researchers working with *Nothobranchius* species use temperatures ranging between 24°C and 27°C (Markofsky and Matias, 1976). It is recommended to establish the best temperature for each species, considering fish growth, reproduction, fertility, and egg quality and not just survival ratio. In this sense, partial studies have been done to observe temperature effects. *Austrolebias nigrofasciatus* growth was studied by Volcan et al. (2012), who concluded that higher temperatures (22°C) are suitable for the first 6 weeks to favor rapid growth, whereas lower temperatures are necessary from puberty onward. The same conditions were observed for other species of *Austrolebias* (N.G. Papa et al., unpublished data). Higher temperature (25°C) can result in reduction of egg quantity and fecundity in *A. nigrofaciatus* females (Volcan et al., 2013). Laboratory observations of *A. charrua* adults kept in 25°C aquaria suggest that males are more sensitive to high temperatures and die earlier than females (N.G. Papa et al., unpublished data). These observations can be related to the decreased sex ratio (male/female) found for this species in temporary ponds at the end of the wet season, when environmental temperature is higher (Arezo, 2012; Passos et al., 2014; *A. reicherti*, N. Papa personal observation).

Growth and reproduction of annual fishes require live food (Laufer et al., 2009; Wourms, 1967). The feeding regimen must be adapted to the experimental design and availability of food. Daily feeding is required when the objective is to stimulate mating and enhance embryo production, although this may vary depending on the quality of the food. On the other hand, feeding three times a week is adequate for fish maintenance while individuals are not engaged in courtship and repro-duction groups. We routinely include live *Tubifex* sp. (Annelida, Oligochaeta) in the feed because its high nutritional value gives the best results (Arezo et al., 2005; García et al., 2009). Given that commercial *Tubifex* worms are collected from sewage drains, caution is recommended when per-forming physiological or developmental research (Denslow and Sepúlveda, 2007). Purging of the worms' gut by daily rinsing with synthetic freshwater, leaving the water level in the recipient at 2 cm afterwards, is recommended. Another possibility is to cultivate *Tubifex* sp. under controlled conditions (Bettinetti and Provini, 2002; Das et al., 2012; Marian et al., 1989). *Daphnia pulex* or *D. magna* can be used as the supplementary food but not as the primary or only a source of nourish-ment. In our experience, it is not enough to obtain good quality eggs. To improve nutritional qual-ity, we recommend using a varied diet. Other authors have had success with live annelids (*Tubifex* sp., *Lombriculus variegatus*, or different species of earthworms), crustaceans (*Artemia*, *Daphnia*, *Gammarus*), and insect larvae (*Chaoborus*, *Drosophila*, *Chironomus*) in different combinations (Genade et al., 2005; Podrabsky, 1999; Reichard and Polačik, 2010). Chopped frozen *Chironomus* larvae and frozen silversides have also been successfully used (Podrabsky, 1999). The food is offered in the area of the aquarium free of stones and plants to facilitate its capture by fish and removal of remains. Species vary in mouth size, and this has to be considered when choosing food. Given the overall small size of most annual fish species used in research, worms should be chopped immediately before feeding. In this case, rinse chopped worms before introducing them into aquaria and siphon the remnants afterward, to preserve the cleanliness of the water.

7.4 REPRODUCTION AND EMBRYO PRODUCTION

Depending on the experimental design, breeding groups could consist of one male and one or two females. A spawning pot with blunt edges is placed at the bottom of the 20-L aquarium (Figure 7.2). The size of the pot should be adapted to specimens' size and space requirements for courtship and spawning behavior of each species. *Austrolebias* courtship is complex and includes partial or total burial of the pair in the substrate (García et al., 2008; Passos et al., 2013; Vaz-Ferreira et al., 1964; Vaz Ferreira and Sierra, 1972). There should be enough substrate to allow fish to display their natural behavior (at least 2-cm deep). There are different options for substrates, such as very fine sand (Markofsky and Matias, 1976), xaxim fibers (arborescent fern trunk; Shibatta, 2005), coconut fibers (Volcan et al., 2012), black acrylic yarn (Volcan et al., 2013), or peat moss or glass beads between 200- and 500-μm diameter (Berois et al., 2012; Podrabsky, 1999). Thomas Scientific beads (200–500 μm in diameter) yielded good results in our experiments. The beads can be steril-ized by boiling them in water or soaking them in 2% sodium hypochlorite. Alternatively, for the initial use, the beads can be soaked in concentrated nitric acid (Podrabsky, 1999). Vegetal substrates have to be sterilized by boiling for an hour. All substrates must be rinsed thoroughly with synthetic freshwater before use.

Egg collection from the spawning pots depends on the substrate chosen. When fine sand or glass beads are used, the embryos can be recovered with a sieve with 0.5-mm openings. Egg sizes vary among species and specimens, and early embryos' chorion is flexible (Wourms, 1972). Thus, the openings in the sieve should be considerably smaller than the eggs to avoid passage through the holes. Peat moss or xaxim fibers are the preferred substrates of the annual fish species with which we have worked, but collection is more labor-intense. It is difficult to gather the embryos because they are translucent and have to be collected one by one with a plastic pipette. To allow fish to rest and to improve embryos' viability, remove the spawning pots for 1 or 2 days once a week or twice a month for 3 days.

The collected embryos can be immediately used in research or can be kept to start the next generation. For the latter purpose, we keep eggs in reserve by mimicking the natural environment. Pots with peat moss are left in reproduction aquaria for more than a week (to ensure they contain enough embryos) and are afterwards withdrawn and left outside until the substrate is almost dry at aquarium room temperature. However, retention of some substrate humidity is necessary for the survival of the embryos. The peat moss is placed inside a labeled, sealed polystyrene bag with a larger volume of air than the volume of the stored substrate. Species, locality data, date of sealing, and other specimen identification are recorded on the bag's label. The bags are kept in the dark at 19–20°C for 3–7 months depending on the humidity of the peat moss. The viability of the embryos decreases when the bags are kept for longer periods of time.

7.5 EMBRYO CULTURE AND DEVELOPMENTAL MONITORING

The collected embryos are cleaned by rolling them with a soft paintbrush over filter paper dampened with synthetic water; they are then placed in Yamamoto's solution (NaCl, 0.75%, KCl, 0.02%, $CaCl_2$, 0.02%; Yamamoto, 1967) in Petri dishes and kept at 25°C in the dark. We determine and follow the embryonic developmental stages by observation under a light microscope (Arezo et al., 2005). We place each embryo and a drop of Yamamoto's solution in a rectangular handmade chamber carved on strips of PVC tape arranged in layers over a microscope slide and covered with a cover glass. The chambers are usually 1.7-mm deep, but with individual and species size variation in eggs, it is sometimes necessary to use chambers with different depths. During observation, the cover glass must touch the top of the egg to make it possible to rotate the egg by gently moving the cover glass (Berois et al., 2012). Developing embryos vary in their fragility at different stages. In our experience, early A. charrua blastula embryos are fragile and must be manipulated with care, but from epiboly to prehatching stages the embryos are resistant to mechanical pressure. Hatching is initiated with the thinning of the chorion from the inside by means of enzymes produced by a hatching gland (Schoots et al., 1982; Yamagami et al., 1992). A "ready-to-hatch" embryo lacks a perivitelline space and has a very thin chorion with a fluffy appearance; these embryos are as fragile as the early blastula embryos, and pressure on them must be avoided. Sudden temperature changes (increases as well as decreases) sometimes act as a stimulus for hatching.

Culture of A. limnaeus eggs in medium (NaCl 10 mM, KCl 0.1424 mM, $MgCl_2 \cdot 6H_2O$ 2.15 mM, $MgSO_4 \cdot 7H_2O$ 0.0013 mM, $CaCl_2 \cdot 2H_2O$ 0.792 mM) with methylene blue (0.0001%) for 4 days without cleaning retards fungal growth and avoids the use of gentamicin (Podrabsky, 1999). Later, on day 4, the eggs are treated with sodium hypochlorite (0.003%) and transferred to the culture medium supplemented with gentamicin sulfate (10 mg/L). The avoidance of gentamicin in early developmental stages is necessary, because it may cause abnormal development (Podrabsky, 1999). In our experience, fungi only colonize dead or nearly dead Austrolebias embryos.

7.6 HATCHING AND FRY CARE

A laboratory population of Austrolebias can be started in two ways, either by collection of adults or juveniles from a natural habitat or by hatching stored embryos from previously bred specimens. Peat moss and embryos are placed in a layer about 1-cm deep in a small glass aquarium with a capacity of 1.5–3 l and are hydrated with synthetic freshwater for at least 30 minutes. Following this, the aquarium is filled to a depth of 2 cm with synthetic freshwater at a temperature of 19–20°C and with high bubble aeration. Most Austrolebias embryos will hatch in less than 24 hours. From the next day on, the intensity of bubbling by aeration should be decreased, and the fry should be fed two times a day with newly hatched Artemia nauplii or with microworms (a free-living nematode, Panagrellus *redivivus*, easily cultivated in laboratory conditions; Biedenbach et al., 1989; Watanabe and Kiron, 1994). Two hours after feeding, the volume of water must be doubled to dilute the concentration of compounds from degradation of food remnants. Once the aquarium is filled, instead of adding water, the best procedure is to change half of volume of water 2 hours after feeding. The water is changed with a siphon pipe; the use of a filter in the starting end of the pipe is advisable to avoid siphoning out the young fish. It is recommended that the fry be transferred to bigger aquaria after 4 days with the objective of avoiding fungus growing on dead *Artemia*. At the same time, it is then possible to split the specimens into two aquaria, reducing juvenile density, which favors rapid growth until hierarchies are established (Volcan et al., 2012; Chapter 12). From the beginning, adult *Daphnia* spp. can be added to help with keeping the water clean, mainly when feeding with unwashed microworms. A positive side effect from using *Daphnia* is that they reproduce in aquaria, with the newly born crustaceans providing the fry with another source of food. Once the young fish have big enough mouths, *Daphnia* could be the main food. We recommend following this regimen for at least 20 days in order to decrease mortality. Genade et al. (2005) complement *Austrofundulus* fry diet with frozen *Chironomus* larvae. To avoid overcrowding, change some of the fish, as needed, to a new aquarium consistent with the size of the fish and space requirements. The times to include finely chopped *Tubifex* sp. in the diet will depend on the growth rate, especially the mouth size. As the fish grow and can be sexed, we separate males from females, as explained in section 7.3, "General Conditions of Maintenance."

Newly born fish unable to float ("belly sliders") is a common problem. Obtaining a high hatching percentage with few "belly sliders" depends on time left in reserve, humidity levels, temperature, and kind of substrate used. *Austrolebias charrua* embryos need to be kept a minimum of 3 months before hatching. Insufficient reserve time results in low hatching, no hatching at all, or a high percentage of belly-slider fry. Although the causes and mechanisms of belly-slider occurrence need thorough study, it has been suggested that exercise during hatching and immediately after is necessary to fill the gas bladder (Calviño et al., 2007). We argue that the belly-slider occurrence is multifactorial. Developmental time, amount of yolk remaining in the egg, reserve time, temperature during reserve time, humidity of the substrate, level of oxygen dissolved in the water, and exercise during and immediately after hatching are some of the factors that could affect belly-slider percentages in laboratory conditions. We also noticed that high percentages of substrate humidity require longer periods in reserve, probably to allow embryos to complete development. One explanation is the possible entrance into diapauses that extends developmental time. The humidity percentage of the substrate and the time the eggs are reserved also depend on the species.

7.7 CONSTRUCTION OF THE LABORATORY BREEDING LINES

Different researchers have successfully established laboratory lines of annual fish and some aquarium hobbyists have been able to maintain populations in captivity. We attempted to obtain a full-sibling mating (FS) endogamic line in *A. charrua* starting with a pair (one male and one female; F0), pairing off brothers in every generation. The F3 embryos show a decrease in viability, and viability decreases abruptly in the next generation (N.G. Papa et al., unpublished data). Starting with two breeding groups consisting of one male and two females each, one (F0) half-sibling mating (HS) was performed. In this case, the initial decrease in the offspring viability was delayed by one generation, but it happened nonetheless (N.G. Papa et al., unpublished data). This would reveal barriers against inbreeding, which would make the generation of stable reproduction lines in this species difficult. On the other hand, the establishment of a laboratory line could shed light about the mechanisms acting in isolated ponds with low gene flow.

The robustness of *A. charrua* for husbandry under laboratory conditions has been the base of our work for more than 10 years. Nevertheless, we found the disadvantages that come with the great genetic variability of a species. This variability is inherent to a species in the middle of a process of speciation that possibly has an ancient hybrid origin (García, 2006). A review of the populations' genetic structure in the selected species has to be considered to establish a stable laboratory line. The focus should be on the choice of the founding reproductive group, which should be chosen based on the locality of the pond and the number of initial breeding groups. An exhaustive mating plan is necessary that takes into account the requirements of facilities, personnel, and food supply. The genetic characterization of the line and continuous monitoring will allow monitoring of the evolution of the line in captivity.

7.8 FINAL CONSIDERATIONS

The Neotropical genus *Austrolebias* is highly variable in its morphology (Costa, 1990, 1995; Loureiro and de Sá, 1998; Vaz-Ferreira and Sierra, 1972) and behavior (García et al., 2008; Vaz-Ferreira and Sierra, 1973). Multiple karyotypic and phylogenetic analysis show the divergence between species from Uruguay, Argentina, and Rio Grande do Sul in Brazil, including differences within and among local populations (García et al., 1993, 1995, 2000, 2001, 2002). When a species is selected for research, we only breed specimens from the same pond or from ponds that can be connected by seasonal flooding to reduce the among-pond variability. As annual fish constitute endemics species, we aim to protect them by only collecting the minimum number of specimens needed for research purposes.

Little is known about the conservation status of most species and populations of annual fish. However, historical records about locations compared with present locations show that a great number of ponds have been destroyed by human activities (agriculture, roads, etc.). Moreover, given that several species inhabit seasonal flood areas, they are highly susceptible to rainfall and current climate change alterations and, above all, to a lack of or shortening of the dry season (Chapter 10). Populations maintained by hobbyists and laboratories will probably be the unique reserve for some of the endemic species of annual fish if the current trend of human activity continues.

ACKNOWLEDGMENTS

The authors would like to thank the editors for giving us the opportunity to share our experience in this matter. We thank Martín Astesiano for illustrations and Asociación Uruguaya de Acuaristas (AUDA) members for their ever-helpful advice. In addition, we thank Dr. Nibia Berois for her guidance and support.

REFERENCES

Arezo, M.J. 2012. Análisis de los mecanismos de determinación y diferenciación del sexo en *Austrolebias charrua* (Cyprinodontiformes: Rivulidae) (Unpublished doctoral dissertation thesis). PEDECIBA, Facultad de Ciencias, Uruguay.

Arezo, M.J., L. Pereiro, and N. Berois. 2005. Early development in the annual fish *Cynolebias viarius*. *Journal of Fish Biology*, 66: 1357–1370. doi: 10.1111/j.0022-1112.2005.00688.x

Berois, N., M.J. Arezo, N.G. Papa, and G.A. Clivio. 2012. Annual fish: Developmental adaptations for an extreme environment. *Wiley Interdisciplinary Reviews Developmental Biology*, 1: 595–602. doi: 10.1002/wdev.39

Bettinetti, R., and A. Provini. 2002. Toxicity of 4-nonylphenol to *Tubifex tubifex* and *Chironomus riparius* in 28-day whole-sediment tests. *Ecotoxicology Environmental Safety*, 53: 113–121. doi: 10.1006/eesa.2002.2206

Biedenbach, J.M., L.L. Smith, T.K. Thomsen, and A.L. Lawrence. 1989. Use of the nematode *Panagrellus redivivus* as an *Artemia* replacement in a larval penaeid diet. *Journal of the World Aquaculture Society*, 20: 61–71. doi: 10.1111/j.1749-7345.1989.tb00525.x

Calviño, P.A., F. Alonso, and J. Sanjuán de Torres. 2007. Llenado de gas de la vejiga natatoria de post-larvas de peces anuales sudamericanos (Cyprinodontiformes; Rivulidae). *Boletín del Killi Club Argentino* 13: 18–39. doi: 10.13140/2.1.2318.5288

Costa, W.J.E.M. 1990. Análise filogenética da família Rivulidae (Cyprinodontiformes, Aplocheiloidei). *Revista Brasileira de Biología*, 50: 65–82.

Costa, W.J.E.M. 1995. *Pearl Killifishes of the Cynolebiatinae: Systematics and Biogeography of a Neotropical Annual Fish Subfamily (Cyprinodontiformes, Rivulidae)*. Neptune City, NJ: TFH Publications.

Das P., S.C. Mandal, S.K. Bhagabati, M.S. Akhtar, and S.K. Singh. 2012. Important live food organisms and their role in aquaculture. In M. Sukham (ed.), *Frontiers in Aquaculture* (pp. 69–86). Delhi, India: Narendra Publishing House.

Denslow, N., and M. Sepúlveda. 2007. Ecotoxicological effects of endocrine disrupting compounds on fish reproduction. In E. Babin, P.J. Cerdà, and J. Lubzens (eds.) *The Fish Oocyte: From Basic Studies to Biotechnological Applications* (pp. 255–322). The Netherlands: Springer. doi: 10.1007/978-1-4020-6235-3_10

García D., M. Loureiro, and B. Tassino. 2008. Reproductive behavior in the annual fish *Austrolebias reicherti* Loureiro and García 2004 (Cyprinodontiformes: Rivulidae). *Neotropical Ichthyology,* 6: 243–248. doi: 10.1590/S1679-62252008000200012

García, G. 2006. Multiple simultaneous speciation in killifishes of the *Cynolebias adloffi* species complex (Cyprinodontiformes, Rivulidae) from phylogeography and chromosome data. *Journal of Zoological Systematics and Evolutionary Research*, 44: 75–87. doi: 10.1111/j.1439-0469.2005.00346.x

García, G., F. Alvarez-Valin, and N. Gómez. 2002. Mitochondrial genes: Signals and noise in phylogenetic reconstruction within killifish genus *Cynolebias* (Cyprinodontiformes, Rivulidae). *Biological Journal of the Linnean Society*, 76: 49–59. doi: 10.1111/j.1095-8312.2002.tb01713.x

García, G., A.I. Lalanne, G. Aguirre, and M. Cappetta. 2001. Chromosome evolution in annual killifish genus *Cynolebias* and mitochondrial phylogenetic analysis. *Chromosome Research*, 9: 93–100. doi: 10.1023/A:1011664009509

García, G., M. Loureiro, N. Berois et al. 2009. Pattern of differentiation in the annual killifish genus *Austrolebias* (Cyprinodontiformes: Rivulidae) from a biosphere reserve site in South America: A multidisciplinary approach. *Biological Journal of the Linnean Society*, 98: 620–635. doi: 10.1111/j.1095-8312.2009.01303.x

García, G., E. Scvortzoff, and A. Hernández. 1995. Karyotypic heterogeneity in South American annual killifishes of the genus *Cynolebias* (Pisces, Cyprinodontiformes, Rivulidae). *Cytologia*, 60: 103–110. doi: 10.1508/cytologia.60.103

García, G., E. Scvortzoff, M.C. Máspoli, and R. Vaz-ferreira. 1993. Analysis of karyotypic evolution in natural populations of *Cynolebias* (Pisces, Cyprinodontiformes, Rivulidae) using banding techniques. *Cytologia*, 58: 85–94. doi: 10.1508/cytologia.58.85

García, G., G. Wlasiuk, and E. P. Lessa. 2000. High levels of mitochondrial cytochrome B divergence in the annual killifishes of the genus Cynolebias (Cyprinodontiformes, Rivulidae). *Zoological Journal of the Linnean Society*, 129: 93–110. doi: 10.1006/zjls.1999.0202

Genade, T., M. Benedetti, E. Terzibasi et al. 2005. Annual fishes of the genus *Nothobranchius* as a model system for aging research. *Aging Cell*, 4: 223–233. doi: 10.1111/j.1474-9726.2005.00165.x

Laufer, G., M. Arim, M. Loureiro, J.M. Piñeiro-Guerra, S. Clavijo-Baquet, and C. Fagúndez. 2009. Diet of four annual killifishes: An intra and interspecific comparison. *Neotropical Ichthyology*, 7: 77–86. doi: 10.1590/S1679-62252009000100010

Loureiro, M., and R. O. de Sá. 1998. Osteological analysis of the killifish genus *Cynolebias (Cyprinodont iformes: Rivulidae). Journal of Morphology*, 238: 245–262. doi: 10.1002/(SICI)1097-4687(199811)238:2<245::AID-JMOR5>3.0.CO;2-F

Marian, M.P., S. Chandran, and T.J. Pandian. 1989. A rack culture system for *Tubifex tubifex. Aquacultural Engineering*, 8: 329–337. doi: 10.1016/0144-8609(89)90039-3

Markofsky, J., and J. R. Matias. 1976. The effects of temperature and season of collection on the onset and duration of diapause in embryos of the annual fish *Nothobranchius guentheri. Journal of Experimental Zoology*, 11: 49–56. doi: 10.1002/jez.1402040209

Nijman, V., and B.A. Heuts. 2000. Effect of environmental enrichment upon resource holding power in fish in prior residence situations. *Behavioural Processes*, 49: 77–83. doi: 10.1016/S0376-6357(00)00078-4

Passos, C., B. Tassino, M. Loureiro, and G.G. Rosenthal. 2013. Intra and inter-sexual selection on male body size in the annual killifish *Austrolebias charrua*. *Behavioural Processes*, 96: 20–26. doi: 10.1016/j.beproc.2013.01.008

Passos, C., B. Tassino, F. Reyes, and G.G. Rosenthal. 2014. Seasonal variation in female mate choice and operational sex ratio in wild populations of an annual fish, *Austrolebias reicherti*. *PLoS ONE*, 9: e101649. doi:10.1371/journal.pone.0101649

Peres da Fonseca A., M.V. Volcan, L.A. Sampaio, L.A. Romano, and R.B. Robaldo. 2013. Growth of critically endangered annual fish *Austrolebias wolterstorffi* (Cyprinodontiformes: Rivulidae) at different temperatures. *Neotropical Ichthyology*, 11: 837–844. doi: 10.1590/S1679-62252013000400012

Podrabsky, J.E. 1999. Husbandry of the annual killifish *Austrofundulus limnaeus* with special emphasis on the collection and rearing of embryos. *Environmental Biology of Fishes*, 54: 421–431. doi: 10.1023/A:1007598320759

Pri-Tal, B.M., S. Blue, F.K.Y. Pau, and J.E. Podrabsky. 2011. Hormonal components of altered developmental pathways in the annual killifish, *Austrofundulus limnaeus*. *General and Comparative Endocrinology*, 174: 166–174. doi: 10.1016/j.ygcen.2011.08.016

Reichard, M., and M. Polačik. 2010. Reproductive isolating barriers between colour-differentiated populations of an African annual killifish, *Nothobranchius korthausae* (Cyprinodontiformes). *Biological Journal of the Linnean Society of London*, 100: 62–72. doi: 10.1111/j.1095-8312.2010.01406.x

Schoots, A.F.M., J.J.M. Stikkelbroeck, J.F. Bekhuis, and J.M. Denucé.1982. Hatching in teleostean fishes: Fine structural changes in the egg envelope during enzymatic breakdown *in vivo* and *in vitro*. *Journal of Ultrastructural Research*, 80: 185–196. doi: 10.1016/S0022-5320(82)90017-X

Shibatta, O.A. 2005. Reprodução do pirá-brasília, *Simpsonichthys boitonei* Carvalho (Cyprinodontiformes, Rivulidae), e caracterização de seu habitat na reserva ecológica do instituto brasileiro de geografia e estatística, Brasília, distrito federal, Brasil. *Revista Brasileira de Zoologia*, 22: 1146–1151. doi: 10.1590/S0101-81752005000400048

Vaz-Ferreira, R., and B. Sierra. 1972. Caracteres etológicos genéricos y específicos en los peces del género Cynolebias Steindachner. *Boletín de la Sociedad Zoológica del Uruguay*, 2: 22–35.

Vaz-Ferreira, R., and B. Sierra. 1973. Los géneros de Cyprinodontidae de aguas temporales Sudamericanas. *Boletín de la Sociedad Zoológica del Uruguay*, 2: 36–42.

Vaz-Ferreira, R., B. Sierra, and S. Scaglia. 1964. Eco-etología de la reproducción en los peces del género *Cynolebias Steindachner*, 1876. *Apartados de los Archivos de la Sociedad de Biología de Montevideo*, 26: 44–49.

Volcan, M.V., A. Peres da Fonseca, M.R.C. Figueiredo, L.A. Sampaio, and R.B. Robaldo. 2012. Effect of temperature on growth of the threatened annual fish *Austrolebias nigrofasciatus* Costa & Cheffe 2001. *Biota Neotropica*, 12: 1–6.

Volcan, M.V., L.A. Sampaio, D.C. Bongalhardo and R.B. Robaldo. 2013. Reproduction of the annual fish *Austrolebias nigrofasciatus* (Rivulidae) maintained at different temperatures. *Journal of Applied Ichthyology*, 29: 648–652. doi: 10.1111/jai.12013

Watanabe, T., and V. Kiron. 1994. Prospects in larval fish dietetics. *Aquaculture*, 124: 223–251. doi: 10.1016/0044-8486(94)90386-7

Williams, T.D., G.D. Readman, and S.F. Owen. 2009. Key issues concerning environmental enrichment for laboratory-held fish species. *Laboratory Animals*, 43: 107–120. doi: 10.1258/la.2007.007023

Wourms, J.P. 1967. Annual fishes. In F.H. Wilt and N. Wessells (eds.), *Methods in Developmental Biology* (pp. 123–137). New York: Thomas & Crowell Company.

Wourms, J.P. 1972. Developmental biology of annual fishes I. stages in the normal development of *Austrofundulus myersi* Dahl. *Journal of Experimental Zoology*, 182: 143–168. doi: 10.1002/jez.1401820202

Yamagami, K., T.S. Hamazaki, S. Yasumasu, K. Masudat, and I. Iuchi. 1992. Molecular and cellular basis of formation, hardening, and breakdown of the egg envelope in fish. *International Review of Cytology*, 136: 2–10. doi: 10.1016/S0074-7696(08)62050-1

Yamamoto, T. 1967. Medaka. In F.H. Wilt and N. Wessels (eds.), *Methods in Developmental Biology* (pp. 101–111). New York: Thomas & Crowell Company.

PART II

Ecology and Conservation

CHAPTER **8**

Austrolebias in Space
Scaling from Ponds to Biogeographical Regions

**Marcelo Loureiro, Ana Borthagaray, Daniel Hernández,
Alejandro Duarte, Verónica Pinelli, and Matías Arim**

CONTENTS

8.1 INTRODUCTION

The unique life cycle of annual killifishes is increasingly attracting the interest of researchers in developmental biology (Arezo et al., 2005; Podrabsky et al., 2010; Berois et al., 2014), genetics (García et al., 2014), senescence (Genade et al., 2005; Terzibasi et al., 2008), trophic ecology (Laufer et al., 2009; Arim et al., 2010; Polacik and Reichard, 2010; Polacik et al., 2014), life-history strategies (Blažek et al., 2013; Reichard et al., 2009; Polacik et al., 2011, 2014; Lanés et al., 2014), and evolution (Reichard and Polacik, 2010; Costa, 2010; Dorn et al., 2011; Sedlácek et al., 2014). The main focuses have been the developmental mechanisms in embryos and the structure and composition of the egg chorion related to the survival under the stressful environmental conditions of complete drought. This characteristic biology of annual fishes plus their unique habitat (i.e., fragmented, discrete, and seasonal) have a strong influence on how distribution and diversity patterns are generated (Kiflawi et al., 2003).

Seasonal patches strongly push individuals to accommodate spatially to optimize food supply, metabolic efficiency, mating opportunities, and predator avoidance, all in few months (Wilbur, 1997). In addition, the flow of individuals among ponds determines metapopulation and metacommunity dynamics from the local to the regional scale and plays a main role in connectivity of the spatial trends in biodiversity. Furthermore, historical processes that facilitate or restrict dispersion and gene flow generate topographic heterogeneity that may promote the processes of speciation, as well as the patterns and limits of endemic areas and ecoregions (Albert and Reis, 2011). All these processes related to the interaction between killifishes and their environment strongly shape their evolution and their role in the structure of their communities and ecosystems.

In this chapter, we present a review of the spatial patterns of annual killifishes, scaling from individuals to higher clades and from local ponds to biogeographical regions. We focus on the genus *Austrolebias* as a model for the study of the spatial-dependent ecological and evolutionary processes that shape biodiversity.

Austrolebias is a relatively rich genus of the family Rivulidae, with ~42 currently recognized species, extensively distributed in the La Plata-Paraná and Patos-Merín basins (Calviño, 2005; Costa, 2006, 2014; Ferrer et al., 2008; Loureiro et al., 2011; Volcan et al., 2014). Species size ranges from small (max. 40 mm of standard length) to large (max. 150 mm of standard length). Most large species are considered to form a monophyletic group of top predators (García et al., 2000; Costa, 2006, 2009, 2010), with three species developing elongated jaws to different degrees (Loureiro, 2004; Costa, 2006, 2009). The small- and medium-sized species correspond to five or six clades, depending on the topologies recovered in different analyses (Loureiro, 2004; Costa, 2006, 2010; García et al., 2014); overall, these clades consist of generalist predator species without or with less conspicuous jaw specializations (Costa, 2009; Laufer et al., 2009). In the Patos-Merín basin, *Austrolebias* may be syntopic with species of the annual genus *Cynopoecilus*, a small-sized generalist predator that differs from *Austrolebias* in adult and egg morphology, sexual behavior, and life history strategies (Arenzon et al., 2001, 2002; Costa, 2002; Ferrer et al., 2014; Keppeler et al., 2014). In the Chaco region (Paraná-Paraguay river basins), *Austrolebias* species can co-occur with other rivulids, such as *Neofundulus paraguayensis*, *Papiliolebias bitteri*, *Trigonectes aplocheiloides*, *Pterolebias longipinnis*, and *Melanorivulus punctatus*, and in the lower Paraná basin with *M. punctatus* (P. Calviño, personal communication).

8.2 LIVING IN A POND: COEXISTENCE IN SMALL CONTAINERS

Seasonal ponds inhabited by *Austrolebias* range from rounded to irregularly shaped shallow depressions in grasslands and seasonal flooded areas (max. depth around 40 cm) adjacent to rivers or large permanent wetlands (Figure 8.1). The necessary environmental condition for a pond to be inhabited by *Austrolebias* species is for it to undergo seasonal droughts. Throughout the *Austrolebias* range, the wet season extends from early fall to mid-spring (March to October), with some yearly variation. However, summer rains, if intense, may accelerate the onset of the wet season. In areas without a defined rainy season, the presence of water in temporary ponds is mediated by the balance between precipitation, temperature, and evaporation (Williams, 2006).

Ponds are heterogeneous environments that cover a large range of areas, have variable depths, contain organic matter, are rich in invertebrates and plant diversity and biomass, and usually occur in isolation from other ponds (Laufer et al., 2009; Arim et al., 2011; Piñeiro-Guerra et al., 2014; Borthagaray et al., in press). In addition, important sources of heterogeneity among ponds are areas of emerged patches of soil that result in a reticulate configuration (Figure 8.1).

Under a niche theory context, species should differ in their functional traits and intertrait compromises should occur in order to achieve coexistence (Mayfield and Levine, 2010). Up to five species of annual fishes (*Austrolebias* and *Cynopoecilus*) may cohabit the same pond (Calviño, 2003; Laufer et al., 2009; Gonçalves et al., 2011; Keppeler et al., 2014; Lanés et al., 2014). These species belong to different clades and correspond to five different morphs (Figure 8.2). One of the clades consists of large-sized and top predator species (including predation over other annuals) characterized by an elongated and robust body shape; these species are less abundant in the ponds (Costa, 2009). The other clades correspond to small- and medium-sized *Austrolebias* with compressed and

(a)

(b)

Figure 8.1   (a) Grassland pond, in Barra Grande (Laguna de Castillos basin). (b) Floodplain pond in Yacaré creek (Río Uruguay basin).

ellipsoidal to elongated bodies that function as overall generalist predators on invertebrates, without significant differences in diet among them (Laufer et al., 2009; Keppeler et al., 2014). Nonetheless, it is possible that body shape differences among species are related to alternative strategies or differences in foraging areas in heterogeneous environments. In spite of this, diet similarity suggests that body shape differences, at least among small- to medium-sized species, may be influenced by additional factors beyond prey identity (Keppeler et al., 2014), that is, spatial preferences in pond depth or vegetation density.

    This pattern is congruent with a similar analysis performed in annual killifishes from central and east African savannahs (Polacik and Reichard, 2010). The relatively low interspecific differences in diet contrast with the systematic trends in individuals' diet along a body size gradient (Arim et al., 2010; Keppeler et al., 2014). Independently of the *Austrolebias* species identity, when

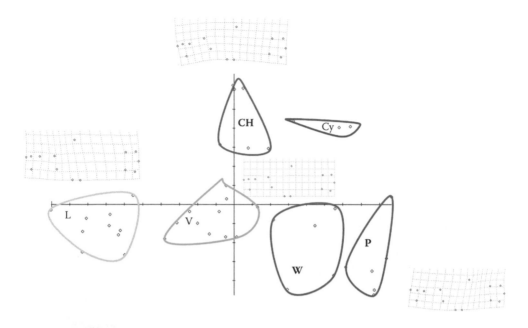

Figure 8.2   Relative warp analyses of body shape variation in sympatric annual fishes from southern Laguna Merín basin. Digitized Landmark data analyzed with TPS relw software (Rohlf 2014). Ch = *A. cheradophilus*; Cy = *Cynopoecilus melanoatenia*; L = *A. luteoflammulatus*; P = *A. prognathus*; V = *A. viarius*; W = *A. wolterstorffi*. Large predator species in **bold**.

individuals become larger, they systematically increase their prey diversity (both the richness and evenness), trophic position, and the number of energy sources indirectly consumed—for example, phytoplankton, macrophytes, detritus, and terrestrial prey (Arim et al., 2010).

Seven species from different lineages of Rivulidae were reported to cohabit the same area in central Brazil (Costa, 1998); these species probably occupy different microhabitats in terms of macrophyte presence, distance from the edge of the pond, and depth. A similar pattern was reported for *Nothobranchius* species, with different habitat preferences, at least at a regional level (Reichard et al., 2009). Cooccurrence analysis is a powerful tool for detecting species segregation (Gotelli and McCabe, 2002). However, it does not give information about the local structure of this segregation—for example, about the spatial autocorrelation in species dominance, as reported for Brazilian *Rivulidae* and Africans *Nothobranchius*.

Spatial segregation within local ponds has been observed among *Austrolebias* species. In this context, a pond in which two species of *Austrolebias* co-occur in the Cebollatí River basin (Patos-Merín basin) was sampled using 10 transects (100-m long, separated 10 m from each other), taking samples each 10 m. We found that the two species were significantly segregated in space within the pond ($p < 0.001$). Furthermore, *A. luteoflammulatus* (elongated body) preferred shallower places than *A. reicherti* (deep body) (Figure 8.3a; t-test = −6.84; df = 170; $p < 0.001$), suggesting that spatial segregation was promoted by some biotic or abiotic environmental factor related to pond depth and *Austrolebias* body shape. However, we found no significant differences in depth preferences in samples from several ponds in Laguna Castillos basin (Atlantic coastal basin), where *A. luteoflammulatus* coexist with up to three other annuals (*A. cheradophilus*, *A. viarius*, and *Cynopoecilus melanotaenia*) (Figure 8.3b–e).

Moreover, in our analyses, coexistence analyses in the community did not differ from neutrality, since the observed spatial arrangement of species in the pond can be expected by chance (Table 8.1). A recent study used the same pond system in Laguna Castillos and suggested that vegetation density favored small-sized *A. luteoflammulatus* but has no effect on *A. viarius* (Hernández, 2013).

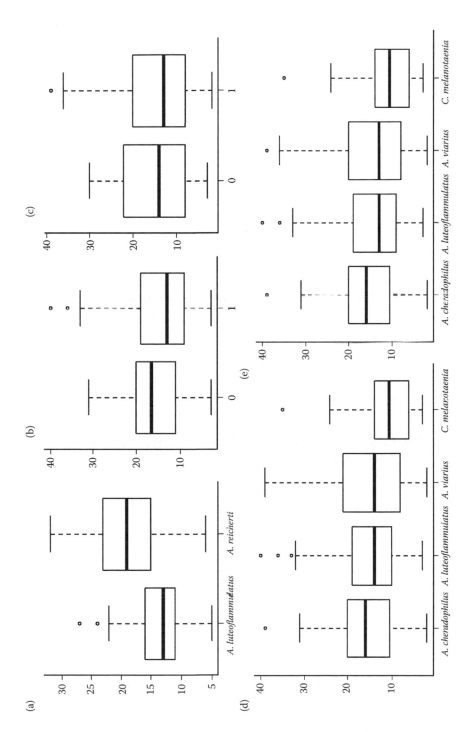

Figure 8.3 Figure 8.3. Species presence in relation to pond depth (cm). (a) Parao (Cebollatí River basin). (b) Barra Grande (Laguna de Castilos basin). *A. luteoflammulatus* alone (0) or with *A. viarius* (1). (c) Barra Grande (Laguna de Castilos basin). *A. viarius* alone (0) or with *A. luteoflammulatus* (1). (d) Barra Grande (Laguna de Castilos basin) all species. (e) Barra Grande (Laguna de Castilos basin), only coexistent communities.

Table 8.1 Null Models Observed and Expected Coexistence Frequencies by Each Sampling Unit (1000 Simulations)

sp/Sample	Observed	Random Individuals		Random Samples	
		Mean	IC 95%	Mean	IC 95%
1	0.78	0.71	0.2–1	0.6	0–1
2	0.2	0.26	0–0.7	0.37	0–1
3	0.017	0.029	0–0.25	0.036	0–0.3

Note: All observed frequencies are within the 95% confidence intervals of null models, except the observation of the four species of the system in the same sampling unit ($N = 1$ in 640, not shown). This implies that the distribution of coexisting species in a community does not differentiate from random. Random individuals: the model randomizes the sampling units where each individual appears. Random samples: the model randomizes the number of individuals by sample, keeping the amount and specific identity in each pond constant.

These data suggest differences in microhabitat selection between both species, at least for some size classes. Furthermore, this author found that species in high-density ponds tend to minimize size ranges and thus diminish prey size overlap (Hernández, 2013).

The structure of any community always has some degree of dependence on its structure in previous times (Fukami, 2010; Fukami and Nakajima, 2011). This is particularly important in temporary ponds, assembled by the balance between system memory—recruitment from resistance structures—and the novo assembly of local communities from a sample of the species pool (Padisák, 1992; Belmonti and Rossi, 1998). The role of immigration versus local recruitment is related to the relative location of a community in the landscape in comparison with all others. Isolated communities are prone to being shaped by local conditions rather than by the flow of immigrants, while the opposite is true for centrally located communities (Borthagaray et al., in press).

8.3 THE LANDSCAPE SCALE: ROLE OF CONNECTIVITY IN THE COHABITATION AND SUBSTITUTION PATTERNS

Species distributions are shaped by processes that operate at large temporal and spatial scales, determining the species pool that can potentially colonize local communities (Losos and Ricklefs, 2009). Migration, birth, and mortality rates are related to the area of the distributional range analyzed (Brown, 1995; Holt and Keitt, 2005). Although the geographical range of a species can still be viewed as reflection of the niche, other factors such as metapopulation and metacommunity dynamics, influenced by connectivity and stability of patches, are of critical importance (Rickets, 2001; Holt and Keitt, 2005; Filipe et al., 2010; Fernandes et al., 2014). Thus, in a matrix of interconnected seasonal habitats, the role of local interactions is balanced by processes related to species migration through the metacommunity and from the regional species pool (Chesson, 2000; Williams, 2006; Borthagaray et al., in press).

Austrolebias ponds can be found in large plains humid grassland where they may form a reticulate and have a dynamic pattern of connectivity (García et al., 2009, 2012; Keppeler et al., 2014). However, ponds may also be found in isolated margins of streams where connectivity among ponds is water flow-oriented, with a strong downstream component in dispersion (see also Humphries and Ruxton, 2002). In this wide range of scenarios, it is expected that movement of individuals among ponds has to exert a strong influence on metapopulation dynamics and hence on distribution patterns (Hanski, 1999). However, explicit analyses of metapopulation or metacommunity ecological and distributional dynamics in killifishes have not been considered. A recent study showed that the spatial distribution of organisms in a metacommunity was strongly connected with individuals' body sizes rather than with their taxonomic identity (Canavero et al., 2014). A large and significant spatial segregation of size classes was detected, indicating that different-sized organisms occupy different spatial locations. However, it should be noted that spatial segregation of species was also an important source of guild structuring among killifishes (Canavero et al., 2014). In summary, the pattern of both the change in diet with body size and the size-dependent spatial occurrence of organisms indicated that killifish assemblages are probably shaped by processes operating through individuals' traits associated not only with species identity but also with those traits associated with body size (Arim et al., 2010; A. I. Borthagaray et al., unpublished data; Canavero et al., 2014). The significant spatial segregation of individuals could attenuate intra and interspecific competition, fostering species coexistence (Canavero et al., 2014). A study on temporal ponds adjacent to streams in Central Amazonia found a hierarchical structure composed by nested subsets (Pazin et al., 2006). Smaller ponds had subsets of species found in larger ponds. However, some of the characteristics (e.g., species life cycles, climate, connection to streams, and available energy) of their ponds are very different from those of *Austrolebias*.

8.4 BIOGEOGRAPHIC PATTERNS IN LA PLATA-PARANÁ AND THE PATOS-MERÍN BASINS

At a regional scale, historical processes increase their influence on killifish populations and assemblage structure, and their integration with previous factors may promote differentiation and eventually speciation processes and patterns within lineages. In addition, considering whole communities together with global climatic factors, ecoregions and areas of endemism could be generated (Linder, 2001).

At the assemblage level, an interesting example of this interaction is the distribution pattern found in the Cebollati River basin. Nine species of annual fishes (eight *Austrolebias* and one *Cynopoecilus*) are distributed across this basin of ~30,000 km². However, not all species coexist in the same pond or area. A coexistence analysis of a presence–absence matrix of these species across this basin showed a checkerboard-shaped spatial pattern for the species different from the pattern expected if these species were randomly combined in the matrix ($p < 0.01$) (Figure 8.4). The checkerboard pattern emerges when there are pairs of species that tend to co-occur in the same pond or area or both, more or less than what is expected by chance (Diamond, 1975; Stone and Roberts, 1990; Gotelli and McCabe, 2002). Some pairs of species were never found together in the same site

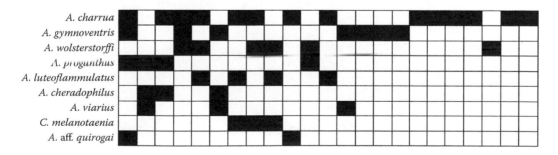

Figure 8.4   Incidence matrix of species coexistence in the Cebollatí River wetlands (Laguna Merín basin), showing a significant checkerboard pattern. In this matrix, each row corresponds to an annual fish species, and each column to a pond in a different location in the basin (data obtained from Ichthological Fish Collection of Facultad de Ciencias; institutional code ZVC-P). The black and white-filled cells indicate the presence and absence of the species, respectively.

within this basin (e.g., *A. wolterstorffi* and *A. viarius*; *A. charrua* and *A. viarius*; *A.* aff. *quirogai* and *A. luteoflammulatus*; *A.* aff. *quirogai* and *Cynopoecilus melanotaenia*), contributing to the observed pattern (Figure 8.4).

Three of the species (*A. prognathus*, *A. cheradophilus*, and *A. gymnoventris*) are distributed along the entire basin (including other areas of Laguna Merín), while the others show different ranges associated with sharp distribution limits (Figure 8.5). In this basin, *A. viarius* is endemic to an area located 65 m above sea level and abruptly disappearing downstream after a gorge of the river (Figure 8.5). Below that gorge three new species appear and extend to other areas of the Laguna Merín basin (*A. wolterstorffi*, *A. charrua*, and *A.* aff. *quirogai*). However, the latter species abruptly disappears in a similar gorge located downstream, 35 m above sea level. From this point, two species are recorded (*A. luteoflammulatus* and *C. melanotaenia*) that also are widely distributed in the rest of Laguna Merín (Figure 8.5). These sharp limits may correspond to ancient sea level fluctuations affecting wetlands levels in the area, because the most recent Pleistocene–Holocene fluctuations did not reach such altitudes (Bossi and Ortiz, 2011).

At the basin scale in the Cebollatí River, both local and historical forces could interact simultaneously to shape distribution patterns of annual fishes. The sharp limits of some of the species indicate low vagility and perhaps competitive exclusion acting together. Another example of sharp limits occurs in the lower Río Negro basin. The downstream limit of *A. arachan* occurs close to Mercedes city (Figure 8.6a), and just a few kilometers downstream it is abruptly replaced by the Uruguay River basin species (*A. bellottii*, *A. nigripinnis*, *A. alexandri*, and *A. elongatus*).

Alternatively, phylogeographic patterns of cytochrome *b* sequences of two *Austrolebias* species inhabiting high-connectivity ponds or wetlands (*A. bellottii* in the lower La Plata-Paraná basin and *A. charrua* in southern Laguna Merín basin) show a scenario of differentiation in allopatry and dispersion events mediated by sea level fluctuations during the late Pliocene and Pleistocene (García et al., 2009, 2012; Chapter 15). *Austrolebias bellottii* has one of the largest range distributions covering ~360,000 km². Three haplotype groups of *A. bellottii* were identified corresponding

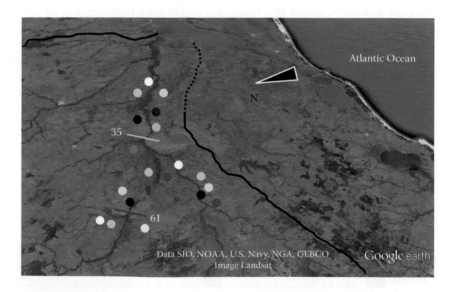

Figure 8.5 Species distribution boundaries in Cebollatí River basin: yellow dot = *A. viarius*; white dots = *A. cheradophilus*; red dots = *A. prognathus*; light blue dots = *A. gymnoventris*; black dots = *A. charrua*; green dots = *A. wolterstorffi*; orange dots = *A.* aff. *quirogai*; purple dot = *A. luteoflammulatus*; blue dot = *C. melanotaenia*. Numbers in white = meters above sea level; black continuous line = Cebollati River basin border; black discontinuous line = basin limit diffuse due to topographic flatness.

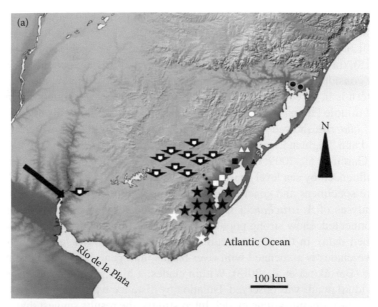

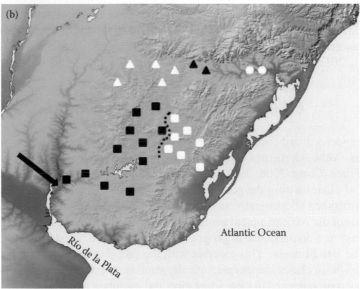

Figure 8.6   (a) Distribution of the *A. adloffi* species group; arrows with white box = *A. arachan*; white stars = *A. viarius*; white squares = *A. reicherti*; white squares = *A. nachtigalli*; black triangles = *A. minuano*; black squares = *A. nigrofasciatus*; white circles = *A. bagual*; black circles = *A. adloffi*; black stars = *A. charrua*. (b) Distribution of the *A. affinis* species group: black squares = *A. affinis*; white squares = *A. juanlangi*; white triangles = *A. periodicus*; black traingles = *A. litzi*; white circles = *A. cyaneus*. Black arrow = Mercedes city, downstream limit of *A. arachan* and *A. affinis*. Dotted lines = parapatric limits. (Distribution data taken from: Colección de Peces Facultad de Ciencias, Montevideo (ZVC-P), Colección de Peces Museo Nacional de Historia Natural, Montevideo (MNHNM), Colección de Peces Universidad Federal de Río Grande do Sul (UFRGS), and Costa, W.J.E.M. 2006. *Zootaxa*, 1213: 1–162. (Map modified from Shuttle Radar Topography Mission (SRTM), Courtesy NASA/JPL-Caltech.)

to three different areas (middle-low Paraná, Paraná-La Plata, and Uruguay River); these haplotype networks are connected by a central haplotype found on the west bank of the lower Uruguay River (García et al., 2012). Although originally this was interpreted as a possible source area (García et al., 2012; Chapter 15), the opposite scenario is also possible, and this area could be considered a sink area, especially considering that it is located at the same altitude or below the other localities analyzed. This pattern is in agreement with sea level fluctuations that generated marine environments in the lower and middle Paraná River, acting as barrier to those partially differentiated populations. Interestingly, *A. charrua* showed a "star" configuration of haplotypes in which the central haplotype was located at a "highland" locality (35 m asl) that was not affected by the Pleistocene marine transgressions (García et al., 2009; Chapter 15). All other localities included in that study are in an area strongly influenced by sea level changes over the last 10,000 years (Bossi and Ortiz, 2011). In both cases, more specimens and genetic markers are needed to further evaluate these hypotheses. Preliminary analyses of *A. arachan*, a species that includes many populations or ponds that are more loosely connected, show strong population structure, suggesting low dispersal (M. Loureiro et al., unpublished data). In the east African annual *Nothobranchius furzeri*, genetic differentiation between two clades is associated with river basins, and gene flow may have been interrupted by a larger river (Bartáková et al., 2013). Within clades, a high genetic structuring of populations inhabiting individual ponds was observed. Isolation by distance within genetic groups suggests that dispersal may follow a stepping-stone model. Interestingly, the results suggest that dispersion in this species may occur upstream, following expansion from past humid climate refuges. Populations of this species are subject to strong fluctuations in size, extinctions, and recolonizations, congruent with a metapopulational dynamic of their populations, at least in an evolutionary timescale (Bartáková et al., 2013).

8.5 BIOGEOGRAPHY AND SPECIATION

Contrasting species distribution ranges with phylogenetic information might shed light onto speciation processes. Although *Austrolebias* phylogeny is not yet stable and tree topologies vary using different sets of data (Costa, 2006, 2010; García et al., 2014), some clear patterns emerge regarding the monophyly of clades within the genus. In most *Austrolebias* species complexes, the resulting subclades are distributed allopatrically, suggesting a general pattern of allopatric speciation. A phylogenetic analysis of the African annual genus *Nothobranchius* (Dorn et al., 2014) recovered similar results, in this case as a consequence of geographical isolation promoted by aridification of western Africa during the late Miocene. This together with the fact that species of both *Austrolebias* and *Nothobranchius* within clades are morphologically similar (see above) indicates that niche conservationism has a strong influence in their adaptation and evolution (Wiens and Graham, 2005).

One example of an allopatric pattern is observed in the *A. adloffi* species group, recovered monophyletically both in morphological and molecular phylogenies (Costa, 2006, 2010; García et al., 2014; Chapter 15). Species of this clade are distributed in the Patos-Merín and Río Negro basins (Figure 8.6a). All species endemic to Patos-Merín are distributed in association with different drainages. However, the basal species in the clade, *A. arachan*, is distributed across most of the Río Negro basin, but is also found in the headwater wetlands of Laguna Merín effluents. Furthermore, this clade shows an example of parapatry between putative sister taxa, *A. charrua* and *A. reicherti*. In the lower Cebollati River basin, they have an area of secondary contact (García et al., 2009; Chapter 15), and intermediate phenotypes are found, suggesting that parental species do breed in the contact zone. Recently, a study showed that both species interbred under laboratory conditions, and the resulting hybrids have similar morphology and coloration patterns to those found in the wild (Pasos, 2013; Chapter 17). However, hybrid viability decreased notoriously after three generations.

The *A. affinis* species group apparently represents another case of allopatric distribution (Figure 8.6b). In this case, species are morphologically extremely similar (Costa, 2006; Clavijo-Baquet, 2008) and can be differentiated only by subtle differences in pigmentation patterns; this raises the question about true species identity and limits. This species group is distributed across all basins that originate in the Río Grande del Sur Central Depression (headwaters of the Negro, Ibicuy, and Jacui rivers) and the upper portions of Yaguarón and Tacuarí rivers (Laguna Merín basin), with one population in the Cuareim River (Uruguay River basin). In this clade, a parapatric distribution between *A. affinis* and *A. juanlangi* occurs, with a very diffuse limit in the middle Río Negro basin. However, detailed phylogeographic patterns are needed to determine the true species limits and speciation processes in this clade.

The allopatric pattern within clades is broken into three species of the large top predators that are sympatric and even syntopic in the southwest Laguna Merín. These species can be found in the same pond in different combinations, but all three never occur together. Interestingly, the three species have a well-differentiated (and possibly trophically specialized) jaw morphology: *A. cheradophilus* with a blunt snout (similar to other clades), *A. wolterstorffi* with elongated jaws, and *A. prognathus* with extremely elongated jaws. *Austrolebias wolterstorffi* has one of the largest distribution ranges in the genus (~50,000 km^2), from the northern Patos to the southern Merín lagoons. It is possible that its presence in the area is the result of dispersion from other regions. On the other hand, *A. cheradophilus* and *A. prognathous* have highly overlapping ranges restricted to the southwestern Laguna Merín wetlands, suggesting the possibility of sympatric or ecological speciation in this area. Unfortunately, current alternative phylogenetic hypotheses within the group do not allow for further tests.

In African aplocheilids, Agnèse et al. (2006) found that in a monophyletic clade (sub gen. *Chromaphyosemion*) within the nonannual *Aphyosemion*, species are also distributed allopatrically (or parapatrically with few poorly documented exceptions, Agnèse et al., 2006); but species ranges do not correspond to river basins. Indeed, large rivers may sometime act as barriers to their dispersion (Agnèse et al., 2006). However, as they are not annuals and live in small streams within highly forested areas, dispersion may be higher than in annuals. Amiet (1987) observed that species distributions in this genus form an "intricate parapatry." According to Agnèse (2006), the major factor affecting species distribution is interspecific competition. The simplest hypothesis to explain speciation is to consider vicariance events generated by the dry phases of the climate that have been occurring in their range in the last 800,000 years, due to small refuges spread across the original forest area (Agnèse, 2006). Collier et al. (2009) found a similar pattern in the genus *Epiplatys*, with an allopatric differentiation generated in this case by marine transgressions in the Dahomey Gap (95–53 mya) and with two posterior pulses of dispersion when sea level decreased, one by coastal lowlands and other through the savannahs. The geographic restriction of the other (five) genera of West African aplocheiloids suggests they are ecologically restricted to their respective rainforest blocks and lack the ability to disperse out of the regions where they diversified (Collier, 2009). Shidlovskiy et al. (2010) and Ng'oma et al. (2013) found that in the African annual *Nothobranchius*, related species are distributed allopatrically associated with river basins. In another African genus, *Fenerbahce* species are also in allopatry in different sections of the Congo basin (Sonnenberg et al., 2011). A phylogenetic analysis of the African annual *Nothobranchius* also shows the allopatric pattern within subclades (Dorn et al., 2011). However, an interesting difference arises between the African *Nothobranchius* and the Neotropical *Austrolebias*. In *Nothobranchius* there are few range overlaps of related subclades (Dorn et al., 2014), while in *Austrolebias* subclades overlap their distribution across the genus range. One explanation of this may be differences in the geological, geomorphological, and climatic histories between the areas of distribution of these genera in South America (Lundberg et al., 1998; Ribeiro, 2006) and Africa, respectively. Another possibility could be the age of clades. However, García et al. (2014) in *Austrolebias* and Dorn et al. (2014) in *Nothobranchius*, suggest similar dates for the origin of these genera (between 8 and 20 mya) and also for the most recent speciation origins (Quaternary).

Working at the Neotropical scale, Costa (2010) analyzed patterns of distribution of the tribe of annuals Cynolebiasini (which includes *Austrolebias*), with a dispersal-vicariance approach based on a morphological phylogeny. He found that distribution patterns have been shaped both by vicariance and dispersal events. However, since phylogenies of Rivulidae are still controversial and differences between morphological and molecular analysis are evident, these results need to be confirmed with more robust phylogenetic trees.

8.6 AREAS OF DISTRIBUTION AND SPECIES DIVERSITY

Annual fish species richness greatly varies through the different areas. The Paraguay River basin only has three *Austrolebias* species, and the middle Paraná has five species. However, both areas also have the presence of nonannual rivulids (see above). The upper and middle Uruguay River, and its largest effluent, the Ibicuy River, are also species-poor areas with four species; the Jacui River (Patos lagoon effluent) only presents three species; the lower Paraná-La Plata basin has seven species; the Río Negro basin has six species; the Laguna de los Patos lowlands have five species.

On the other hand, the Laguna Merín basin presents 15 species, 8 of them in the Cebollatí River wetlands. Why such a diversity? One possible answer may be a combination of the extent of suitable habitats for annual fishes, from large lowland wetlands to relatively large flat areas with wetlands 50 m above sea level, and the high connectivity of this area to others (see Biogeographical Modules). Some of the species or clades present here are shared with Laguna de los Patos: *A. wolterstorffi*, *A. gymnoventris*, *A. adloffi* species group; and *A. prognathus* putatively closer relatives (Loureiro, 2004; Costa, 2006, 2010). *A. elongatus* and *A. monstruosus* are found in the Paraná basin. Sea level in the area has been changing at least over the last 100,000 years (Ayup-Zouain, 2006). In this context, an interesting event was the sea level regression that occurred ~18,000 years ago, when sea levels dropped ~140 m. In this scenario, the La Plata estuary did not exist, and it is proposed that the Paraná River flowed directly into the Atlantic Ocean (Ayup-Zouain, 2006), with plausible connections with current Laguna Merín effluents. Lambeck et al. (2002) also mention a severe global sea level drop ~150,000 years ago, so biogeographic connections between both basins may be older than the Holocene. The most important source of species, however, seems to be the Río Negro basin. Six of the species found in the Laguna Merín basin (*A. arachan*, *A. juanlangi*, *A. quirogai*, *A. vazferreirai*, *A.* aff. *melanoorus*, and *A.* aff. *luteoflammulatus*) have their largest ranges in the middle and upper Río Negro. This pattern has probably been generated by rearrangements of the drainage system between the Río Negro, Tacuarí, and Yaguarón rivers (both Laguna Merín effluents; Loureiro et al., 2011). All these species are present in the wetlands associated with these rivers and located above 100 m asl, in very close proximity to Río Negro effluents. Furthermore, some parts of the river configurations show elbows of capture (Loureiro et al., 2011), which are considered a source of evidence regarding rearrangements in this drainage (Bishop, 1995). If the other annual species in the Patos-Merín drainage, *Cynopoecilus melanotaenia*, whose more closely related genera are distributed in the northern Atlantic coastal basins in southeastern Brazil (Costa, 1998; Ferrer et al., 2014), is taken into account, the Laguna Merín basin can be considered an endemic area built by high connectivity with different sources.

8.7 BIOGEOGRAPHICAL MODULES

The definition of biogeographical regions and their boundaries represent an important challenge in biogeography studies (Cumming et al. 2010; Thébault, 2013; Kougioumoutzis et al., 2014). Environmental factors, physical barriers, dominant biota, or target taxa are considered the main criteria for classifying species into biogeographic units (Quinn and Harrison, 1998; Presley et al., 2010).

Recently, the network approach has emerged as a promising perspective within biogeography, providing new insights into the identification of ecologically meaningful units (Cumming et al. 2010; Carstensen and Olesen, 2009; Carstensen et al., 2012; Theébault, 2013; Trøjelsgaard et al., 2013; Dalsgaard et al., 2014; Kougioumoutzis et al., 2014). Biogeographic networks are defined on the basis of species occurrences among habitat patches, where species and habitat patches are the nodes and occurrences among them represent the links (Carstensen et al., 2012, 2013). This kind of network is termed a bipartite or two-mode network and consists of two classes of nodes—species and habitat patches—wherein links occur only among nodes of different classes—that is, species–habitat patches. Therefore, the connections among habitat patches through a species could be attributed to an interchange of individuals via dispersal (Carstensen et al., 2012), while species connected to the same habitat patch represent overlap in their distributions (Carstensen et al., 2013).

The increasing explosion of studies on biogeography from a network approach is due to its contribution to organizing geographic patterns of species distribution across habitat patches into objectively delimited compartments (e.g., Carstensen and Olesen, 2009; Carstensen et al., 2012, 2013; Dalsgaard et al., 2014; Kougioumoutzis et al., 2014). Particularly, modularity analysis offers a robust tool for identifying groups of species and habitat patches more mutually linked than with the rest of the network, termed *biogeographic modules* (Carstensen et al., 2012). In this sense, modules have been suggested as appropriate biogeographical study units delimited by areas of low species interchange (Carstensen and Olesen, 2009). The importance of this analysis is that it provides a clear division of habitat patches and species into modules at intermediate scales not evident from other techniques (see Box 8.1; Borthagaray et al., 2014b). The concept of modularity was suggested for ecological systems in general in May 1972, but was only recently detected after the development of strong algorithms (Newman and Girvan, 2004; Guimerà and Amaral, 2005; Marquitti et al., 2014).

Based on the species and habitat patches connections within modules and across them, this analysis provides information about the spatial importance—or *biogeographical role*—of species and habitat patches in the network's structure (Carstensen et al., 2012). Specifically, biogeographical roles give information on how individual species and patches contribute to the connectivity of the system. Four biogeographical roles are defined according their linkage pattern: peripherals, connectors, module hubs, and network hubs (see Box 8.1; Olesen et al., 2007; Carstensen et al., 2012, 2013). These topological roles have been associated with biological (e.g., body size and trophic level) and physical (e.g., area, altitude, and isolation) attributes to determine which aspects of species and locations are involved in the connectivity of the whole network (Borthagaray et al., 2014a).

In order to advance in the understanding of the regional assembly of the genus *Austrolebias* in the La Plata basin and Patos-Merín drainage system in South America, we identified biogeographical modules, that is, highly connected subgroups of basins and killifish species. To this end, first using a presence–absence matrix of 41 killifish species and 8 basins, we constructed a bipartite network wherein killifish species and basins are nodes and links represent occurrences of killifish species in the basins. That is, if a killifish species occurs in a basin, a link is established between them. Then a modularity analysis was applied to the bipartite network and, based on the linkage pattern of species and basins, biogeographical roles were assigned (see Box 8.2 for methodological details on modularity analysis).

The La Plata basin and Patos-Merín drainage system present a significantly modular structure ($M = 0.59$; $p = 0.02$) composed of five biogeographic modules (Figure 8.7). Specifically, modules were composed of one basin and several species, except in two cases integrated by two and three basins located in contiguous geographic regions (Table 8.2, Figure 8.7). For example, one module comprised the Laguna Merín and coastal Atlantic basins. These results likely reflect similar habitat conditions beyond the physical border of the individual basins or past connections between them due to the documented sea level fluctuations in the area (Bossi and Ortiz, 2011). A transition zone, composed of the Río Negro basin, corresponds to only one biogeographic module and, interestingly, is the area with the lowest percentage of endemic species (Table 8.2).

BOX 8.1   MODULARITY IN BIOGEOGRAPHY

Modularity describes the degree to which a network is arranged into clearly defined groups of nodes—i.e. modules—with more connections within them than among them (Olesen et al., 2007). Several modularity metrics and optimization methods have been proposed (see Marquitti et al., 2014; Thébault, 2013). In ecology, the Newman and Girvan's modularity index (2004) for unipartite network and the optimization technique called simulated annealing (SA) (Guimerà and Amaral, 2005) are frequently applied.

Once a significant modular organization is detected, the topological role of nodes are estimated based on their membership in a module (Olesen et al., 2007). The role of each node is quantified by two parameters. The first one is defined as the number of links that a node has to other node in the same module—standardized within module degree, z.

The second parameter is a measure of the number of links of a node with nodes in others modules—among module connectivity, c. Finally, the zc-parameter space is divided in four regions following the criteria defined by Olesen et al. (2007) and nodes classified as peripherals, module hubs, connectors, and network hubs (see Figure B8.1).

In biogeographical networks, based on species occurrences among patches, modularity analysis is used to identify spatial aggregations of species (S) and habitat (H) patches. In this sense, biogeographic modules may correspond to species that have similar habitat requirements and locations that share species. Then, species and habitat patches are classified in four topological roles, defined according Carstensen et al. (2012) by their local (z) and regional (c) topological linkage patterns (see Figure B8.1).

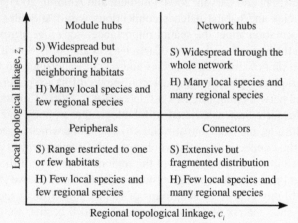

Figure B8.1 zc parameter space showing the distribution of the biogeographical roles. The threshold value of z and c that define the four roles are presented in Olesen et al. (2007). (Modified from Carstensen, D.W. et al. 2012. *Journal of Biogeography*, 39: 739–749; Carstensen, D.W. et al. 2013. *Ecography*, 36: 1097–1105.)

In contrast, modules located at both ends of the environmental gradient present high values of endemism. This pattern could be attributed to a high flow of individuals between the transition zone and their neighbors' modules. Consequently, the Río Negro basin could be considered as a link area for species dispersal from more distant basins.

Biogeographical roles are defined according to the pattern of local and regional topological linkage that species and basins present (Box 8.1). Therefore, all species were classified as peripherals. However, two species subgroups could be considered. On one hand, species restricted to one or two basins but belonging to the same module—for example, *A. elongatus* (Figure 8.7), could be

BOX 8.2   METHODS TO IDENTIFY THE MODULAR
STRUCTURE IN A BIOGEOGRAPHIC NETWORK

Aim. To identify biogeographical units based on the distributions of killifishes-species of the genus *Austrolebias.*

Data. A presence–absence matrix of killifish-species x basins to construct a bipartite network. That is, links only connect species and basins—i.e. species never connect with species and basins never connect with basins.

Analysis. The modularity analysis was performed with the software MODULAR (Marquitti et al., 2014), which used an algorithm based on simulated annealing (SA) to found the partition that maximizes a modularity metric. In this case, the modularity metric is the one proposed by Barber (2007) for bipartite networks. This metric has been suggested as the one with the best performance to identify modules in bipartite networks (Thébault et al., 2013). To test its significance, the observed modularity was compared with an expected modularity distribution estimated on random networks (N = 2000) that conserve the original distribution of links per node (see Marquitti et al., 2014).

To determine the biogeographical roles of species and basins in the bipartite network, the vector of their memberships was imported to the free software R, wherein the two parameters, z—the standardized within module degree and c—the among module connectivity, were estimated according to Olesen et al. (2007). Finally, from a zc parameter space killifish-species and basins were classified as peripherals, connectors, module hubs, or network hub.

classified as endemic to their modules and represent 76% of the regional species pool. Congruent with module composition, these species occupy the more isolated modules in the studied system. On the other hand, 25% of species inhabit few basins across different modules, and so they present an extensive distribution compared with the first subgroup. Congruently, these species present high values of regional topological linkage (Box 8.1, Figure 8.8) but not enough to be classified as connector species. However, they may represent important connections among biogeographical modules and so to the whole network. For example, *A. nigripinnis* or *A. bellottii* could be key species to connect two different biogeographic modules to the transition zone (Río Negro).

Similarly, basins were also classified as peripherals but four basins were also classified as module hubs. Peripheral basins are composed of a few local or short-dispersal species and belong to biogeographic modules at the extremes of the environmental gradient. Interestingly, all modules, except the one composed of the Paraguay River basin, present a basin classified as module hub and so are important to maintaining the internal flow within their module. In this sense, module hubs have been suggested as source habitats for their modules, while peripherals are sink habitats, receiving species from sources within their modules (Carstensen et al., 2012).

Finally, the results found herein illustrate the benefits of using a network approach to detect biogeographic modules. Specifically, they emphasize the relevance of modularity analysis to identify areas of endemism and also to capture intermediate levels of geographic aggregation not evident from other analysis (Box 8.2).

8.8 DISTRIBUTION AND CONSERVATION

Most *Austrolebias* species are endemic to very restricted areas (see Figure 5.6). The extension of their presence and area of occupancy, two parameters used by IUCN in the Red List Assessments, together with the severe land-use transformation that is taking place in the area (deforestation, rice and soy crops, open-sky minery, and tourism) place them in the Endangered category. As a matter

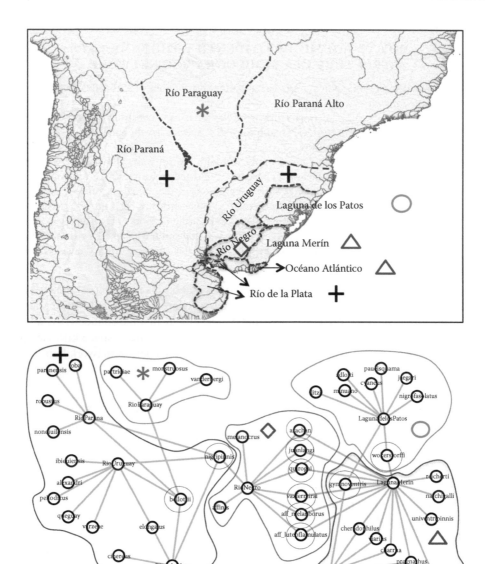

Figure 8.7   The upper panel shows the La Plata basin and Patos-Merín drainage systems. In the lower panel, the biogeographical network showing the modular structure of the study system (41 killifish species and 8 basins) is organized in five modules. Species that presented high values of regional topological linkage (parameter c in Box 8.1) are highlighted with a red circle (see Figure 8.8). In both the cases, the symbols correspond to the five modules detected.

of fact, two *Austrolebias* species have already been placed on the IUCN Red List; *A. cinereus* as Critically Endangered (Loureiro et al., 2007) and *A. affinis* as Vulnerable (Reis and Lima, 2009). However, the other species have not been formally evaluated yet.

8.9 CONCLUSIONS AND FUTURE WORK

Throughout this chapter we reviewed what it is known about distribution patterns in the annual killifish genus *Austrolebias* at different spatial scales. At the pond scale, coexisting species seem

Table 8.2 Description of Biogeographical Modules Detected for the La Plata Basin and Patos-Merín Drainage System in Terms of Number of Basins, Number of Species, Number of Species Per Basin, and Percentage of Module-Endemic Species

Module	No. of Basins	No. of Species	No. of Species Per Basin	Module Endemics (%)
Paraná Uruguay Río de la Plata	3	16	5.3	0.69
Paraguay	1	4	4	0.75
Negro River	1	9	9	0.22
Laguna Merín Atlantic Ocean	2	11	5.5	0.73
Laguna de los Patos	1	9	9	0.78

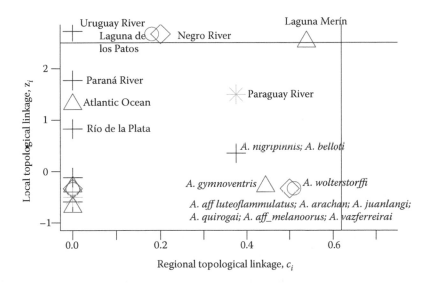

Figure 8.8 *zc* parameter space showing the distribution of killifish species and basins according to their network roles. The different symbols and colors correspond to the five modules found (see Figure 8.7).

to relax competitive pressures by adjusting their body shape (evolutionary timescale) and size (ecologic scale). Coexistence also seems to be favored by processes occurring at the landscape scale through the metapopulation and metacommunity dynamics mediated, among other factors, by connectivity. Connectivity and species vagility also have a strong influence on the large-scale biogeographical patterns, on species' genetic structure and clade diversity, as well as on the richness of endemic areas. We also proposed modularity analysis as an alternative method to determine endemic areas. Spatial analyses in annual fishes are incipient; more information and research at all scales are still needed, even from other fields, such as systematics. Finally, many of the patterns observed should be reconsidered in the light of robust phylogenies, which are currently lacking.

ACKNOWLEDGMENTS

We would like to thank all the undergraduate and graduate students who participated in the fieldwork at all scales. MA was partially funded by Fondo Clemente Estable 2011-2-7117 and 2007-054. All the authors acknowledge funding from PEDECIBA and ANII.

REFERENCES

Agnèse, J.F., F. Zentz, O. Legros, and D. Sellos. 2006. Phylogenetic relationships and phylogeography of the killifish species of the subgenus *Chromaphyosemion* (Radda 1971) in West Africa, inferred from mitochondrial DNA sequences. *Molecular Phylogenetics and Evolution*, 40: 332–346. doi: 10.1016/j.ympev.2006.03.018

Albert, J.S., and R.E. Reis. 2011. Major biogeographic and phylogenetic patterns. In J.S. Albert and R.E. Reis (eds.), *Historical Biogeography of Neotropical Freshwater Fishes* (pp. 3–19). Oakland, CA: University of California Press.

Amiet, J.L. 1987. *Le Genre Aphyosemion Myers (Pisces, Teleostei, Cyprinodontiformes). Faune du Cameroun (tome 2)*. Paris: Sciences Nat. Compiègne.

Arenzon, A., C.A. Lemos, and M.B.C. Bohrer. 2002. The influence of temperature on the embryonic development of the annual fish *Cynopoecilus melanotaenia* (Cyprinodontiformes, Rivulidae). *Brazilian Journal of Biology*, 62: 743–747. doi: 10.1590/S1519-69842002000500002

Arenzon, A., A.C. Peret, and M.B.C. Bohrer. 2001. Growth of the annual fish *Cynopoecilus melanotaenia* (Regan 1912) based in a temporary water body population in Río Grande Do Sul State, Brazil (Cyprinodontiformes, Rivulidae). *Revista Brasileira de Biologia*, 61: 117–123. doi: 10.1590/S0034-7108 2001000100015

Arezo, M.J., L. Pereiro, and N. Berois. 2005. Early development in the annual fish *Cynolebias viarius*. *Journal of Fish Biology*, 66: 1357–1370. doi: 10.1111/j.0022-1112.2005.00688.x

Arim, M., S. Abades, G. Laufer, M. Loureiro, and P.A. Marquet. 2010. Food web structure and body size: Trophic position and resource acquisition. *Oikos*, 119: 147–153. doi: 10.1111/j.1600-0706.2009.17768.x

Arim, M., M. Berazategui, J.M. Barreneche, L. Ziegler, M. Zarucki, and S.R. Abades. 2011. Determinants of density-body size scaling within food webs and tools for their detection. *Advances in Ecological Research*, 45: 1–43. doi: 10.1016/B978-0-12-386475-8.00001-0

Ayup-Zouain, R.N. 2006. Evolución paleogeográfica y dispersión de los sedimentos del Río de la Plata. In R. Menafra, L. Rodriguez-Gallego, F. Scarabino, and D. Conde (eds.), *Bases para la conservación y el manejo de la costa uruguaya* (pp. 1–8). Montevideo: Vida Silvestre.

Bartáková, V., M. Reichard, K. Janko et al. 2013. Strong population genetic structuring in an annual fish, *Nothobranchius furzeri*, suggests multiple savannah refugia in southern Mozambique. *BMC Evolutionary Biology*, 13: 196. doi: 10.1186/1471-2148-13-196

Belmonte, G., and V. Rossi. 1998. Resurrection and time travelling: Diapause in crustaceans (and others). *Trends in Ecology and Evolution*, 13: 4–5. doi: 10.1016/S0169-5347(97)01234-2

Berois, N., M.J. Arezo, and R.O. de Sá. 2014. The neotropical genus *Austrolebias*: An emerging model of annual killifishes. *Cell and Developmental Biology*, 3: 2. doi: 10.4172/2168-9296.1000136

Bishop, P. 1995. Drainage rearrangement by river capture, beheading, and diversion. *Progress in Physical Geography*, 19: 449–473. doi: 10.1177/030913339501900402

Blažek, V., M. Polačik, and M. Reichard. 2013. Rapid growth, early maturation and short generation time in African annual fishes. *EvoDevo*, 4: 24. doi: 10.1186/2041-9139-4-24

Borthagaray, A.I., M. Arim, and P. Marquet. 2014a. Inferring species roles in metacommunity structure from species coexistence networks. *Proceedings of the Royal Society B*, 281: 1792. doi: 10.1098/rspb.2014.1425

Borthagaray, A.I., J.M. Barreneche, S. Abades, and M. Arim. 2014b. Modularity along organism dispersal gradients challenges a prevailing view of abrupt transitions in animal landscape perception. *Ecography*, 37: 564–571. doi: 10.1111/j.1600-0587.2013.00366.x

Borthagaray, A.I., V. Pinelli, M. Berazategui, L. Rodriguez-Tricot, and M. Arim. in press. Effects of metacommunity network on local communities structure: From theoretical predictions to empirical evaluations. In A. Belgrano, G. Woodward, and U. Jacob (eds.), *Aquatic Functional Biodiversity: An Eco-Evolutionary Approach*. USA: Elsevier.

Bossi, J., and A. Ortiz. 2011. Geología del Holoceno. In F. García Rodriguez (ed.), *El Holoceno en la zona costera de Uruguay* (pp. 13–48). Montevideo: Universidad de la República.

Brown, J.H. 1995. *Macroecology*. Chicago, IL: The University of Chicago Press.

Calviño, P.A. 2003. Redescubriendo *Austrolebias robustus* (Günther 1883). *KCA—Suplemento 1*, Buenos Aires: KCA.

Calviño, P.A. 2005. *Austrolebias toba* (Cyprinodontiformes: Rivulidae), una especie nueva de pez anual de la Argentina. *Revista del Museo Argentino de Ciencias Naturales*, 7: 183–190.

Canavero, A., D. Hernández, M. Zaruki, and M. Arim. 2014. Patterns of co-occurrences in a killifish meta-community are more related with body size than with species identity. *Austral Ecology*, 39: 455–461. doi: 10.1111/aec.12103

Carstensen, D.W., B. Dalsgaard, J.C. Svenning et al. 2012. Biogeographical modules and island roles: A comparison of Wallacea and the West Indies. *Journal of Biogeography*, 39: 739–749. doi: 10.1111/j.1365-2699.2011.02628.x

Carstensen, D.W., B. Dalsgaard, J.C. Svenning et al. 2013. The functional biogeography of species: Biogeographical species roles of birds in Wallacea and the West Indies. *Ecography*, 36: 1097–1105. doi: 10.1111/j.1600-0587.2012.00223.x

Carstensen, D.W., and J.M. Olesen. 2009. Wallacea and its nectarivorous birds: Nestedness and modules. *Journal of Biogeography*, 36: 1540–1550. doi: 10.1111/j.1365-2699.2009.02098.x

Chesson, P. 2000. Mechanisms of maintenance of species diversity. *Annual Review of Ecology and Systematics*, 31: 343–366. doi: 10.1146/annurev.ecolsys.31.1.343

Clavijo-Baquet, S. 2008. Variación morfológica en el grupo de especies de peces anuales: *Austrolebias affinis* (Cyprinodontiformes; Rivulidae): Implicancias en la sistemática del grupo. Unpublished master's thesis. PEDECIBA-Facultad de Ciencias, Uruguay.

Collier, G.E., W.J. Murphy, and M. Espinoza. 2009. Phylogeography of the genus *Epiplatys* (Aplocheiloidea: Cyprinodontiformes). *Molecular Phylogenetics and Evolution*, 50: 190–196. doi: 10.1016/j.ympev.2008.10.006

Costa, W.J.E.M. 1998. Phylogeny and classification of Rivulidae revisited: Origin and evolution of annualism and miniaturization in rivulid fishes (Cyprinodontiformes: Aplocheiloidei). *Journal Comparative Biology* 3:33–94.

Costa, W.J.E.M. 2002. The annual fish genus *Cynopoecilus* (Cyprinodontiformes: Rivulidae): Taxonomic revision, with descriptions of four new species. *Ichthyological Exploration of Freshwaters*, 13: 11–24.

Costa, W.J.E.M. 2006. The South American annual killifish genus *Austrolebias* (Teleostei: Cyprinodontiformes: Rivulidae): Phylogenetic relationships, descriptive morphology and taxonomic revisión. *Zootaxa*, 1213: 1–162.

Costa, W.J.E.M. 2009. Trophic radiation in the South American annual killifish genus *Austrolebias* (Cyprinodontiformes: Rivulidae). *Ichthyological Exploration of Freshwaters*, 20: 179–191.

Costa, W.J.E.M. 2010. Historical biogeography of Cynolebiasine annual killifishes inferred from dispersal–vicariance analysis. *Journal of Biogeography*, 37: 1995–2004. doi: 10.1111/j.1365-2699.2010.02339.x

Costa, W.J.E.M. 2014. *Austrolebias araucarianus*, a new seasonal killifish from the Iguacu river drainage, southern Brazilian Araucarian Plateau Forest (Cyprinodontiformes: Rivulidae). *Ichthyological Exploration of Freshwaters*, 25: 97–101.

Dalsgaard, B., D.W. Carstensen, J. Fjeldsa et al. 2014. Determinants of bird species richness, endemism, and island network roles in Wallacea and the West Indies: Is geography sufficient or does current and historical climate matter? *Ecology and Evolution*, 4: 4019–4031. doi: 10.1002/ece3.1276

Diamond, M. 1975. Assembly of species communities. In M.L. Cody and J.M. Diamond (eds.), *Ecology and Evolution of Communities* (pp. 342–444). Cambridge, MA: Belknap Press, 342–444.

Dorn, A., E. Ng'oma, K. Janko et al. 2011. Phylogeny, genetic variability and colour polymorphism of an emerging animal model: The short-lived annual *Nothobranchius* fishes from southern Mozambique. *Molecular Phylogenetics and Evolution*, 61: 739–749. doi: 10.1016/j.ympev.2011.06.010

Dorn, A., Z. Musilová, M. Platzer, K. Reichwald, and A. Cellerino. 2014. The strange case of East African annual fishes: Aridification correlates with diversification for a savannah aquatic group? *BMC Evolutionary Biology*, 14: 210. doi: 10.1186/s12862-014-0210-3

Fernandes, I.M, R. Henriques-Silva, J. Penha, J. Zuanon, and P.R. Peres-Neto. 2014. Spatiotemporal dynamics in a seasonal metacommunity structure is predictable: The case of floodplain-fish communities. *Ecography*, 37: 464–475. doi: 10.1111/j.1600-0587.2013.00527.x

Ferrer, J., L.R. Malabarba, and W.J.E.M. Costa. 2008. *Austrolebias paucisquama* (Cyprinodontiformes: Rivulidae), a new species of annual killifish from southern Brazil. *Neotropical Ichthyology*, 6: 175–180. doi: 10.1590/S1679-62252008000200004

Ferrer, J., J.M. Wingert, and L.R. Malabarba. 2014. Description of a new species and phylogenetic analysis of the subtribe Cynopoecilina, including continuous characters without discretization (Cyprinodontiformes: Rivulidae). *Zoological Journal of the Linnean Society*, 172: 846–866. doi: 10.1111/zoj.12190

Filipe, A.F., M.F. Magalhaes, and M.J. Collares-Pereira. 2010. Native and introduced fish species richness in Mediterranean streams: The role of multiple landscape influences. *Diversity and Distributions,* 16: 773–785. doi: 10.1111/j.1472-4642.2010.00678.x

Fukami, T. 2010. Community assembly dynamics in space. In H.A. Verhoef and P.J. Morin (eds.), *Community Ecology: Processes, Models, and Applications* (pp. 45–54). Oxford: Oxford University Press. doi: 10.1093/acprof:oso/9780199228973.003.0005

Fukami, T., and M. Nakajima. 2011. Community assembly: Alternative stable states or alternative transient states? *Ecology Letters,* 14: 973–984. doi: 10.1111/j.1461-0248.2011.01663.x

García, G., V. Gutiérrez, N. Ríos et al. 2014. Burst speciation processes and genomic expansion in the neotropical annual killifish genus *Austrolebias* (Cyprinodontiformes, Rivulidae). *Genetica,* 142: 87–98. doi: 10.1007/s10709-014-9756-7

García, G., V. Gutiérrez, J. Vergara, P. Calviño, A. Duarte, and M. Loureiro. 2012. Patterns of population differentiation in annual killifishes from the Paraná–Uruguay–La Plata Basin: The role of vicariance and dispersal. *Journal of Biogeography,* 39: 1707–1719. doi: 10.1111/j.1365-2699.2012.02722.x

García, G., M. Loureiro, N. Berois et al. 2009. Pattern of differentiation in the annual killifish genus *Austrolebias* (Cyprinodontiformes: Rivulidae) from a biosphere reserve site in South America: A multidisciplinary approach. *Biological Journal of the Linnean Society,* 98: 620–635. doi: 10.1111/j.1095-8312.2009.01303.x

García, G., G. Wlasiuk, and E. Lessa. 2000. High levels of mitochondrial cytochrome b divergence in annual killifishes of the genus *Cynolebias* (Cyprinodontformes, Rivulidae). *Zoological Journal of the Linnean Society,* 129: 93–110. doi: 10.1006/zjls.1999.0202

Genade, T., M. Benedetti, E. Terzibasi et al. 2005. Annual fishes of the genus *Nothobranchius* as a model system for aging research. *Aging Cell,* 4: 223–233. doi: 10.1111/j.1474-9726.2005.00165.x

Gonçalves, C., U. Souza, and M. Volcan. 2011. The opportunistic feeding and reproduction strategies of the annual fish *Cynopoecilus melanotaenia* (Cyprinodontiformes: Rivulidae) inhabiting ephemeral habitats on southern Brazil. *Neotropical Ichthyology,* 9: 191–200. doi: 10.1590/S1679-62252011000100019

Gotelli, N.J., and D.J. McCabe. 2002. Species co-occurrence: A meta-analysis of J.M. Diamond's assembly rules model. *Ecology,* 83: 2091–2096. doi: org/10.2307/3072040

Guimerà R., and L.A.N. Amaral. 2005. Functional cartography of complex metabolic networks. *Nature,* 433: 895–900. doi.org/10.1038/nature03288

Hanski, I. 1999. *Metapopulation Ecology.* New York: Oxford University Press.

Hernández, D. 2014. Determinantes de la distribución de tamaños corporales en comunidades de charcos temporales. MSc Thesis Dissertation. Montevideo, PEDECIBA: Facultad de Ciencias.

Holt, R.D., and T.H. Keitt. 2005. Species' borders: A unifying theme in ecology. *Oikos,* 108: 3–6. doi: 10.1111/j.0030-1299.2005.13145.x

Humphries, S., and G.D. Ruxton. 2002. Is there really a drift paradox? *Journal of Animal Ecology,* 71: 151–154. doi: 10.1046/j.0021-8790.2001.00579.x

Keppeler, F.W., L.E.K. Lanes, A.S. Rolon et al. 2014. The morphology–diet relationship and its role in the coexistence of two species of annual fishes. *Ecology of Freshwater Fish,* 24: 77–90. doi: 10.1111/eff.12127

Kiflawi, M., A. Eitam, and L. Blaustein. 2003. The relative impact of local and regional processes on macroinvertebrate species richness in temporary pools. *Journal of Animal Ecology,* 72: 447–452. doi: 10.1046/j.1365-2656.2003.00712.x

Kougioumoutzis K., S.M. Simaiakis, and A. Tiniakou. 2014. Network biogeographical analysis of the central Aegean archipelago. *Journal of Biogeography,* 41: 1848–1858. doi: 10.1111/jbi.12342

Lambeck, K., T.M. Esat, and E.K. Potter. 2002. Links between climate and sea levels for the past three million years. *Nature,* 419: 199–206. doi: 10.1038/nature01089

Lanés, L.E.K., F.W. Keppeler, and L. Maltchik. 2014. Abundance variations and life history traits of two sympatric species of neotropical annual fish (Cyprinodontiformes: Rivulidae) in temporary ponds of southern Brazil. *Journal of Natural History,* 48: 1971–1088. doi: 10.1080/00222933.2013.862577

Laufer, G., M. Arim, M. Loureiro, J.M. Piñeiro-Guerra, S. Clavijo-Baquet, and C. Fagúndez. 2009. Diet of four annual killifishes: An intra and interspecific comparison. *Neotropical Ichthyology,* 7: 77–86. doi: 10.1590/S1679-62252009000100010

Linder, H.P. 2001. On areas of endemism, with an example from African Restionaceae. *Systematic Biology,* 50: 892–912. doi: 10.1080/106351501753462867

Losos, J.B., and R.E. Ricklefs. 2009. Adaptation and diversification on islands. *Nature*, 457: 830–836. doi: 10.1038/nature07893

Loureiro, M. 2004. Sistemática y biogeografía de los peces anuales de la subtribu Cynolebiatina (Cyprinodontiformes: Rivulidae: Cynolebiatinae). PhD dissertation. PEDECIBA: Facultad de Ciencias.

Loureiro, M., A. Duarte, and M. Zarucki. 2011. A new species of *Austrolebias* Costa (Cyprinodontiformes: Rivulidae) from northeastern Uruguay, with comments on distribution patterns. *Neotropical Ichthyology*, 9: 335–342.

Loureiro, M., F. Teixeira de Mello, A. D'Anatro, and V. Cardozo. 2007. *Austrolebias cinereus*. In *IUCN 2010. IUCN Red List of Threatened Species*. Version 2010.4. www.iucnredlist.org.

Lundberg, J.G., L.G. Marshall, J. Guerrero, B. Horton, M.C.S.L. Malabarba, and F. Wesselingh. 1998. The stage for neotropical fish diversification: A history of tropical South American rivers. In L.R. Malabarba, R.E. Reis, R.P. Vari, Z.M.S. Lucena, and C.A.S. Lucena (eds.), *Phylogeny and Classification of Neotropical Fishes* (pp. 13–48). Porto Alegre: EDIPUCRS.

Marquitti, F.M.D., P.R. Guimarães, M.M. Pires, and L.F. Bittencourt. 2014. MODULAR: Software for the autonomous computation of modularity in large network sets. *Ecography*, 37: 221–224. doi: 10.1111/j.1600-0587.2013.00506.x

May, R.M. 1972. Will a large complex system be stable? *Nature*, 238: 413–414. doi: 10.1038/238413a0

Mayfield, M.M., and J.M. Levine. 2010. Opposing effects of competitive exclusion on the phylogenetic structure of communities. *Ecology Letters*, 13: 1085–1093. doi: 10.1111/j.1461-0248.2010.01509.x

Newman, M.E.J., and M. Girvan. 2004. Finding and evaluating community structure in networks. *Physical Review*, 69: 26–113. doi: 10.1103/PhysRevE.69.026113

Ng'oma, E., S. Valdesalici, K. Reichwald, and A. Cellerino. 2013. Genetic and morphological studies of *Nothobranchius* (Cyprinodontiformes) from Malawi with description of *Nothobranchius wattersi* sp. nov. *Journal of Fish Biology*, 82: 165–188. doi: 10.1111/jfb.12001

Olesen, J.M., J. Bascompte, Y.L. Dupont, and P. Jordano. 2007. The modularity of pollination networks. *Proceedings of the National Academy of Sciences of the United States of America*, 104: 19891–19896. doi: 10.1073/pnas.0706375104

Padisák, K.J. 1992. Seasonal succession of phytoplankton in a large shallow lake (Balaton, Hungary)—A dynamic approach to ecological memory, its possible role and mechanisms. *Journal of Ecology*, 80: 217–230. doi: 10.2307/2261008

Pasos, C. 2013. Austrolebias: Un modelo para explorar la selección sexual. PhD dissertation. Montevideo, PEDECIBA: Facultad de Ciencias.

Pazin, V.F.V., W.E. Magnusson, J. Zuanon, and F.P. Mendonca. 2006. Fish assemblages in temporary ponds adjacent to "terra-firme" streams in Central Amazonia. *Freshwater Biology*, 51: 1025–1037. doi: 10.1111/j.1365-2427.2006.01552.x

Piñeiro-Guerra, J.M., C. Fagúndez-Pachón, M. Oesterheld, and M. Arim. 2014. Biodiversity–productivity relationship in ponds: Community and metacommunity patterns along time and environmental gradients. *Austral Ecology*, 39: 808–818. doi: 10.1111/aec.12149

Podrabsky, J., A. Tingaud-Sequeira, and J. Cerdà. 2010. Metabolic dormancy and responses to environmental desiccation in fish embryos. In E. Lubzens, J. Cerdà, and M. Clark (eds.), *Dormancy and Resistance in Harsh Environments* (pp. 203–226). Topics in Current Genetics, The Netherlands: Springer. doi: 10.1007/978-3-642-12422-8_12

Polacik, M., M.T. Donner, and M. Reichard. 2011. Age structure of annual *Nothobranchius* fishes in Mozambique: Is there a hatching synchrony? *Journal of Fish Biology*, 78: 796–809. doi: 10.1111/j.1095-8649.2010.02893.x

Polacik, M., C. Harrod, R. Blazek, and M. Reichard. 2014. Trophic niche partitioning in communities of African annual fish: Evidence from stable isotopes. *Hydrobiologia*, 721: 99–106. doi: 10.1007/s10750-013-1652-0

Polacik, M., and M. Reichard. 2010. Diet overlap among three sympatric African annual killifish species (*Nothobranchius* spp.) from Mozambique. *Journal of Fish Biology*, 77: 754–768. doi: 10.1111/j.1095-8649.2010.02717.x

Presley S.J., L.M. Cisneros, B.D. Patterson, and M.R. Willig. 2010. Vertebrate metacommunity structure along an extensive elevational gradient in the tropics: A comparison of bats, rodents and birds. *Global Ecology and Biogeography*, 21: 968–976. doi: 10.1111/j.1466-8238.2011.00738.x

Quinn, J.F., and S.P. Harrison. 1998. Effects of habitat fragmentation and isolation on species richness: Evidence from biogeographic patterns. *Oecologia*, 75: 132–140. doi: 10.1007/BF00378826

Reichard, M., and M. Polacik. 2010. Reproductive isolating barriers between colour-differentiated populations of an African annual killifish, *Nothobranchius korthausae* (Cyprinodontiformes). *Biological Journal of the Linnean Society*, 100: 62–72. doi: 10.1111/j.1095-8312.2010.01406.x

Reichard, M., M. Polacik, and O. Sedlacek. 2009. Distribution, colour polymorphism and habitat use of the African killifish *Nothobranchius furzeri*, the vertebrate with the shortest life span. *Journal of Fish Biology*, 74: 198–212. doi: 10.1111/j.1095-8649.2008.02129.x

Reis, R., and F. Lima. 2009. *Austrolebias affinis*. In *IUCN 2011. IUCN Red List of Threatened Species*. Version 2011.2. www.iucnredlist.org.

Ribeiro, A.C. 2006. Tectonic history and the biogeography of the freshwater fishes from the coastal drainages of eastern Brazil: An example of faunal evolution associated with a divergent continental margin. *Neotropical Ichthyology*, 4: 225–246. doi: 10.1590/S1679-62252006000200009

Rickets, T.H. 2001. The matrix matters: Effective isolation in fragmented landscapes. *The American Naturalist*, 158: 87–99. doi: 10.1086/320863

Rohlf, F.J. 2014. *Relative Warps 1.54*. Stony Brook: Department of Ecology and Evolution, State University of New York. http://life.bio.sunysb.edu/morph/.

Sedlácek, O., B. Baciaková, and L. Kratochvíl. 2014. Evolution of body colouration in killifishes (Cyprinodontiformes: Aplocheilidae, Nothobranchiidae, Rivulidae): Is male ornamentation constrained by intersexual genetic correlation? *Zoologischer Anzeiger A Journal of Comparative Zoology*, 253: 207–215. doi: 10.1016/j.jcz.2013.12.004

Shidlovskiy, K.M., B.R. Watters, and R.H. Wildekamp. 2010. Notes on the annual killifish species *Nothobranchius rachovii* (Cyprinodontiformes; Nothobranchiidae) with the description of two new species. *Zootaxa*, 2724: 37–57.

Sonnenberg, R., T. Woeltjes, and J.R. Van Der Zee. 2011. Description of *Fenerbahce devosi* (Cyprinodontiformes: Nothobranchiidae), a new species of dwarf killifish from the eastern Congo Basin in the Democratic Republic of Congo. *Zootaxa*, 2966: 1–12.

Stone, L., and A. Roberts. 1990. The checkerboard score and species distributions. *Oecología*, 85: 74–79. doi: 10.1007/BF00317345

Terzibasi, E., D.R. Valenzano, M. Benedetti et al. 2008. Large differences in aging phenotype between strains of the short-lived annual fish *Nothobranchius furzeri*. *PLoS ONE*, 3: 1–13. doi: 10.1371/journal.pone.0003866

Thébault, E. 2013. Identifying compartments in presence–absence matrices and bipartite networks: Insights into modularity measures. *Journal of Biogeography*, 40: 759–768. doi: 10.1111/jbi.12015

Trøjelsgaard, K., M. Báez, X. Espadaler et al. 2013. Island biogeography of mutualistic interaction networks. *Journal of Biogeography*, 40: 2020–2031. doi: 10.1111/jbi.12165

Volcan, M.V., L.E.K. Lanés, and Â.C. Gonçalves. 2014. *Austrolebias bagual*, a new species of annual fish (Cyprinodontiformes: Rivulidae) from southern Brazil. *Aqua*, 20: 4–15.

Wiens, J.J., and C.H. Graham. 2005. Niche conservatism: Integrating evolution, ecology, and conservation biology. *Annual Review of Ecology and Systematics*, 36: 519–39. doi:10.1146/annurev.ecolsys.36.102803.095431

Wilbur, H.M. 1997. Experimental ecology of food webs: Complex systems in temporary ponds. *Ecology*, 78: 2279–2302. doi: 10.2307/2265892

Williams, D.D. 2006. *The Biology of Temporary Waters*. New York: Oxford University Press. doi:10.1093/acprof:oso/9780198528128.001.0001

CHAPTER 9

The Evolutionary Ecology of African Annual Fishes

Martin Reichard

CONTENTS

9.1 DISTRIBUTION AND BIOGEOGRAPHY

African annual fishes form two clades within the aplocheiloid killifishes, both belonging to the family Nothobranchiidae (Murphy and Collier, 1997; see Chapter 1). The first clade comprises *Nothobranchius*, *Pronothobranchius*, and *Fundulosoma*. The genus *Nothobranchius* currently contains more than 60 described species that inhabit temporary pools in East African savannah habitats, ranging from the Nuba Mountains in Sudan to the northeastern part of South Africa. The closely related *Pronothobranchius* consists of four described and poorly known species recorded from a few sites. *Pronothobranchius* inhabit temporary pools in savannahs across the Sahel region between the Gambia and Central African Republic. *Fundulosoma* (a monotypic genus related to *Pronothobranchius*) is a facultatively annual species. The second clade of annual African killifishes consists of *Callopanchax* from the forested region of West Africa (with five described species from Guinea, Liberia, Sierra Leone) and *Fundulopanchax*. At least some species of *Fundulopanchax* (29 described species in total) from West Africa are facultatively annual. They produce eggs capable of surviving desiccation, but the pools where they are found do not dry out routinely, and their typical development does not involve habitat desiccation. Most information on African annual fishes comes from studies on *Nothobranchius*; this chapter primarily considers the ecology of that genus. However, whenever possible, I also report information on other annual taxa of African killifishes.

9.1.1 Habitat Types

Within the vast geographic range of *Nothobranchius*, the location of suitable habitats is extremely patchy. *Nothobranchius* pools are limited to grassland and woodland savannah, typically within alluvia of streams and rivers but sometimes completely isolated from permanent waters. For example, in the dry part of *Nothobranchius'* range in southwestern Mozambique, many pools inhabited by *N. furzeri* and *N. orthonotus* appear isolated and are filled exclusively by rainwater during intensive precipitation (Figure 9.1a and b). Such pools, however, may often be partly connected, as they represent depressions in the savannah where water drained from a larger area accumulates

Figure 9.1 *Nothobranchius* habitats. The same pool in the dry part of the gradient in southern Mozambique during wet (a) and dry (b) phases. *N. furzeri* and *N. orthonotus* were collected in the pool. (c) Extensive marsh in relatively humid coastal region of Central Mozambique, with *N. krysanovi* and *N. orthonotus* populations. (d) Remnant pool within a stream channel that is either annually colonized by three *Nothobranchius* species or supports their stable populations. (Photographs by Martin Reichard.)

after heavy rains. Populations of *Nothobranchius* can also regularly be found in isolated remnant pools within the channels of temporary streams (Figure 9.1d; e.g., Valdesalici and Wildekamp, 2005; Reichard et al., 2009; Valdesalici, 2012; Valdesalici et al., 2012). It is not clear whether these populations are stable and complete the full life cycle within the stream pools or simply comprise individuals flushed from pools adjacent to the stream channel during annual flooding and are not capable of sustaining a viable population over several generations (i.e., sink populations). Such habitats may be disproportionally targeted during sampling due to their convenient locations near fords or bridges and may not represent typical *Nothobranchius* habitats. However, at some of such sites, the same three species were recorded regularly over several years (Reichard et al., 2014). Their population genetic characteristics were comparable with those of populations in savannah pools, though their genetic diversity was relatively high (Bartáková, 2013); it is therefore possible that they represent sustainable populations but with an elevated immigration rate through connectivity.

In the humid parts of *Nothobranchius'* range, such as the coastal area of central and northern Mozambique, the shores of Lake Malawi, Zanzibar Island, and the Tanzanian coastal plains, *Nothobranchius* inhabit vast swampy areas. Likely many of these swamps are permanent, or at least connected to permanent streams, as they frequently harbor a diverse community of nonannual freshwater fishes. The occurrence of *Nothobranchius* is supported by the existence of annually desiccating areas of the grassland matrix that meet the standard requirements for the development of their eggs. After hatching, however, *Nothobranchius* in these habitats frequently coexist with nonannual teleost species and are able to sustain long-term viable populations (Valdesalici and Wildekamp, 2004; Watters et al., 2007).

The critical prerequisite of *Nothobranchius* occurrence in a particular pool is the specific composition of the substrate (Watters, 2009). Soil conditions are the primary drivers of habitat suitability for *Nothobranchius*, as the eggs can only survive the embryonic period and develop successfully on Quaternary vertisol and calcimorph soils. Specific alkaline clay minerals (smectites) are prerequisites for suitable conditions during embryonic development in desiccated pool substrates (Watters, 2009). The mud-rich layer in such pools has low permeability, enabling water to remain in the pool after the surrounding water table has receded. Without this impermeable layer, the pool rapidly desiccates (Watters, 2009). Visually, this substrate is dark brown to black, often forming a thick layer of soft mud on the bottom of the pool. Organic material aggregates in the pool in the form of dead aquatic and terrestrial vegetation but does not cover the large part of the bottom, as is typical of water bodies in forested areas with leaf litter. Despite the presence of rapidly decaying material, the water stays alkaline due to the high buffering capacity of the alkaline clay in the sediment. The water quality is also affected by the action of large mammals. Currently, throughout most of the *Nothobranchius* range, domestic cattle visit these pools, frequently urinating and defecating into the pools (Reichard et al., 2009). The pools hold water for several months after the rains and thus serve as drinking reservoirs. Historically, large African herbivores probably frequented such pools as they still do where they occur. *Nothobranchius* sometimes inhabit pools with sandy substrates (especially pools within active or former streambeds), but such pools always contain at least a small area of accumulated vertisol substrate. *Nothobranchius* never inhabit pools consisting only of orange-colored laterite soils (Reichard et al., 2009; Watters, 2009). Although these pools are very common in the African savannah, especially after heavy precipitation, they are characterized by kaoline-type clay minerals and are slightly acidic, and their substrate is not suitable for *Nothobranchius* embryo survival during the dry period. Clawed frogs (*Xenopus* spp.) and anostracan shrimps frequently occur in these pools and are rarely syntopic (co-occurring in the same pool, i.e., interacting ecologically) with *Nothobranchius* despite their broad sympatry (co-occurrence at the geographic scale). The soil type can clearly be distinguished during the dry season; vertisol mud substrates form deep cracks when desiccated, sometimes covered by a thin layer of white calcite salts, and can be readily distinguished from adjacent sandy and laterite substrates. Soil characteristics and their effect on *Nothobranchius* embryo survival have been previously reviewed (Watters, 2009).

Callopanchax occurs in slow-flowing streams and adjacent forest pools, though their presence in streams is likely secondary (Sonnenberg and Busch, 2010). They were also recorded from a rubber tree plantation, where small depressions around tree trunks formed shallow (2–5 cm) pools inhabited by *Callopanchax*, and from forest pools apparently filled exclusively by rainwater (Dinesen, 2006). *Callopanchax* pools were always shaded within secondary forest plantations, their bottoms covered by leaf litter with sandy and stony substrates. The water was acidic (pH 5.2–6.7, with pH below 6 prevalent). *Pronothobranchius* were also recorded from partly forested areas, but their typical habitat appears to be woodland savannah (Blažek et al., 2012; Valdesalici, 2013). In Niokolo Koba National Park, southeastern Senegal, the habitat of *Pronothobranchius gambiensis* was reminiscent of typical *Nothobranchius* habitats (small temporary savannah pools in sandy soils, but with a layer of black mud). At the same time, a few individuals were collected in adjacent temporary streams, draining a large flooded area into an oxbow lake of the Gambia River. However, *P. gambiensis* was never found in the oxbow lake despite frequent sampling with more than 10,000 individual fish collected over numerous visits (White et al., 2012).

9.1.2 Species Distribution and Range Size

The ranges of individual species vary greatly in size. This may be partly an artifact of incomplete knowledge about the extent of their distribution, of actual differences in species ranges, and of the different species concepts applied for the delineation of new species during recent species descriptions. Typically, however, species ranges span several river drainages. *Nothobranchius orthonotus*, whose range is relatively well studied, inhabits a vast range of at least 1200 km north-south, from the northern part of South Africa (Kwa-Zulu Natal) to southern Malawi and part of Mozambique north of the Zambezi River. In contrast, *N. guentheri* is endemic to Zanzibar Island, only 80 km along its longest axis. Zanzibar Island is also inhabited by *N. melanospilus*, a species otherwise widespread along the Tanzanian coast. Mafia Island, about 150 km south of Zanzibar, also harbors two *Nothobranchius* species, both of which are also common on the adjacent mainland (*N. korthausae* and *N. lucius*). These two islands are located on a shallow continental shelf and were largely connected to the mainland during most of the Quaternary. The third larger island in the area, Pemba, has no record of *Nothobranchius*.

Pronothobranchius ranges are geographically extensive but always contain only a few populations sparsely scattered across a large area (Figure 9.2; Valdesalici, 2013). This may be a real phenomenon wherein existing populations are relicts from a wider historical distribution during a more climatically favorable period. Alternatively, the seemingly patchy distribution may be only an artifact of poor sampling effort. It is unlikely, however, that sampling effort has been equally poor throughout the entire range of the genus. *Callopanchax* has one widely distributed species (*C. occidentalis*) across Sierra Leone, with all other species being more geographically restricted. They all occur in coastal areas drained by small streams and rivers and are not associated with major river basins (Sonnenberg and Busch, 2010).

9.1.3 Climatic Conditions

The distribution of *Nothobranchius* includes areas with either two or only a single rainy season. Two rainy seasons in the equatorial area are associated with more humid conditions than those found in areas with a single rainy season further away from the equator. However, local climatic and geomorphological conditions largely modify the effect of overall seasonality. For example, a relatively steep gradient of aridity has been described in southern Mozambique along the east-west axis. The humid coastal areas receive annual rainfall of up to 1000 mm, but the inland region in Gaza and Inhambane provinces receives only 300 mm annually. As an extension of the Mozambican gradient, the Sazale pan in Gonarezhou National Park in Zimbabwe (the type locality of *N. furzeri*

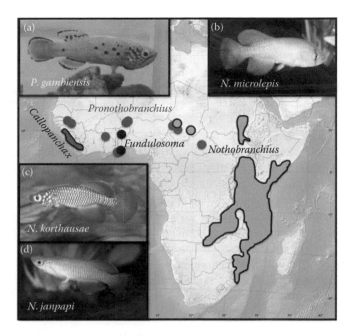

Figure 9.2 Distribution of African annual fishes and representative variation in *Nothobranchius* phenotypic variation. *Callopanchax* range represented by the blue area, *Nothobranchius* by the gray area (and circles for isolated populations), *Pronothobranchius* by red circles, and *Fundulosoma* by black circles. Note circles may represent several adjacent populations. The insets represent males of (a) *Pronothobranchius gambiensis* (Niokolo Koba National Park, Senegal), (b) a robust species from Northern clade (*N. microlepis*, population Mnazini KEN08-8), (c) an example of a coastal species phenotype (*N. korthausae,* Mafia Island TZN 08-4), and (d) a relatively delicate phenotype characteristic of the subgenus *Aphyobranchius* (*N. janpapi*, Ruvu River TZN 09-4). (Photograph (a) by Martin Reichard, photographs (b), (c), and (d) by Béla Nagy.)

at an altitude of 422 m) likely receives even less rainfall. Maybe even more importantly, rainfall is less predictable in the dry regions within this gradient, and some *Nothobranchius* pools may not be filled every year (Terzibasi Tozzini et al., 2013). In the northern extremes of *Nothobranchius'* distribution (Sudan, Somalia, and Lake Chad region) the habitat is apparently even drier and rains perhaps even more erratic (Valdesalici, 2014). In southeastern Senegal, *P. gambiensis* was collected in only one of the five study years (Blažek et al., 2012; White et al., 2012) despite annual visits to the same sites, suggesting that in the driest regions hatching may not occur every year. However, the temporary pools were inundated only in two years during the five years of fieldwork; the other three visits were conducted some time after the rains, when the pools already desiccated (White et al., 2012).

Ambient temperature is seasonally relatively stable in the equatorial areas (25–28°C) but variable in the southern part of the distribution of *Nothobranchius*. We have recorded an annual amplitude ranging from 38°C in summer to 12°C in winter in southern Mozambique. While water has the capacity to buffer the most extreme fluctuations in air temperature, direct measurements using data loggers revealed that daily fluctuations may be similarly large, ranging from a high of 35°C in late afternoon to a low of 20°C before sunrise in a shallow savannah pool in February (Reichard et al., 2009), that is, when *Nothobranchius* in the area were approximately 6 weeks old (Polačik et al., 2011). The deeper parts of the pools (when available) are less prone to daily temperature fluctuations (Figure 9.3a and b). Over six years of fieldwork in Mozambique, we have recorded water temperature (in pools containing *Nothobranchius* at the time of measurement) from 12.8°C (July, winter when most—but not all—pools were dry) to 38.5°C (February, summer, with abundant occurrence of *Nothobranchius*). Between February and April (i.e., the highest period of *Nothobranchius* occurrence in southern Mozambique), the water temperature rarely fell below 22°C. Even in the equatorial region, *Nothobranchius* often occur at a higher altitude (e.g., 1200 m on Lake Victoria plateau) where ambient temperature also decreases rapidly at night. In coastal areas, daily temperature fluctuation is much lower, but the open savannah habitat is generally more prone to early morning decreases and afternoon peaks in water temperature than the forested areas inhabited by other (nonannual) killifishes.

The duration of the wet season varies among regions, among pools within a region (due to variation in their size and morphology), and among years (Reichard et al., unpublished data). Long-term deployment of temperature loggers set in the upper level of the pool bottom over an entire annual cycle can identify periods of habitat desiccation through daily changes in temperature amplitude. Low amplitude indicates the presence of water, due to its buffering effect on sudden temperature changes. Abrupt changes between day and night indicate a period of desiccation (Figure 9.3c). A long-term data set showed that, at least in southern and central Mozambique, repeated pool desiccation within a season and the existence of secondary pools is relatively common (Polačik et al., 2014a). At the same time, several pools remained dry for the entire rainy season, despite evidence of *Nothobranchius* (i.e., *N. furzeri*, *N. kadleci*) presence in the preceding and following years. Similar conditions may also be common in Sudan, the northern limit of the range of *Nothobranchius*, where

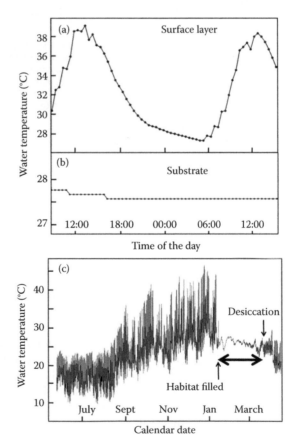

Figure 9.3 Temperature fluctuations recorded by data loggers. Daily variation in water temperature in the upper layer (a) and in the bottom (b) of the pool (water depth 80 cm, lower logger 12 cm inside the substrate) in relatively humid region in southern Mozambique. Seasonal variation in ambient temperature illustrates the duration of habitat inundation (c).

some sites known to host a population of *N. occultus* were dry for almost 10 years. However, it is not clear whether this population still persists (Valdesalici, 2014). In more humid areas of the *Nothobranchius* range, pools are likely to fill reliably every year.

9.1.4 Biogeography

The sister clade to East African *Nothobranchius* consists of *Pronothobranchius* and *Fundulosoma* in West Africa. Currently, the ancestral range of *Nothobranchius* is unknown. The species-rich area along the Tanzanian coast (where 10 different *Nothobranchius* species occur in a radius of less than 100 km) has been assumed as the origin of the genus radiation (Wildekamp, 2004). However, a recent molecular phylogeny of the genus suggested a Nilo-Sudanian origin (Dorn et al., 2014); this is compatible with its sister clade relationship to the West African taxa. This phylogeny is only partly consistent with the former division of *Nothobranchius* into subgenera and species groups (Wildekamp, 2004); furthermore, it suggests convergent evolution of characteristic phenotypic morphs in the genus. Still, these new findings need to be tested with a richer taxon sampling for molecular studies.

Four major clades of *Nothobranchius* have been clearly identified (Dorn et al., 2014). They suggest an entirely allopatric mode of diversification and, even today, their distribution shows very limited secondary overlap. The boundaries of their distribution are largely consistent with the East African Rift system. The dating of their divergence, on the basis of a molecular clock calibrated with the mutation rate of other cyprinodontiform taxa (6–11 mya, mean estimate 8.3 mya), coincides with the aridification of East Africa following its uplifting and with the establishment of savannah habitats between 8 and 5 mya. Alternative dating, using the upsurge of the Ethiopian plateau as the first vicariance event in the genus (approximately 20 mya), suggests a more ancient origin of the genus and its basal diversification in the more humid epoch of the Cenozoic. Each clade went through a period of burst diversification likely driven by climatic oscillations, and extant species are likely of Quaternary origin (Dorn et al., 2014).

Under this recent phylogenic scenario, the basal clade (called the Northern clade) is distributed across arid areas of northeastern Africa, including Sudan, Somalia, and northeastern Kenya. The Southern clade occurs in the southern periphery of the range of *Nothobranchius*, south of the main rift valley, and the lower Zambezi River, with only a few populations recorded on the northern bank of the Zambezi and along the Shire River, corresponding to the southernmost extension of the rift fault containing Lake Malawi. Phylogeographic analyses suggest that all three species complexes forming the Southern clade colonized their range from north to south (Bartáková et al., 2015). The Northern and Southern clades, unless large extinctions have taken place, have diversified poorly (5–10 species known) compared with the other two clades, especially considering the size of their range. The Inland clade is distributed from Uganda in the north to southern Zambia across the East African plains located at an altitude of about 1000 m. The Coastal clade, the most species-rich clade, is distributed in low-altitude plains between southern Kenya and northern Mozambique (Dorn et al., 2014).

Herein, I propose that species diversity in the Coastal clade may be associated with more rapid geomorphological changes such as river captures, changes in the position of the main channel, and repeated isolation and expansion following the more frequent and dynamic occurrence of forested habitats in this more humid region. All these events create ample opportunities for allopatric diversification as a result of frequent isolation of small populations and consequent mechanisms of diversification such as genetic drift, founder effect, sexual selection, or their combination. It is in this region where the highest species richness per unit area is found.

The intraspecific population structure of *Nothobranchius*, where known, is deep. In the three species complexes of the Southern clade, the distributions of intraspecific lineages and, in some cases, allopatric sister species share similar boundaries across all three lineages. The main river channels frequently form a barrier to dispersal, and species and intraspecific lineages are confined within areas bordered by two major rivers. In some cases, populations were able to disperse across the main channel and co-occur with other lineages in secondary sympatry. These cases appear more frequent in the lower parts of the rivers, where the topography is flat and suggestive of more frequent changes in the main channel position and overall river dynamics (Bartáková et al., 2015).

9.1.5 Dispersal and Colonization

The typical mode of *Nothobranchius* dispersal and colonization of new habitats is unknown, and different scenarios are possible. First, adult (or juvenile) fish may (actively or passively) enter different water bodies during major catastrophic floods, which irregularly connect adjacent pools (termed *flood dispersal hypothesis, FDH*). This hypothesis predicts that the dispersers will settle successfully in a suitable new habitat. Fishes are sensitive to minor changes in water velocity and are capable of navigating in desiccating flooded areas (Lucas et al., 2001), increasing the chance that adult *N. furzeri* can colonize depressions where water accumulates on a suitable substrate. Second, the eggs may be transported between savannah pools (termed *egg dispersal hypothesis, EDH*). These pools are frequented by large herbivores, including elephants, rhinoceroses, and buffalo, for drinking and mud bathing. The eggs of macroinvertebrates are known to attach on the skin of mud-bathing African megafauna (Vanschoenwinkel et al., 2011) and the same process may apply to *Nothobranchius* eggs, given that a muddy substrate is characteristic of their habitat. A similar scenario is feasible for *water-bird-related dispersal* (WRD), where the fish eggs are entrained in the mud on birds' feet. The ornamented chorion of annual fishes may facilitate or play role in the latter two dispersal mechanisms, as suggested for Neotropical annual fishes (Loureiro and de Sá, 1996; Fava and Toledo-Piza, 2007).

These three hypotheses make different predictions about the genetic structure of *Nothobranchius* populations. In the case of the FDH, populations should be structured according to river basins (drainages), with no dispersal across major river channels. *Nothobranchius* are poor swimmers, and only passive dispersal would be possible in torrential currents, with little chance of leaving the river channel once entrained. Thus, migration would only be in a downstream direction and genetic diversity would decline with increasing altitude. Under EDH and WRD, dispersal to a higher altitude (upstream dispersal) would be possible, and boundaries between river basins would not form a significant barrier to dispersal. The two hypotheses generate different predictions for the role of river channels. Large rivers would likely wash the *Nothobranchius* eggs off large animals even during the dry season and would thus remain significant barriers to dispersal. In contrast, river channels are less likely to play a major role in structuring *Nothobranchius* populations if their eggs are dispersed by waterbirds. Importantly, these three hypotheses are not mutually exclusive, and fish may disperse via any combination of them.

Given the long history of clade divergences, it is possible that what may appear retrospectively to be dispersal across river channels may actually result from changes in river geomorphology over time (*geomorphological vicariance hypothesis, GVH*). Historical climatic changes in Africa were dramatic (Cohen et al., 2007), with large lakes desiccating and reappearing repeatedly. Similarly, the size and position of the major river channels will have altered in conjunction with changes in climate. The lower reaches on flat plains are particularly prone to changes in river channel morphology, and the majority of apparent channel crossing is expected there. Importantly, if the GVH plays a role, several sympatric annual fish species should disperse concurrently across the channel. Abrupt changes in the position of river channels are a possible explanation for a more recent intraspecific divergence via apparent channel crossing. At an interspecific level, the timing of divergence can coincide with geological events such as uplifting and other tectonic processes associated with rifting that may divide or connect river basins (Watters, 2009).

The basic outlines, predictions, and current evidence for the various dispersal hypotheses are summarized in Table 9.1. The data currently available are not exclusively consistent with alternative

Table 9.1 An Overview of the Main Hypotheses Explaining the Dispersal of African Annual Fishes, Together with Their Mechanisms, Predictions, Current Evidence, and Weakness

Acronym	Hypothesis	Mechanism	Predictions for Genetic Diversity and Its Structuring	Expected Founder Effect	Evidence	Weakness
FDH	Flood dispersal hypothesis	Flood-related displacement of juvenile or adult fish	Higher at lower altitude, importance of river basins, genetic similarity higher in populations along a stream	Strong	Genetic similarity in some adjacent populations[a]	Significant dispersal to a higher altitude inferred from population genetic data[a]
EDH	Embryo dispersal hypothesis	Eggs transported in mud attached to large mammals	Isolation by distance, no role of altitude, no role of river basins, importance of river channels	Strong	Significant dispersal to a higher altitude inferred from population genetic data[a]	Tight suture zone without geographic barrier[b]
WRD	Waterbird-related dispersal hypothesis	Eggs transported on waterbirds' feet	Weak isolation by distance, no role of altitude, no role of river basins, no role of river channels	Strong	Significant dispersal to a higher altitude inferred from population genetic data[a]	Tight suture zone without geographic barrier[b]
GVH	Geomorphological vicariance hypothesis	Changes in river morphology alter the relative position of the pool	Spatially and temporally synchronous estimates of the channel crossing consistent among sympatric lineages	Weak	Synchronous apparent river crossing[b]	Cannot explain dispersal at fine scale, across pools

Note: More details are given in the main text.
[a] Bartáková et al. (2013).
[b] Bartáková et al. (2015).

hypotheses, and it is likely that the dispersal of African annual fishes resulted from a combination of these processes. In three species complexes in southern and central Mozambique, both river channels and river basins have played important roles in structuring *Nothobranchius* communities (Dorn et al., 2011; Bartáková et al., 2015). Strong evidence of a combination of migration to higher and lower altitudes, calculated from genetic data for *N. furzeri* (Bartáková et al., 2013), lends some support for the EDH. However, a strong suture zone between two intraspecific genetic clades without a clear geographic barrier is not compatible with the EDH (Bartáková et al., 2013). It is also notable that barriers to dispersal are frequently shared among sympatric species (Dorn et al., 2011), including apparent river channel crossing, thus supporting the GDH (Bartáková et al., 2015). In conclusion, I agree with Watters (2006) that a combination of rare adult dispersal and geomorphological processes leading to changes in river drainages appear the most likely processes involved in *Nothobranchius* dispersal at both small and large spatiotemporal scales. While some genetic data indicate upstream dispersal consistent with the EDH, additional data are needed to exclude the possibility of a false population-genetic signal indicating apparent upstream dispersal within a metapopulation.

9.2 SPECIES COEXISTENCE

9.2.1 Community Assembly

Several species of *Nothobranchius* often co-occur syntopically, in the same pool. This contrasts with *Pronothobranchius* and *Callopanchax*, which consist of strictly allopatric species. Up to five species of *Nothobranchius* have been recorded in a single pool, though coexistence of two species is most common (Wildekamp, 2004; Watters, 2006). In the most intensively researched area of *Nothobranchius* in southern Mozambique, three species (i.e., *N. furzeri*, *N. orthonotus*, and *N. pienaari*) co-occurred in 19 cases (24%), *N. furzeri* co-occurred with *N. orthonotus* in another 19 cases (24%) and with *N. pienaari* only in three cases (4%). *Nothobranchius furzeri* was the only species in the pool in 31 cases (39%), *N. orthonotus* in two cases (3%), and *N. pienaari* in three cases (4%). *Nothobranchius orthonotus* and *N. pienaari* co-occurred three times (4%). Notably, the absence of *N. furzeri* in pools with other *Nothobranchius* species was always recorded at the coastal periphery of its range, where conditions were more humid; so it is possible that these pools were actually outside the range of *N. furzeri*. The distribution of the three species was significantly nested, but the nested pattern was not evident outside the range of *N. furzeri*, where this species is replaced by its sister species and ecological vicariant, *N. kadleci*.

In the lower Rufiji River system in coastal Tanzania, five sympatric species co-occur in pools in various combinations (Watters, 2006); however, no quantitative information is available from this or any other region. There is also interannual variability in the abundance of each species, and hence the probability of capturing all species during a single survey is likewise variable, especially if the number of individuals collected is relatively low. The apparent absence of a certain species (e.g., *N. pienaari* or *N. orthonotus* in the case of southern Mozambique) from a community in any given year may be the result of the species failure to hatch or survive to adulthood in that year or it may be that the species was present only at very low numbers. There is also variation in community structure within a year that is unlikely to result from a sampling bias (Nagy and Horváth Kis, 2010). In pools along the Tanzanian coast with up to four *Nothobranchius* species, several species disappeared from the community over a short period of 3 weeks (between January and February). This region has two rainy seasons, the long one lasting from March to May and the shorter one from October to mid-December. Sampling therefore likely coincided with the seasonal loss of habitat and, indeed, the water level in most pools had decreased and some had already dried out. However, at least one pool had a higher water level during the second visit, and cichlids were abundant in the pool despite the disappearance of all *Nothobranchius*, indicating that the pool had been flooded from the adjacent river between the two sampling occasions. During the second visit, the number (and general abundance) of *Nothobranchius* had decreased to a single species (*N. luekei* in the case of species-rich original communities) or disappeared entirely. It is not clear whether environmental conditions had deteriorated to the level where *Nothobranchius* existence was not sustainable (high water temperature and associated factors) or whether they were largely exterminated by predation (Nagy and Horváth Kis, 2010). In at least one case, flooding from an adjacent stream was associated with *Nothobranchius* disappearance. Notwithstanding the cause of annual fishes mortality, this simple case demonstrates how inferences from a single visit to a pool can result in misleading conclusions regarding community structure and population density.

Species that are largely sympatric may nevertheless differ in their specific habitat preferences. In Uganda, sympatric *N. robustus* and *N. ugandensis* were recorded from pools that varied in their water chemistry and location (Nagy, 2010). *Nothobranchius robustus* inhabited relatively cooler (25°C or less), stagnant sections of streams with slightly acidic water, while *N. ugandensis* lived in typical *Nothobranchius* pools with alkaline water and a mean temperature of 28.5°C. The two species co-occurred only in a single pool out of 20 pools containing *Nothobranchius* (Nagy, 2010). Variation in habitat preference at a landscape scale is also apparent in the Mozambican clade. *Nothobranchius furzeri* is a dominant species within *Nothobranchius* communities in the drier region further from the coast. *Nothobranchius pienaari* may inhabit pools in this dry part of the region, but its populations are relatively scarce. In contrast, the species is more abundant in the humid coastal areas where *N. furzeri* is absent. *Nothobranchius orthonotus* is generally the least abundant species in the community, but its frequency is relatively stable across dry and humid regions (Reichard et al., 2009). Unlike the Ugandan species, the Mozambican species are largely syntopic.

Nothobranchius may regularly co-occur with nonannual fishes. While *Nothobranchius* have stable populations in annually desiccating habitats, other freshwater fishes annually colonize these habitats from adjacent permanent water bodies. Lungfishes (see below) and several small species of '*Barbus*' (i.e., evolutionary diploid group of *Barbus* sensu lato) syntopic with *Nothobranchius* are frequently captured during sampling (e.g., Valdesalici and Wildekamp, 2004; Watters et al., 2007; Reichard et al., 2009; Reichard, 2010; Valdesalici, 2012; Valdesalici et al., 2012). *Clarias gariepinus* also commonly co-occurs with *Nothobranchius* (Reichard, 2010; Valdesalici et al., 2012) and is presumably their predator. Other teleosts collected with *Nothobranchius* include a mormyrid *Petrocephalus* sp. (Valdesalici et al., 2012), juvenile tilapias (Reichard, 2010), anabantids *Ctenopoma* spp., and nonannual killifishes *Aplocheilichthys* spp. (Larsen, 1999). *Pronothobranchius gambiensis* was recorded co-occurring with '*Barbus*' *leonensis*, '*B.*' *pobeguini*, *Clarias* sp., mormyrids *Marcusenius senegalensis* and *Petrocephalus bovei*, and *Schilbe intermedius*, though sometimes *P. gambiensis* was the only species in the habitat (M. Reichard, unpublished observation). *Callopanchax sidibeorum* co-occurred with the nonannual killifishes *Epiplatys fasciolatus* and *Sciptaphyosemion geryi* (Sonnenberg and Busch, 2010) and *C. monroviae* shared the habitat with *E. fasciolatus* and *Epiplatys dageti monroviae* (Dinesen, 2006).

Lungfishes (*Protopterus* spp.) are the only other African fish species able to have viable populations in annually desiccating habitats. *Protopterus annectens* frequently coexist with *Nothobranchius* in Mozambique (Reichard et al., 2009) and likely elsewhere. Lungfishes are larger (up to 80 cm) and have a different strategy for surviving annual desiccation from killifishes. They survive the dry period as large juveniles and adults buried in the sediment and spawn soon after the pool is inundated. *Nothobranchius*, at least as adults, are not common prey of *Protopterus*, though we detected one individual (out of approximately 10 dissected) with *Nothobranchius* in its stomach (M. Poláčik, unpublished data). Lungfishes' diet consists of crabs, mollusks, and debris (Reichard et al., 2014), avoiding any niche competition with *Nothobranchius*.

9.2.2 Habitat Use

There is little quantitative evidence of habitat separation at the microhabitat scale. This may reflect the inherent difficulty of collecting data on undisturbed fish distribution within a pool because of their small size. Indirect data from sampling in southern Mozambique suggests that *Nothobranchius pienaari* may be more common in heavily vegetated shallow areas (mainly flooded grass at pool margins), whereas *N. orthonotus* may be more common in deeper, unvegetated sections of the pool, and *N. furzeri* is intermediate and occurs in vegetated areas (often associated with water lilies *Nymphaea* spp.) away from the margins of the pool (M. Reichard, R. Blažek, M. Polačik, unpublished observation). However, it is important to note that this observation is not based on a quantitative estimate. Watters (2006) also remarked that co-occurring *Nothobranchius* species inhabit different parts of the pool.

9.2.3 Morphology and Diet

Nothobranchius species have a relatively uniform body shape, and morphological differentiation among species is low. The general body shape and its variation are depicted in Figure 9.2. Typical adult size ranges between 3 and 7 cm. Some species (i.e., *N. ocellatus* and *N. orthonotus*) have the dorsal and anal fin shifted posteriorly, giving the appearance of a lurking predatory fish. Indeed, their morphology is likely associated with their prevalent mode of feeding, the posterior position of the unpaired fins enabling rapid-burst swimming. A group of species, classified in the subgenus *Aphyobranchius* (Wildekamp, 2004), have a more delicate appearance, a larger anal than dorsal fin, and the base of the dorsal fin shifted posteriorly (Figure 9.2). Species differ also in the position of the mouth, being either terminal or superior (upward pointing) to various degrees. This is apparently associated with their diet and dominant mode of feeding. No *Nothobranchius* species possess the inferior (downward pointing) mouths of specialized benthic feeders.

Most *Nothobranchius* are generalized carnivores, feeding on a range of planktonic and benthic invertebrates (Polačik and Reichard, 2010; Reichard et al., 2010). *Nothobranchius ocellatus* is uniquely a large predator (body size 15 cm), and it may be piscivorous (Wildekamp, 2004). Another specialized mode of feeding has been reported for *N. microlepis*, which uses its specially adapted gill rakers to feed on small planktonic crustaceans and their nauplii (Wildekamp and Haas, 1992). *Nothobranchius jubbi*, a species with a generalized body shape, preys on larger planktonic crustaceans, *Coryxa* nymphs (Hemiptera), and mosquito larvae (Wildekamp, 1983). This was largely confirmed by a quantitative study on other species with a generalized body shape (i.e., *N. furzeri*, *N. kadleci*, *N. pienaari*, and *N. orthonotus*), with crustaceans and coarse insect larvae (Odonata, Ephemeroptera, and Coleoptera) being their primary prey (Polačik and Reichard, 2010).

Co-occurring *Nothobranchius* species demonstrate morphological differentiation (within the limits of low overall variability in the genus). In the Mozambican group, *N. piennari* has a relatively small superior mouth, and *N. furzeri* is a robust species (5–8 cm) with a larger head and terminal mouth, while *N. orthonotus* can reach the largest body size (5–10 cm) and has a relatively large mouth. Niche separation in resource use among the three species has been confirmed *via* analysis of their gut contents and stable isotopic signatures (Polačik and Reichard, 2010; Polačik et al., 2014b). *Nothobranchius orthonotus* had the most distinct diet, which included small tadpoles and lungfish, although no teleost fishes were identified. The other two species feed predominantly on small crustaceans (Cladocera, Copepoda, Ostracoda, and Conchostraca). Mosquito larvae formed a negligible part of the diet but were also uncommon in the pools. The diet differentiation between *N. furzeri* and *N. pienaari* was more evident in their isotopic signatures than their stomach contents, demonstrating that ontogenetic changes in the diet play an important role in resource partitioning. It should be emphasized that there was considerable overlap in the diet, and differentiation among species was only evident when resource abundance and diversity was high. In resource-poor habitats, all species consumed the same resources (Polačik and Reichard, 2010; Polačik et al., 2014b). This confirms that most *Nothobranchius* are generalized predators of small aquatic invertebrates.

9.3 POPULATION ECOLOGY

9.3.1 Population Genetic Structure

Many *Nothobranchius* populations are discrete units, with little regular (i.e., annual) dispersal between adjacent populations. In more humid regions, it is likely that populations from adjacent pools form a metapopulation, with a common exchange of individuals when the water level is high and several pools are connected within a larger floodplain. Such metapopulation dynamics can be indirectly inferred by the presence of nonannual species in a habitat. Population genetic data are consistent with this assumption. Most populations (defined as individuals inhabiting one savannah pool) are genetically distinct, that is, demonstrating significant differences in F_{ST} estimates. This includes spatially adjacent populations (Bartáková et al., 2013). In *N. furzeri*, significant genetic differentiation was found between populations located only 1.45 km ($F_{ST} = 0.015$, $P > 0.05$) and 1.25 km apart ($F_{ST} = 0.048$, $P > 0.05$) and even 0.25 km ($F_{ST} = 0.009$, $P > 0.05$) in *N. kadleci*.

In a larger-scale study across the entire range of the three Mozambican species groups, out of all possible pairwise differences among 26, 29, and 41 populations (i.e., 325–703 pairwise values for each species group), only 2–6 population pairs had nonsignificant F_{ST} estimates, indicating strong genetic structuring among most populations (Bartáková et al., 2015). It should be noted that adjacent populations (pools <5 km apart) were typically omitted from the sampling (with the exceptions described above), as the main aim of the analysis was to provide an overview of the population genetics across the range. The groups of populations with nonsignificant F_{ST} values were located either in flat plains near the coast (altitude <10 m) or in the floodplain of the same stream, despite a relatively large geographic distance (up to 75 km, populations from altitudes of 126–74 m) in the latter case. This suggests that downstream dispersal during major floods can maintain metapopulation dynamics even in savannah pools outside the alluvia of major rivers and along a relatively steep gradient. However, in this case, only unidirectional downstream dispersal is possible. There are no data on the population genetic structures of other African annual fishes.

9.3.2 Age Structure

The presence of a single cohort and hence a very simple age structure among *Nothobranchius* populations is another distinctive feature of African annual fishes. Exact birth date can be read from otoliths, that is, mineralized aragonite structures found in the inner ears of fishes, in which daily increments are deposited. In our years of research in southern Mozambique, *Nothobranchius* hatching has coincided with the arrival of heavy monsoonal precipitation. The fish typically hatch between mid-December and mid-January (Polačik et al., 2011). The onset of the rainy season in Mozambique is earlier, with rains beginning in October and peaking in December and January (Figure 9.4). It is possible that the pools are not inundated until the height of the monsoon (with rainfall of up to 80 mm per day; Figure 9.4). However, our data from temperature loggers deposited in the pools suggests the pools are frequently inundated earlier than December but then undergo rapid desiccation, usually within few days. It is not known whether such precipitation induces the hatching of some fish that will not survive until the main inundation of the pool, but so far we have no evidence that this is the case (Polačik et al., 2011). Instead, these initial rains may trigger embryos from diapause through environmental perturbation (e.g., moistened substrate, altered oxygen levels, and partial pressure) and partly dissolve the hardened substratum to enable rapid

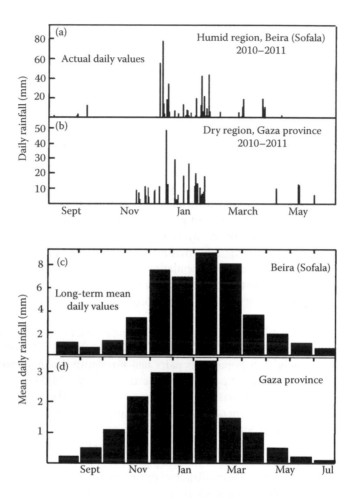

Figure 9.4 Annual precipitation dynamics in humid region (data from the city of Beira) and dry region (Mabalane and Chigubo meteorological stations) of Mozambique. Precipitation pattern expressed as actual daily record over a single annual cycle (a, b) and long-term daily mean values (c, d) illustrate that seasonal precipitation is typically clustered into a few bouts of heavy rains. (Data from National Weather Service. 2014. Climate prediction center. http://www.cpc.noaa.gov/products/fews/AFR_CLIM/GRAPHS/africa.html.)

hatching during the subsequent flooding of the habitat (Watters, 2009). Slight interindividual variation in birth dates can be ascribed to the gradual inundation of a pool, where the egg banks located in different parts of the pool are submerged at different times. The greatest age difference among fishes in the same population was 25 days, with a typical range of 10–14 days (Polačik et al., 2011). It would be extremely interesting to obtain data from different *Nothobranchius* species, particularly from the equatorial region, where two rainy seasons may effectively double the generation time if the habitats are regularly desiccated in the intervening dry seasons. This is more likely to occur in the drier equatorial regions, such as at higher altitudes. Alternatively, given a metapopulation structure, two rainy seasons may result in two overlapping age cohorts (or generations). Indeed, there is at least one observation of *N. melanospilus* from the equatorial region in Tanzania, where two size cohorts, an abundant cohort of 3- to 5-cm large fish and several larger individuals of 8–12 cm, were found coexisting in a single pool (Shidlovskiy, 2010). One population of *Callopanchax sidibeorum* was reported to consist of a single age cohort in one year but a wide size spectrum of juveniles and adults in another year (Sonnenberg and Busch, 2010).

9.3.3 Population Size, Sex Ratio, and Their Temporal Dynamics

The population size (abundance) of *Nothobranchius* varies greatly among populations but also within populations among different years. In our study sites in Mozambique, population sizes were generally congruent among years. Where annual fishes were rare, they were typically proven to be relatively rare across multiple years. In contrast, habitats with high fish abundance tended to support abundant populations across years. Obviously, there was considerable interannual variation, but this can generally be ascribed to environmental conditions and demographic stochasticity. Estimates of total population size for individual pools were made using capture-mark-recapture methods and depletion sampling. The calculated estimates were within the range of 116–216 fishes in a 1800 m² pool (density of 0.1 fishes per m²) to 1750–1956 individuals in a 875 m² pool (density of 2 fishes per m²). Many larger pools conceivably contained larger fish populations, on the order of two to three magnitudes, and there were also pools with a higher density of fishes; recapture-based population estimates in these pools were logistically impossible. These estimates clearly indicate that population sizes of *Nothobranchius* may vary by several orders of magnitude.

From a population genetic perspective, some populations departed from Hardy–Weinberg equilibrium and were likely cases of recent colonization or a severe bottleneck (Bartáková et al., 2013). Indeed, these populations were located in a narrow (15-m wide) strip between a road and railway embankment, strongly suggesting recent human influence on their structure.

Nothobranchius reach sexual maturity at 3–7 weeks after hatching, with a record successful reproduction at 17 days in *N. kadleci* and 18 days in *N. furzeri* (Blažek et al., 2013). Sexually mature individuals of *N. furzeri* approximately 3 weeks old were collected in the field (Polačik et al., 2011), and I anticipate that sexual maturity could be reliably attained earlier than in 17 days under natural conditions with abundant food. Both males and females continue growing rapidly after sexual maturity and only attain the inflexion point in their growth curve later in adulthood. The increase in age and body size is associated with an increase in fecundity, from an initial production of 2–10 eggs to hundreds of eggs produced each day when resources are abundant. In the longer term, both in the wild and in captivity, fecundity of 20–50 eggs per day is observed, at least in the group of Mozambican species (Blažek et al., 2013; Vrtílek and Reichard, 2015; Vrtílek and Reichard, unpublished data). It should be noted that *Nothobranchius* are extreme income breeders (Vrtílek and Reichard, 2015), and a positive association between female size and fecundity is only retained under analogous environmental and ration conditions (see below).

There is striking sex-ratio bias among adult *Nothobranchius* in the wild. At the same time, no such sex bias is observed in the same populations when held in captivity (Reichard et al., 2014). This clearly implies selective male mortality in the wild. Males of *N. guentheri* are more often targets of heron attacks than females under seminatural conditions (Haas, 1976). A recent study found that adult sex-ratio bias is lower at sites lacking giant predatory hemipterans (that are capable of preying on *Nothobranchius*), with clear water and abundant vegetation, and argued that complex vegetation protects males from strong predation (Reichard et al., 2014). However, the environmental predictors of sex bias varied among sympatric species, severely limiting their generalization. Another explanation for increased male mortality stems from male aggression (Reichard et al., 2014; Polačik and Podrabsky, 2015). Male–male competition is fierce in most *Nothobranchius* species and commonly results in the death of the weaker opponent, at least under the spatially restrictive conditions of captivity (Polačik and Reichard, 2011).

The sex-ratio bias becomes apparent soon after sexual maturity suggesting that male-biased mortality may start even before the onset of sexual maturity, possibly due to inherently higher frailty of males. In contrast, in replicated populations of the Neotropical annual fish *Austrolebias reicherti*, equal sex ratios across populations at the start of adulthood was found, with a decrease in male abundance later in the season (Passos et al., 2014). It is clear that male-biased mortality is only expressed under natural conditions and therefore must be mediated by predators, interactions among males, and/or challenging environmental conditions. In Trinidadian guppies, it was shown that there are bouts of increased male-specific mortality in particular periods (in the case of guppies during flood events associated with water turbulence and turbidity) rather than the successive stable disappearance of one sex from the adult population due to its consistently higher mortality (Arendt et al., 2014). Whether a similar process is also applicable to annual fishes is unclear.

The seasonal decline in *Nothobranchius* abundance within populations (in both sexes) could provide an interesting insight into sources of their mortality. The pattern of adult mortality has important consequences for the evolution of aging. This is of great interest, because *Nothobranchius* have become a popular model in aging research (Chapter 6). There are three basic scenarios. First, there may be strong adult mortality due to predation, diseases, or intraspecific competition for scarce resources, and *Nothobranchius* populations may steadily decrease in abundance with increasing age. Second, population abundance may decline only negligibly, mainly due to accidental deaths, and most individuals die only during habitat desiccation. Third, there may be an initial decline in adult abundance due to density-dependent population effects, but little mortality afterward until habitat desiccation (Cellerino et al., in press). This distinction is not trivial, because each scenario sets different predictions for the evolution of aging and corresponding life history traits such as investment in reproduction, growth, and immunity. At present, we only know that male mortality is higher than that of females. This means that each sex has a different life expectancy. Research into the population dynamics of *Nothobranchius* fishes in the wild is needed and will provide an excellent opportunity to advance the fundamental understanding of *Nothobranchius* aging.

9.3.4 Ecology of Embryonic Development

The embryonic stage is a relatively long period of *Nothobranchius'* life, and in some species even the longest. This is a notable feature, particularly given their brief posthatching lifespan. Any evolutionary and ecologically relevant calculation of generation time should take account of the duration of the embryonic period. Embryo development, including characteristic diapauses (i.e., forms of developmental arrest) is strikingly similar between African and Neotropical annual fishes (see Chapter 2). It is notable that the presence of diapauses (and hence the ability to complete the annual life cycle) is considered an ancestral trait in killifish phylogeny (Murphy and Collier, 1997), though a recent analysis suggested that diapauses evolved in at least six killifish lineages in parallel (Furness et al., 2015).

Under standard circumstances, a fertilized *Nothobranchius* egg is deposited in the pool substrate, where it remains until the pool desiccates. The egg's chorion is covered by short, evenly dispersed filaments. Their density, length, and shape are largely species specific (Wildekamp, 2004), and they can be potentially used in species identification from a sample of eggs in the substrate. Functionally, they create a narrow space between the egg and the surrounding sediment, enabling gas exchange during development (Wildekamp, 2004). Environmental conditions change dramatically with the disappearance of the water, and the embryo survives in the egg envelope with a thickened chorion, remaining in diapause over a protracted period of drought until eventually hatching when the habitat is inundated. This simple scenario has several alternative pathways, which are considered further below.

The timespan of several weeks to months when the eggs are deposited in the sediment while water is still present in the pool is certainly much longer that the minimum time needed for completion of embryogenesis and hatching (12–30 days depending on the temperature). At the same time, gradual hatching has never been observed in *Nothobranchius*. After a brief initial development phase, the embryo may be arrested in diapause (diapause I). The presence of adult fish effectively halts embryo development beyond diapause (Inglima et al., 1981). In the laboratory, embryos kept in aqueous aerobic media at high temperatures without the presence of adults (shallow water in Petri dishes) may omit this diapause and continue to develop. *Nothobranchius furzeri* embryos can reach the fully developed stage at 12–18 days at high ambient temperature (28°C; Valenzano et al., 2011). Such rapid development, however, may incur significant cost to posthatching viability (R. Blažek, unpublished data).

During the standard developmental trajectory, embryos can undergo a total of three diapauses (Wourms, 1972; Chapter 2). However, all these are facultative and can be omitted, at least under laboratory conditions (Blažek et al., 2013; Polačik et al., 2014a). The duration of each diapause is not constant among embryos, even if they are full siblings fertilized on the same day and incubated under identical conditions. The variability in the timing of embryo development is enormous and certainly cannot be ascribed solely to nonhomogeneous microenvironments within incubation media. Further, there are some general trends, despite a high level of stochasticity in the process. In *N. furzeri* and the Neotropical *Austrofundulus limnaeus*, young females (more precisely females laying their first eggs) tend to produce more embryos with very rapid development; later in life, the same females (paired with the same male) tend to produce more embryos that enter diapause (Podrabsky et al., 2010; M. Polačik et al., unpublished data). This trend may be apparent even among eggs produced only a few days after the first eggs. M. Polačik et al. (unpublished data) suggest "an internal reproductive effort counter" that females use to epigenetically modify the fate of embryo development. However, environmental cues also likely play a role, and further experiments are needed to fully understand the cues females use for embryo development modification (see also Chapter 4). A tentative conclusion is that embryos enter alternative developmental trajectories and likely often do so under natural conditions.

The process is clearly of epigenetic origin, and an understanding of the mechanisms involved would be of great significance and interest for biomedical research. How is the functional expression of certain genes—and which ones—silenced or promoted by maternal epigenetic marks? Are RNAi, microRNA, histones, alternative splicing, or any other mechanisms involved? To what extent does the process remain stochastic and to what extent is it deterministic? What are the relative roles of environmental factors, epigenetic background, and their interaction? To what extent can environmental conditions affect the process of embryogenesis once it has started? These are questions that we are only beginning to learn how to answer methodologically, despite solid developmental background information (Furness et al., 2015; Chapter 2). Such a research agenda is undoubtedly one of the areas where annual fishes can provide general insights into an enigmatic biological phenomenon.

Environmental conditions certainly play a nonnegligible role in embryo development. Under natural conditions, embryos likely persist under anoxic conditions during the first weeks after the pool becomes desiccated. In an anoxic environment, all embryos are likely in diapause; any nondiapausing embryos would die. Reduced oxygen content causes the embryo to enter diapause I as early as 3 days after fertilization during the dispersed phase of development at the end of epiboly (Wourms, 1972). This diapause is rarely observed under laboratory conditions (Valenzano et al., 2011; Dolfi et al., 2014), but this may be an experimental artifact, since reduced oxygen conditions are not compatible with standard observation in aqueous media under a stereomicroscope. This indirectly suggests that this early embryogenesis may be more deterministic and dependent on environmental conditions. Still, some embryos enter diapause I even under oxygenated laboratory conditions. Given that the substrate in *Nothobranchius* habitats is likely to be mostly anoxic (Watters, 2009), most embryos should enter diapause I under standard natural conditions. Sometimes the eggs are deposited on aerobic substrates such as sand or mud soon after pool inundation (Watters, 2009) and hence may also continue their development without diapause I in the wild.

Embryos can exit diapause I after further environmental perturbation (with the potential role of epigenetic effects). Such perturbation can be related to the aerobic phase of the sediment, when air starts to penetrate desiccated sediment after several weeks or months (Watters, 2009). It takes less than 1 week to complete a phase of active development (somitogenesis, formation of neural keel, functional tubular heart) between diapause I and diapause II (Wourms, 1972). Natural conditions at this stage of development are aerobic but nonaqueous, and it is most likely that such conditions favor entrance into diapause II (Watters, 2009). In this stage, the embryo is most resistant to environmental challenges and can persist up to several years (Podrabsky et al., 2010). Further development is likely triggered by the onset of the rainy season and saturation of the substrate. Embryos may later enter diapause III, in which they have all the structures of juvenile fish but their metabolism is considerably reduced. While this diapause is obligatory in *Austrofundulus* (Berois et al., 2014), its presence in *Nothobranchius* is unclear (Wourms, 1972; Watters, 2009; Cellerino et al., in press). Again, it may be common under natural conditions but difficult to recognize under laboratory conditions.

In summary, *Nothobranchius* embryo development is extremely variable, even within a single clutch of eggs produced by the same parents and deposited on the same substrate. A suite of environmental conditions and epigenetic effects interact to modulate embryo development and ultimately lead to an effective bet-hedging strategy ensuring that the embryos follow divergent developmental trajectories. This is an excellent adaptation to the unpredictability of precipitation and interannual variation in pool existence and duration.

9.4 ALTERNATIVE PHENOTYPES, MALE COLOR MORPHS, AND SEX DIFFERENCES

9.4.1 Alternative Phenotypes

Intriguingly, the alternative developmental trajectories that the embryo can follow (outlined above) are mirrored in its posthatching phenotype. Using a population of *N. furzeri*, Polačik et al. (2014a) provided experimental evidence that a short embryonic period (maximum of 50 days at 22°C, likely no diapause) produced phenotypes with a more rapid life history than those with a long embryonic period (200 days at 22°C, including a substantial period in diapause stages). Individuals that completed their development in 50 days were smaller at hatching but had larger residual yolk stores. The individuals with rapid embryonic development continued to grow more rapidly after hatching and exceeded the size of those individuals with long embryo development as early as six days posthatching. The individuals with rapid embryo development clearly progressed more rapidly through every life history stage. They reached sexual maturity earlier and were shorter lived, and their final body size was smaller. Such ability to produce alternative phenotypes appears unique among vertebrates. It illustrates that the slow to rapid pace-of-life pattern, known from comparisons among taxa and between populations (e.g., Bronikowski, 2008; Jones et al., 2008), can also be recognized at the intrapopulation level.

In nature, any annual fish that omit all diapauses are destined to inhabit secondary pools, emerging in the savannah following an occasional second wave of heavy precipitation after the primary pools have desiccated. There is evidence that this happens in Mozambique (Polačik et al., 2014a), and extremely rapid life history is then adaptive, allowing individuals to maximize reproductive success in secondary pools that are inherently more prone to rapid desiccation because they appear toward the end of the rainy season (Polačik et al., 2014a). Rapidly desiccating pools with relatively young fish were also reported in *N. robustus* in Uganda (Nagy, 2010) and may be more common than currently assumed. The mechanism permitting rapidly developing individuals to grow at a faster rate is unknown. It may be supported by the extra portion of yolk reserve that is found in rapidly developing embryos at hatching that promotes rapid initial growth. Alternatively, the epigenetic mechanism that caused the embryos to depart from the diapause-inclusive trajectory may also act on posthatching phenotypes and modify the expression level of growth-associated genetic pathways.

9.4.2 Sex Differences

Nothobranchius males and females differ in body size, coloration (Wildekamp, 2004), and life expectancy in the wild (Reichard et al., 2014). No differences in the diet or any other niche axis were recorded. Intersexual differences in body size are negligible in some species (e.g., *N. orthonotus*) but reach 30% in others (e.g., *N. furzeri*, *N. kadleci*, and *N. pienaari*) (Polačik et al., 2011, 2014a). From an evolutionary perspective, larger body size in males is likely related to sexual selection and strong male hierarchy where successful competition for access to females is a major determinant of reproductive success. The most striking intersexual difference is in coloration. Females are pale, typically light brown, and in some species possess dark (e.g., *N. melanospilus* species group, *N. kadleci*, and *N. orthonotus*) or blue iridescent (e.g., *N. rachovii* species group) spots. In *N. ocellatus*, females possess a single large spot on the caudal peduncle. In *Pronothobranchius* and *Callopanchax* females are ornamented with red dots dispersed across the body and on the unpaired fins. *Nothobranchius* fishes appear to have escaped genetic linkage between male and female coloration, enabling more intensive male coloration to evolve without any selective constraint on female cryptic coloration (Sedláček et al., 2014).

9.4.3 Male Color Morphs

Males of several *Nothobranchius* and *Callopanchax* species occur in two or more color morphs. Such polymorphism is relatively widespread in African killifishes and occurs in *Fundulopanchax* and *Aphyosemion*. Typically, males are dimorphic, with red and blue or red and yellow morphs. In some cases, these color morphs may be largely sympatric and syntopic (e.g., in *N. furzeri*, *N. orthonotus*, and *N. ugandensis*), in others they appear sympatric but not syntopic despite considerable sampling effort (e.g., *N. eggersi*; Valdesalici, 2010), while no geographic sympatry is recorded in others (Wildekamp, 2004). Sometimes, color morphs represent separate, closely related species (e.g., *N. rachovii* species complex: *N. rachovii*—blue morph, *N. krysanovi*—red morph, and *N. pienaari*—black morph), and the species are indeed effectively reproductively isolated due to chromosomal rearrangements (Shidlovskiy et al., 2010). Yet in other cases, male color morphs are not discrete, and there is intergradation of colors. In *N. korthausae*, populations with pure red males, pure yellow males, and intermediate forms with a mixture of red and yellow forms have been reported (Wildekamp, 2004; Nagy, 2008; Reichard and Polačik, 2010). Likely, the apparent lack of male polymorphism in some species may be a consequence of low sampling effort in their respective ranges, and male polymorphism may be more widespread in *Nothobranchius* than currently understood. Given that species diagnosis is often based solely on male coloration, it is imperative to consider intraspecific male polymorphism during species descriptions.

It is intriguing to view male coloration as either directly adaptive or as a trait linked to some adaptive phenotypic feature. There is ample evidence from other fish taxa as well as from all other vertebrate groups, insects and other invertebrates animals, and even plants that color polymorphism is often related to traits critical to fitness (McLean and Stuart-Fox, 2014). The coexistence of two or more color morphs is then related to variation in environmental conditions or complex demographic processes (Sinervo and Lively, 1996). In other cases, color polymorphism may have evolved due to genetic drift or geographic isolation and may not be associated with adaptation (McLean and Stuart Fox, 2014). It remains to be investigated whether male coloration in *Nothobranchius* is linked to other phenotypic traits and whether male polymorphism is adaptive.

9.5 LIFE HISTORY ADAPTATIONS

9.5.1 Rapid Growth and Sexual Maturation

The environment of *Nothobranchius* and other African annual fishes imposes strong selective pressure on a network of adaptations promoting extremely rapid life history. Rapid growth and sexual maturation is perhaps the most apparent and is exemplified in species from the drier parts of the distribution. In the south, *N. kadleci* and *N. furzeri* were reported to reach sexual maturity in captivity as early as 17 and 18 days, respectively (Blažek et al., 2013). The youngest wild fish with an age estimate (from daily increments on otoliths) were 19 days old (Polačik et al., 2011) and possessed the coloration of sexually mature individuals. No younger fish were collected for age estimation. While histological confirmation of sexual maturity (i.e., the presence of ripe gametes) was not made in this sample, I suggest it is likely that sexual maturation in the wild can be attained even earlier than 17 days when resources are abundant. Furthermore, it is very likely that species from the northern periphery of the range of *Nothobranchius*, for example, *N. rubroreticulatus*, *N. bellemansi*, *N. occultus*, *N. microlepis*, and *N. bojiensis*, may be capable of even more rapid sexual maturation. This suggestion is based on the fact that these species face even drier and less predictable environmental conditions (Valdesalici, 2014), and consequently there would be a selective advantage in reaching sexual maturity even sooner.

The daily growth rate is faster during the second week of life, where a rate of 2.7 mm per day (constituting an increase of 23% in the body size per day) has been recorded in *N. furzeri* (Blažek et al., 2013). Under the conditions reported in Blažek et al. (2013), the steep growth continues after sexual maturation and levels off later at the age of 35 days. This underscores the extreme demands of *Nothobranchius'* life history. The rate of fish growth normally decreases upon sexual maturity (Wootton, 1990). In *Nothobranchius*, the risk of habitat desiccation provides strong pressure for early reproduction, requiring rapid growth. The resolution of the trade-off between current and future fecundity apparently results in a compromise when reproduction starts early, but there is "maturity of maturation" when fecundity increases sharply with age along with body size (Blažek et al., 2013). Females therefore initially produce small clutches of few eggs per day and only reach full fecundity when their rapid growth decelerates. Males also keep growing steadily after sexual maturity, until a break point in their growth is reached. Consequently, there are clear intersexual differences in growth rates. Males start to grow more rapidly than females from the age of approximately 2 weeks (Polačik et al., 2014a), which may be associated with the start of allocation to reproduction. Alternatively, the difference in growth rates can be related to hormonal changes, affecting juvenile development in general.

9.5.2 Immunity and Diseases

Allocation of resources to rapid growth and reproduction is disproportionately high in *Nothobranchius* fishes. Life history theory (Stearns, 1992) predicts that such allocation should come at the cost of self-maintenance. A weaker ability to cope with immune challenges is therefore predicted for annual fishes. This prediction has remained largely untested, and our current knowledge of how *Nothobranchius* cope with immune challenges is very limited. Observational evidence from captivity demonstrates that *Nothobranchius* are very susceptible to infection by *Piscinoodinium* sp. (a dinoflagellate protozoan) and *Glugea* sp. (a microsporidian). While *Piscinoodinium* infection can be effectively treated by a sudden osmotic shock that kills the parasite but leaves *Nothobranchius* fish unharmed, *Glugea* infection is untreatable, decreases host vigor and reproductive ability, and may become fatal. Fish in poor physiological condition are more susceptible to the negative effects of parasitic infections, indirectly indicating that immune response is at least partly efficient (M. Reichard, personal observation).

In the wild, *Nothobranchius* serve as an intermediate host for several species of internal parasite. Examination of adult individuals from 14 populations of *N. furzeri*, *N. kadleci*, *N. orthonotus*, and *N. pienaari* (Michálková et al., unpublished data) yielded a relatively high diversity and abundance of endoparasites (internal parasites). Across all species, the authors identified seven species of metacercariae (larval stage of trematodes), three species of nematodes, four species of larval cestodes, and one species of adult trematode in the host digestive tracts. The only ectoparasites (external parasites) were two individual unidentified leeches. This demonstrates that *Nothobranchius* are common intermediate hosts of flukes (metacercariae), the definitive hosts being waterbirds that need to consume *Nothobranchius* to be infected.

One fluke species (*Apatemon* sp.) is of special interest, because it affects fish behavior. It has only been recorded in some savannah pools inhabited by *N. furzeri*. When present, it was located either inside the brain tissue or between the skull and brain, with an intensity of up to 10 individuals per host. Its presence can easily be predicted at the commencement of sampling, because *Nothobranchius* infected by *Apatemon* metacercariae stay near the water surface and do not escape. Instead, they frequently jump out of the water and occasionally remain exposed on the leaves of water lilies. This makes them an easy target for piscivorous birds. When imported to the lab, infected fish failed to escape in a simulated predator attack, in sharp contrast to noninfected fish. There were further behavioral and routine activity disruptions making infected fishes more susceptible to a simulated bird attack (Michálková et al., unpublished data). Such a suite of behaviors ultimately leads to increased transmission of the parasite between its intermediate (*Nothobranchius*) and definitive (bird) hosts. Such effects of parasites on fish behavior are known from other host–parasite systems, but our understanding of the mechanisms underlying the process remains poor (reviewed in Barber et al., 2000). The major advantage that *Nothobranchius* offers is the availability of a detailed anatomical map of the *N. furzeri* brain (D'Angelo, 2013). With this aid, it may be possible to associate the location of the parasite with the brain region responsible for a particular functional response. This may help to test whether the physical location of the parasite is important in inducing the changes in host behavior, for instance *via* mechanical stimulation. Alternatively, host behavior may be altered *via* secretion of a chemical substance, making the precise position of the parasite within the host brain less important.

Investment in self-maintenance also includes the ability to cope with internal damage and retain homeostasis, maintaining organ and cellular function and dealing with metabolic waste products. A high incidence of neoplasias (liver tumors in particular) has been reported for several *Nothobranchius* species (Di Cicco et al., 2011), despite being generally rare in teleosts. Other damage at the organ level includes various types of lesions in the kidney, liver, and heart (Di Cicco et al., 2011). These signatures of internal damage are generally associated with aging and are discussed in detail in Chapter 6.

9.5.3 Phenotypic Plasticity

The single overarching feature of *Nothobranchius'* life history is its extremely high level of phenotypic plasticity. This is adaptive, as variation in environmental conditions and population density may be enormous, exposing the same genotypes to a potentially wide range of conditions. The flexibility of the genotypes to produce phenotypic traits according to the prevailing environmental conditions (i.e., strong gene by environment interaction) enables fine-tuning of individual traits to achieve optimal values across conditions.

Some phenotypic traits are irreversible and respond to the conditions prior to their expression. Sexual maturation is the simplest example. Sexual maturity can be achieved in less than 3 weeks when abiotic conditions (e.g., ambient temperature) are optimal and resources (e.g., food ration) plentiful (Blažek et al., 2013). In the same species, and conceivably in the same genotype, time to sexual maturity may be doubled and reached as late as in 5–6 weeks posthatching (Graf et al., 2010), when population density is high, food ration low, or environmental conditions suboptimal. There is also variation in sexual maturity (and growth rate) among individuals due to social interactions, likely modulated by stress hormones in subordinate individuals. In captivity, dominant individuals often inhibit the growth of subordinate individuals, and removal of the dominant individual is followed by rapid growth of the fish that resumes dominant rank.

Other phenotypic traits remain flexible throughout life. Most notably, in *N. furzeri*, fecundity responds strongly to ration manipulation. Actual fecundity was tightly associated with actual ration, and females undergoing changes in ration level tracked current conditions (Vrtílek and Reichard, 2015). Surprisingly, there was no apparent cost of compensatory growth on female fecundity, at least in a time span of 6 weeks. Females fully compensated for their smaller body size (and mass) after 6 weeks on a high ration and produced the same number of eggs as females fed a high ration throughout their lives (Vrtílek and Reichard, 2015). The fecundity of *N. furzeri* females was predicted to vary among populations across an aridity gradient, with a higher allocation to reproduction in populations from the drier parts of the range (where life expectancy is lower due to shorter habitat existence). However, this prediction was not confirmed in wild populations (Vrtílek and Reichard, unpublished data). The two most likely explanations are (1) interannual variation overrides any adaptive value of differential allocation between populations by habitat and (2) high importance of resource availability due to population density leads to high flexibility in fecundity traits overriding any interpopulation differences. Laboratory experiments in which females from several populations are housed in common environmental conditions may illuminate interpopulation differences in fecundity. These experiments are being undertaken.

Egg size is variable across *Nothobranchius* species. In *N. janpapi*, the egg diameter is only 0.65 mm; in *N. furzeri*, it is typically 1.2 mm; and in *N. ocellatus*, it is 2.5–3 mm (Larsen, 1999; Wildekamp, 2004; Vrtílek and Reichard, 2015). Egg size is related to the size of adult fishes rather than environmental conditions and may retain a phylogenetic signal, with closely related species having similar egg sizes. The size of the egg is apparently unrelated to its ability to survive habitat desiccation, at least at an interspecific level. Species with the largest (*N. ocellatus*) and smallest egg sizes (*N. luekei*, only 25% of the egg diameter of *N. ocellatus*) coexist in the same pools. Despite this, there is significant variation in egg size at the interpopulation level in *N. furzeri*. Wild females from populations in the drier part of the range (with a shorter life expectancy) had smaller eggs than females from populations in the humid part of the range (Vrtílek and Reichard, unpublished data). This variation is perhaps associated with the trade-off between egg size and number rather than being directly adaptive.

Egg size is also flexible within a population. Captive *N. furzeri* females receiving a low ration (i.e., under less favorable conditions) produced relatively larger eggs, at the expense of their number (Vrtílek and Reichard, 2015). This is in accordance with adaptive maternal effects but was not expected in annual fishes with unpredictable interannual variability in conditions. Whether this variability reflects *Nothobranchius*' phylogenetic history (retained ancestral trait) or indeed represents an adaptation is unclear.

9.6 SUMMARY AND CONCLUSIONS

Most information on the ecology of African annual fishes comes from research on the genus *Nothobranchius*. There is a mix of reports from collecting trips by hobbyists and aid workers (often themselves non-professional scientists), scientific field expeditions, and experimental work in the laboratory. All three avenues have provided unique insights and are important sources of our current understanding of *Nothobranchius* ecology. Given the popularity of collecting *Nothobranchius* populations in the wild, there is a considerable scope for involvement in citizen science by sharing data on habitat conditions, community assembly, and population parameters (species co-occurrence, sex ratio, body size), and by accumulating samples for phylogenetic and phylogeographic analyses. Such collaboration between scientists and hobbyists may prove instrumental in acquiring a comprehensive data set and considerably advance our knowledge of annual fishes ecology and evolution.

I hope that our ongoing research on the Southern clade of *Nothobranchius* (*sensu* Dorn et al., 2014) will ultimately be extended to include a larger geographic area. Most of our current inferences relate to this clade of annual fishes and to *N. furzeri* in particular. This is because this species has become a model for the biology of aging and several related disciplines (Chapter 6). The methodological resources developed for this species have made it relatively easy to utilize state-of-the-art tools for research on other annual fish species.

The most characteristic features of African annual fishes ecology are linked to their habitat. *Nothobranchius* have rapid life history and short lifespan, both in the wild and in captivity.

Nothobranchius kadleci and *N. furzeri* have shown the most rapid sexual maturation and generation time of any vertebrate species recorded (Blažek et al., 2013). One captive population of *N. furzeri* (GRZ strain) was reported to have a median lifespan of 9 weeks and a maximum lifespan as short as 12 weeks (Valdesalici and Cellerino, 2003). While this has subsequently been shown to be particularly short even for *N. furzeri*, a median lifespan in a range of 20–43 weeks and maximum lifespan of 25–67 weeks (Terzibasi et al., 2008; Terzibasi Tozzini et al., 2013; Polačik et al., 2014a) are still remarkable and enable many studies requiring the lifetime observation of individuals to be completed within conveniently short time scales.

There is a paucity of information on many aspects of the ecology of *Nothobranchius* in the wild. Even basic data such as the sources of mortality, competition, dispersal, habitat use within the pool, reproductive behavior, and mating patterns are based on circumstantial evidence at best. A study visiting multiple wild populations throughout the annual cycle would be instrumental in revealing how African annual fishes respond to annual fluctuations in habitat conditions and how populations cope with the extreme environmental conditions of savannah pools.

ACKNOWLEDGMENTS

I would like to thank Matej Polačik, Radim Blažek, and Milan Vrtílek for a number of discussions on annual fish ecology throughout the years of our *Nothobranchius* research. They have undoubtedly shaped the opinions and interpretations outlined in this review. Béla Nagy kindly provided photographs for Figure 9.2. The funding for my current *Nothobranchius* research comes from the Czech Science Foundation (project P506/11/0112).

REFERENCES

Arendt, J. D., D. N. Reznick, and A. Lopez-Sepulcre. 2014. Replicated origin of female-biased adult sex ratio in introduced populations of the Trinidadian guppy (*Poecilia reticulata*). *Evolution* 68:2343–2356. doi: 10.1111/evo.12445

Barber, I., D. Hoare, and J. Krause. 2000. Effects of parasites on fish behaviour: A review and evolutionary perspective. *Reviews in Fish Biology and Fisheries* 10:131–165. doi: 10.1023/A:1016658224470

Bartáková, V. 2013. Genetic structure of short-lived *Nothobranchius* fish in southern Mozambique. MSc Dissertation Masaryk University.

Bartáková, V., M. Reichard, K. Janko et al. 2013. Strong population genetic structuring in an annual fish, *Nothobranchius furzeri*, suggests multiple savannah refugia in southern Mozambique. *BMC Evolutionary Biology* 13:196. doi: 10.1186/1471-2148-13-196

Bartáková, V., M. Reichard, R. Blažek, M. Polačik, and J. Bryja. 2015. Terrestrial fishes: Rivers are barriers to gene flow in annual fishes from the African savanna. *Journal of Biogeography*. doi:10.1111/jbi.12567

Berois, N., M. J. Arezo, and R. O. de Sá. 2014. The Neotropical genus *Austrolebias*: An emerging model of annual killifishes. *Cell & Developmental Biology* 3:136. doi: 10.4172/2168-9296.1000136

Blažek, R., M. Ondračková, B. Bímová Vošlajerová, L. Vetešník, I. Petrášová, and M. Reichard. 2012. Fish diversity in the Niokolo Koba National Park, middle Gambia River basin, Senegal. *Ichthyological Exploration of Freshwaters* 23:263–272.

Blažek, R., M. Polačik, and M. Reichard. 2013. Rapid growth, early maturation and short generation time in African annual fishes. *EvoDevo* 4:24. doi: 10.1186/2041-9139-4-24

Bronikowski, A. M. 2008. The evolution of aging phenotypes in snakes: A review and synthesis with new data. *Age* 30:169–176. doi: 10.1007/s11357-008-9060-5

Cellerino, A., D. R. Valenzano, and M. Reichard. 2015. From the bush to the bench: The annual *Nothobranchius* fishes as a new model system in biology. *Biological Reviews* in press, doi: 10.1111/brv.12183.

Cohen, A. S., J. R. Stone, K. R. Beuning et al. 2007. Ecological consequences of early late-pleistocene mega-droughts in tropical Africa. *Proceedings of the National Academy of Sciences* 104:16422–16427.

D'Angelo, L. 2013. Brain atlas of an emerging teleostean model: *Nothobranchius furzeri*. *The Anatomical Record* 296:681–691. doi: 10.1002/ar.22668

Di Cicco, E., E. T. Tozzini, G. Rossi, and A. Cellerino. 2011. The short-lived annual fish *Nothobranchius furzeri* shows a typical teleost aging process reinforced by high incidence of age-dependent neoplasias. *Experimental Gerontology* 46:249–256. doi: 10.1016/j.exger.2010.10.011

Dinesen, K. 2006. A review of *Callopanchax monroviae* (Rollof & Ladiges, 1972)—With notes on the type locality, distribution, natural life cycle, and phenotypes. *Journal of the American Killifish Association* 39:103–113.

Dolfi, L., R. Ripa, and A. Cellerino. 2014. Transition to annual life history coincides with reduction in cell cycle speed during early cleavage in three independent clades of annual killifish. *EvoDevo* 5:32. doi: 10.1186/2041-9139-5-32

Dorn, A., Z. Musilová, M. Platzer, K. Reichwald, and A. Cellerino. 2014. The strange case of East African annual fish: Aridification correlates with diversification for a savannah aquatic group? *BMC Evolutionary Biology* 14:210. doi: 10.1186/s12862-014-0210-3

Dorn, A., E. Ng'oma, K. Janko et al. 2011. Phylogeny, genetic variability and colour polymorphism of an emerging animal model: The short-lived annual *Nothobranchius* fishes from southern Mozambique. *Molecular Phylogenetics and Evolution* 61:739–749. doi: 10.1016/j.ympev.2011.06.010

Fava, D., and M. Toledo-Piza. 2007. Egg surface structure in the annual fishes *Simpsonichthys* (subgenera *Opthalmolebia* and *Xenurolobias*) and *Nematolebias* (Teleostei: Cyprinodontiformes: Rivulidae): Variability and phylogenetic significance. *Journal of Fish Biology* 71:889–907. doi: 10.1111/j.1095-8649.2007.01572.x

Furness, A. I., D. N. Reznick, M. S. Springer, and R. W. Meredith. 2015. Convergent evolution of alternative developmental trajectories associated with diapause in African and South American killifish. *Proceedings of the Royal Society of London B* 282:20142189. doi: 10.1098/rspb.2014.2189

Graf, M., A. Cellerino, and C. Englert. 2010. Gender separation increases somatic growth in females but does not affect lifespan in *Nothobranchius furzeri*. *PLoS ONE* 5:e11958. doi: 10.1371/journal.pone.0011958

Haas, R. 1976. Sexual selection in *Nothobranchius guentheri* (Pisces-Cyprinodontidae). *Evolution* 30:614–622.

Inglima, K., A. Perlmutter, and J. Markofsky. 1981. Reversible stage specific embryonic inhibition mediated by the presence of adults in the annual fish *Nothobranchius guentheri*. *Journal of Experimental Zoology* 215:23–33.

Jones, O. R., J. -M. Gaillard, S. Tuljapurkar et al. 2008. Senescence rates are determined by ranking on the fast–slow life-history continuum. *Ecology Letters* 11:664–673. doi: 10.1111/j.1461-0248.2008.01187.x

Larsen, F. 1999. *Nothobranchius ocellatus* (Seegers 1985). *Journal of the American Killifish Association* 32:186–192.

Loureiro, M., and R. O. de Sá. 1996. External morphology of the chorion of the annual fishes *Cynolebias* (Cyprinodontiformes: Rivulidae). *Copeia* 1016–1022.

Lucas, M. C., E. Baras, T. J. Thom, A. Duncan, and O. Slavík. 2001. *Migration of Freshwater Fishes*. Oxford: Blackwell Science.

McLean, C. A., and D. Stuart-Fox. 2014. Geographic variation in animal colour polymorphisms and its role in speciation. *Biological Reviews* 89:860–873. doi: 10.1111/brv.12083

Murphy, W. J., and G. E. Collier. 1997. A molecular phylogeny for aplocheiloid fishes (Atherinomorpha, Cyprinodontiformes): The role of vicariance and the origins of annualism. *Molecular Biology and Evolution* 14:790–799.

Nagy, B. 2008. Mafia Island—Notes on the distribution of *Nothobranchius* and my recent collections. *Journal of the American Killifish Association* 41:129–144.

Nagy, B. 2010. In the pearl of Africa. *Nothobranchius Archives* 1:14–30.

Nagy, B., and A. Horváth Kis. 2010. Variation in habitat characteristics and the occurrence of *Nothobranchius* species in seasonal biotopes of Tanzania. *Journal of the American Killifish Association* 43:130–144.

National Weather Service. 2014. Climate prediction center. http://www.cpc.noaa.gov/products/fews/AFR_CLIM/GRAPHS/africa.html.

Passos, C., B. Tassino, F. Reyes, and G. G. Rosenthal. 2014. Seasonal variation in female mate choice and operational sex ratio in wild populations of an annual fish, *Austrolebias reicherti*. *PLoS One* 9:e101649. doi: 10.1371/journal.pone.0101649

Podrabsky, J. E., I. D. Garrett, and Z. F. Kohl. 2010. Alternative developmental pathways associated with diapause regulated by temperature and maternal influences in embryos of the annual killifish *Austrofundulus limnaeus*. *Journal of Experimental Biology* 213:3280–3288. doi: 10.1242/jeb.045906

Polačik, M., R. Blažek, R. Řežucha, M. Vrtílek, E. Terzibasi Tozzini, and M. Reichard. 2014a. Alternative intra-population life history strategies and their trade-offs in an African annual fish. *Journal of Evolutionary Biology* 27:854–865. doi: 10.1111/jeb.12359

Polačik, M., M. Donner, and M. Reichard. 2011. Age structure of annual *Nothobranchius* fishes in Mozambique: Is there a hatching synchrony? *Journal of Fish Biology* 78:796–809. doi: 10.1111/j.1095-8649.2010.02893.x

Polačik, M., C. Harrod, R. Blažek, and M. Reichard. 2014b. Trophic niche partitioning in communities of African annual fish: Evidence from stable isotopes. *Hydrobiologia* 721:99–106. doi: 10.1007/s10750-013-1652-0

Polačik, M., and J. E. Podbrabsky. 2015. Temporary environments. In: *Extremophile Fishes—Ecology and Evolution of Teleosts in Extreme Environments*, eds. R. Riesch, M. Plath, and M. Tobler. New York: Springer.

Polačik, M., and M. Reichard. 2010. Diet overlap among three sympatric African annual killifish species (*Nothobranchius* spp.) from Mozambique. *Journal of Fish Biology* 77:754–768. doi: 10.1111/j.1095-8649.2010.02717.x

Polačik, M., and M. Reichard. 2011. Asymmetric reproductive isolation between two sympatric annual killifish with extremely short lifespans. *PLoS One* 6:e22684. doi: 10.1371/journal.pone.0022684

Reichard, M. 2010. *Nothobranchius kadleci* (Cyprinodontiformes: Nothobranchiidae), a new species of annual killifish from central Mozambique. *Zootaxa* 2332:49–60.

Reichard, M., and M. Polačik. 2010. Reproductive isolating barriers between colour-differentiated populations of an African annual killifish, *Nothobranchius korthausae* (Cyprinodontiformes). *Biological Journal of Linnean Society* 100:62–72. doi: 10.1111/j.1095-8312.2010.01406.x

Reichard, M., M. Polačik, R. Blažek, and M. Vrtílek. 2014. Female bias in the adult sex ratio of African annual fishes: Interspecific differences, seasonal trends and environmental predictors. *Evolutionary Ecology* 28:1105–1120. doi: 10.1007/s10682-014-9732-9

Reichard, M., M. Polačik, and O. Sedláček. 2009. Distribution, colour polymorphism and habitat use of the African killifish, *Nothobranchius furzeri*, the vertebrate with the shortest lifespan. *Journal of Fish Biology* 74:198–212. doi: 10.1111/j.1095-8649.2008.02129.x

Reichard, M., B. R. Watters, R. H. Wildekamp et al. 2010. Potential negative impact and low effectiveness in the use of African annual killifish in the biocontrol of aquatic mosquito larvae in temporary water bodies. *Parasites & Vectors* 3:89. doi: 10.1186/1756-3305-3-89

Sedláček, O., B. Baciaková, and L. Kratochvíl. 2014. Evolution of body colouration in killifishes (Cyprinodontiformes: Aplocheilidae, Nothobranchiidae, Rivulidae): Is male ornamentation constrained by intersexual genetic correlation? *Zoologische Anzeiger* 253:207–215. doi: 10.1016/j.jcz.2013.12.004

Shidlovskiy, K. M. 2010. Collecting *Nothobranchius* in south-western Tanzania. *Nothobranchius Archives* 1:18–27.

Shidlovskiy, K. M., B. R. Watters, and R. H. Wildekamp. 2010. Notes on the annual killifish species *Nothobranchius rachovii* (Cyprinodontiformes; Nothobranchiidae) with the description of two new species. *Zootaxa* 2724:37–57.

Sinervo, B., and C. M. Lively 1996. The rock-paper-scissors game and the evolution of alternative male strategies. *Nature* 380:240–243. doi:10.1038/380240a0

Sonnenberg, R., and E. Busch. 2010. Description of *Callopanchax sidibei* (Nothobranchiidae: Epiplateinae), a new species of killifish from southwestern Guinea, West Africa. *Bonn Zoological Bulletin* 57:3–14.

Stearns, S. C. 1992. *The Evolution of Life Histories*. New York: Oxford University Press.

Terzibasi, E., D. R. Valenzano, M. Benedetti et al. 2008. Large differences in aging phenotype between strains of the short-lived annual fish *Nothobranchius furzeri*. *PLoS One* 3:e3866. doi: 10.1371/journal.pone.0003866

Terzibasi Tozzini, E., A. Dorn, E. Ng'oma et al. 2013. Parallel evolution of senescence in annual fishes in response to extrinsic mortality. *BMC Evolutionary Biology* 13:77. doi: 10.1186/1471-2148-13-77

Valdesalici, S. 2010. *Nothobranchius eggersi* Seegers 1982. *Nothobranchius Archives* 40:25–29.

Valdesalici, S. 2012. *Nothobranchius kardashevi* and *Nothobranchius ivanovae* (Cyprinodontiformes: Nothobranchiidae): Two new annual killifishes from the Katuma River drainage, western Tanzania. *Aqua, Internationl Journal of Ichthyology* 18:191–198.

Valdesalici, S. 2013. *Pronothobranchius chirioi* n. sp. a new annual killifish species from the Niger River drainage, with redescriptions of *P. kiyawensis*, *P. gambiensis* and *P. seymouri* (Cyprinodontiformes: Nothobranchiidae). *Killi-Data Series* 2013:21–41.

Valdesalici, S. 2014. *Nothobranchius bellemansi* and *Nothobranchius occultus* (Cyprinodontiformes: Nothobranchiidae) two new annual killifish from Sudan. *Killi-Data Series* 2014:4–19.

Valdesalici, S., R. Bills, A. Dorn, K. Reichwald, and A. Cellerino 2012. *Nothobranchius niassa* (Cyprinodontiformes: Nothobranchiidae), a new species of annual killifish from northern Mozambique. *Ichthyological Exploration Freshwaters* 23:19–28.

Valdesalici, S., and A. Cellerino 2003. Extremely short lifespan in the annual fish *Nothobranchius furzeri*. *Proceedings of the Royal Society of London B* 270:S189–S191. doi: 10.1098/rsbl.2003.0048

Valdesalici, S., and R. H. Wildekamp. 2004. A new species of the genus *Nothobranchius* Peters, 1868 from the Lufwa River basin, Katanga Province, Democratic Republic of Congo (Pisces, Cyprinodontiformes, Aplocheilidae). *Annali del Museo Civico di Storia Naturale "Giacomo Doria"* 96:241–251.

Valdesalici, S., and R. H. Wildekamp. 2005. A new species of the genus *Nothobranchius* (Cyprinodontiformes, Nothobranchiidae) from Luapula River basin, Zambia. *Aqua, Journal of Ichthyology and Aquatic Biology* 9:89–96.

Valenzano, D. R., S. Sharp, and A. Brunet. 2011. Transposon-mediated transgenesis in the short-lived African killifish *Nothobranchius furzeri*, a vertebrate model for aging. *G3* 1:531–538. doi: 10.1534/g3.111.001271

Vanschoenwinkel, B., A. Waterkeyn, and T. Nhiwatiwa. 2011. Passive external transport of freshwater invertebrates by elephant and other mud-wallowing mammals in an African savannah habitat. *Freshwater Biology* 56:1606–1619. doi: 10.1111/j.1365-2427.2011.02600.x

Vrtílek, M., and M. Reichard. 2015. Highly plastic resource allocation to growth and reproduction in females of an African annual fish. *Ecology of Freshwater Fish*. doi:10.1007/s10682-014-9732-9.

Watters, B. R. 2006. Dispersal of *Nothobranchius* fishes—Fact and fiction. *Journal of the American Killifish Association* 39:137–144.

Watters, B. R. 2009. The ecology and distribution of *Nothobranchius* fishes. *Journal of the American Killifish Association* 42:37–76.

Watters, B. W., B. J. Cooper, and R. H. Wildekamp. 2007. Description of *Nothobranchius cardinalis* spec. nov. (Cyprinodontiformes: Aplocheilidae), an annual fish from the Mbwemkuru River basin, Tanzania. *Journal of the American Killifish Association* 40:129–145.

White, S. M., M. Ondračková, and M. Reichard. 2012. Hydrologic connectivity affects fish assemblage structure, diversity, and ecological traits in the unregulated Gambia River, West Africa. *Biotropica* 44:521–530. doi: 10.1111/j.1744-7429.2011.00840.x

Wildekamp, R. H. 1983. Assignment report. Preliminary study of the Somalian *Nothobranchius* species of larvivorous fishes. EM/MAL/198 EM/VBC/41 SOM/MPD/001/RB. Rome: World Health Organization.

Wildekamp, R. H. 2004. *A World of Killies: Atlas of the Oviparous Cyprinodontiform Fishes of the World. I.* Elyria: American Killifish Association.

Wildekamp, R. H., and R. Haas. 1992. Redescription of *Nothobranchius microlepis*, description of two new species from northern Kenya and southern Somalia, and note on the status of *Paranothobranchius* (Cyprinodontiformes: Aplocheilidae). *Ichthyological Exploration Freshwater* 3:1–16.

Wootton, R. J. 1990. *Ecology of Teleost Fishes.* London: Chapman & Hall.

Wourms, J. P. 1972. The developmental biology of annual fishes. III. Pre-embryonic and embryonic diapause of variable duration in the eggs of annual fishes. *Journal of Experimental Zoology* 182:389–414.

CHAPTER 10

Tolerance of Environmental Stress

Jason E. Podrabsky, Claire L. Riggs, and Josiah T. Wagner

CONTENTS

10.1 INTRODUCTION

The temporary or marginal aquatic habitats inhabited by annual killifishes are often subject to a high degree of environmental variation. In addition, many of these habitats reach extremes in environmental parameters that approach the very limits of vertebrate survival. Annual killifishes have evolved to exploit their harsh and variable environment as adults and larvae during the rainy season and tolerate or resist environmental stress as embryos during the dry season.

10.1.1 Environmental Conditions Faced by Adults

Rainy season conditions vary across the range of annual killifishes (Table 10.1). For example, fish may be exposed to near-freezing temperatures in the subtropical and temperate zone pools of Argentina, Uruguay, and southern Brazil (Errea and Danulat, 2001; Volcan et al., 2011a), while in the coastal deserts of Venezuela and savannahs of Mozambique they may be exposed to temperatures near their upper lethal limit (Podrabsky et al., 1998; Reichard, 2010). In both cases, the pools are relatively small and shallow and therefore have a limited capacity to buffer environmental changes that occur on a daily or seasonal basis. The water may be highly turbid, or clear and tea-colored in cases where a significant amount of degrading vegetation is present. Thus, annual killifishes must endure a high degree of daily and seasonal variation in important environmental parameters such as temperature, oxygen concentration, pH, salinity, and, of course, water availability (Podrabsky et al., 1998; Reichard, 2010; Volcan et al., 2011a; Volcan et al., 2014; Lanés et al., 2014a,b).

10.1.2 Environmental Conditions Faced by Embryos

The environment experienced by embryos of annual killifishes is less well described compared with that of adults. In principle, the embryos must endure the same rainy season temperature conditions described above for the adults, although they may be buffered somewhat from daily fluctuations when buried in the mud or if deposited deep in the pond or under shelter of some kind. Because embryos are deposited into the pond substrate—they likely experience a very different rainy season environment compared with the adults—that is dictated largely by the properties of the clay muds typically associated with the occurrence of annual killifishes (Watters, 2009). The muddy substrates that are almost always associated with annual killifish ponds are rich in clay minerals and subject to swelling when wet and significant shrinkage when dry. These types of soils are usually classified as vertisols and also have a significant capacity for cation exchange and buffering (especially calcium and magnesium) when hydrated (Watters, 2009). During the rainy season, the high content of clay minerals makes vertisol soils very sticky, soft, and almost gel-like. Water and gases infiltrate these soils very slowly, thus allowing the pond areas to hold standing water for a longer period of time than surrounding areas. These soils, when inundated, become hypoxic (low in oxygen) or anoxic (lacking in oxygen) rather quickly due to microbial activity and limited diffusion of oxygen through the soil matrix. Sediments from ponds in northern Venezuela have been shown to have redox potentials that are consistent with severe hypoxia or anoxia (Podrabsky et al., 1998). During the dry season, vertisol soils tend to dry slowly and hold a significant amount of moisture compared with other soil types, even when they appear dry and cracked due to shrinkage. Therefore, during the early part of the dry season, embryos encased in mud are likely to gain some protection from evaporative water loss and may even be able to attract water from the soil. However, it is important to note that a great deal of the water associated with clay soils is bound too tightly to the soil matrix to be biologically available. It is currently unclear what the relationship between soil moisture and embryo moisture might be during the length of the dry season. Importantly, these soils contract as they lose water, and thus air infiltration is likely minimal during the initial phases of drying, which may lead to prolonged periods of hypoxia or anoxia even after the soils have lost a great deal of their moisture.

Table 10.1 Water Conditions of Annual Killifish Habitats in Africa and South America

Location	Species	Temperature (°C)	pH	D.O. (mg/L)	Cond. (µS/cm)	Month	Reference
Africa							
Mozambique	Nothobranchius kadleci Nothobranchius orthonotus Nothobranchius rachovii	27.7–38.2			68–300		Reichard (2010)
Lower Mozambique	Nothobranchius furzeri Nothobranchius orthonotus Nothobranchius rachovii	20.5–35.2			50–625	Feb	Reichard et al. (2009)
	Nothobranchius furzeri	22–28 12–45 dry					Tozzini et al. (2013)
East Africa	Nothobranchius sp.	17–34	5.8–9.6 7.36		0–531		Watters (2009)
Mweru Wantipa Lake basin, Zambia	Nothobranchius oestergaardi		6.1		207		Valdesalici and Amato (2011)
Kilosa, Tanzania	Nothobranchius guentheri	21–30				Mar–Aug	Bailey (1972)
Malawi	Nothobranchius wattersi Nothobranchius kirki	24–34	5.4–9.6		20–390		Ng'oma et al. (2013)
South America							
Southern Brazil—Patos-Mirim Lagoon System	Austrolebias nigrofasciatus	9–42	6.3–10.5	3.4–8.2	446–518	Sept–Oct	Volcan et al. (2011a)
	Cynopoecilus melanotaenia		6.9 ± 1.3	5.0 ± 2.3	482 ± 51		
	Austrolebias melanoorus	18.9 ± 4.5	6.75 ± 0.3	6.3 ± 3.1	227.2 ± 112.4	Jul–Aug	Volcan et al. (2011b)
	Austrolebias cheradophilus	10.9–21.1	6.85–8.23		19–26	Jul–Sept	Lanés et al. (2014a)
	Austrolebias quirogai Austrolebias juanlangi	27.6 ± 8	5.94 ± 0.35	1.4 ± 2 18–28%sat	136 ± 10	Oct	Volcan et al. (2014)
Southern Brazil—Lagoa do Peixe National Park	Austrolebias minuano Cynopoecilus fulgens	10.3–29.7	6.03–8.88	5.9–12.6	10.4–26	Mar–Nov	Lanés et al. (2014b)
Northern Brazil—Russas, Ceara	Hypsolebias antenori	32	6.9	3.9	141	Jun–Oct	Nascimento et al. (2012)

(Continued)

Table 10.1 (*Continued*) Water Conditions of Annual Killifish Habitats in Africa and South America

Location	Species	Temperature (°C)	pH	D.O. (mg/L)	Cond. (µS/cm)	Month	Reference
Uruguay—Department of Rocha	*Cynolebias viarius*	6–28.8	6.26 ± 0.2	4.4–9.7 44–88%sat	257.9 ± 38.6	Apr–Dec	Errea and Danulat (2001)
	Cynopoecilus melanotaenia						
Venezuela—EstadoPortuguesa—savannah	*Austrofundulus transilis*	25.5–29.5	5.0–6.5	0.7–3.7	20–110	Aug	Nico and Taphorn (1984)
	Pterolebias zonatus						Nico et al. (1987)
	Rachovia maculipinnis						
Venezuela—EstadoPortuguesa—forest	*Pterolebias hoignei*	27–29	5.3–6.5		190		Nico and Taphorn (1984)
	Rachovia maculipinnis						Nico et al. (1987)
	Rivulus stellifer						
	Terranatos dolichopterus						
Venezuela—EstadoPortuguesa—savannah	*Austrofundulus transilis*	25.8–27.3	6.0–6.5	0.2–6.7	27–910	Jun	Podrabsky et al. (1998)
	Pterolebias zonatus			2.5– 82%sat			
	Pterolebias hoignei						
	Rachovia maculipinnis						
Venezuela—Maracaibo basin	*Austrofundulus limnaeus*	27.1–37.5	5.48–9.07	0.2–17.9	23–2350	Jun	Podrabsky et al. (1998)
	Rachovia hummelinki			2–256%sat			
	Rachovia pyropunctata						

Consequently, for embryos, survival of the rainy season may be dependent on enduring long bouts of hypoxia or anoxia (Podrabsky et al., 2012b; Anderson and Podrabsky, 2014), while survival through the dry season is likely invested in both tolerance of hypoxia/anoxia and the ability of the embryos to resist water loss to the environment (Podrabsky et al., 2001; Podrabsky et al., 2010b).

10.1.3 Critical Importance of Environmental Cues to Synchronize Life History

Annual killifishes are highly responsive to their environment and are able to key in on the predictable cycles of the rainy and dry season, while at the same time bet-hedging on embryos that may survive several pond inundations and thus buffer against the loss of entire populations due to unpredictable events. Amazingly, they are able to tolerate and resist rather harsh environments while retaining the ability to sense and respond to critical environmental cues such as temperature and hypoxia.

10.2 TEMPERATURE

Temperature is arguably the most variable and critical environmental parameter with which adult annual killifishes must cope. Temperature has a profound effect at all levels of biological organization. First, temperature influences many other environmental factors such as oxygen availability and pH. Second, temperature has a profound effect on the rate of biological processes such as cellular and organismal metabolism. Due to the high degree of variation in temperatures associated with their temporary pond environments, annual killifishes tend to be eurythermal—tolerant to a wide range of temperatures. For annual killifishes from northern South America and Africa, this translates to the ability to function across a wide range of daily temperatures from the low to mid-20s to low 40s in degrees Celsius (Podrabsky et al., 1998; Podrabsky and Somero, 2003; Reichard, 2010). Fish living in southern South America also experience large temperature fluctuations, although they also must endure temperatures near or below 10°C. For example, habitat temperature ranges of 9–42°C were reported for *Austrolebias nigrofasciatus* in southern Brazil, with temperatures ranging from 11 to 34°C in a single day (Volcan et al., 2011a).

The thermal environment specifically experienced by embryos during the rainy or dry season is relatively unknown for any of the annual killifishes. Embryos likely experience temperatures similar to adults while the ponds are inundated. However, it is highly likely that temperature fluctuations are muted in the pond sediments, and embryos may not experience the full range of temperatures of the overlying water. To our knowledge, the dry season microhabitat conditions experienced by embryos remain to be described in detail for any annual killifish species. It is likely that dry season temperatures exceed rainy season temperatures, at least in the top few centimeters of the soil. Detailed field experiments and sampling will have to be conducted to better understand the thermal environment experienced by embryos during the dry season.

10.2.1 Temperature Variation

Adult *Austrofundulus limnaeus* thrived in laboratory experiments at constant temperatures of 20 and 37°C, as well as daily temperature cycling from 20–37°C that mimics natural variation (Podrabsky and Somero, 2003). These fish appear to be quite well adapted to daily temperature cycling at the cellular level as evaluated by alterations in gene expression of liver tissue. Over 90% of the nearly 5000 elements on a liver tissue-specific cDNA microarray were expressed at constant levels across a 17°C temperature range (Podrabsky and Somero, 2003). For those genes that did respond to temperature cycling, there were distinct patterns of gene expression associated with temperature cycling, as well as constant cold and warm temperatures. However, after a few weeks of temperature cycling, the majority of genes that initially exhibit a strong transcriptional response were expressed at levels similar to control fish held at a constant 26°C (Podrabsky and Somero, 2003). Thus, it appears that these fishes are able to rather quickly adjust to a variety of thermal regimes in a manner that results in stable gene expression patterns. Behaviorally, *A. limnaeus* prefer temperatures near 26°C under laboratory conditions, and fish acclimated to temperature cycling appear to be better at behavioral thermoregulation and seek out their preferred temperature in a thermal gradient more quickly than those acclimated to constant conditions (Podrabsky et al., 2008). While no sex bias in final temperature preferendum was found in this species, males initially chose cooler temperatures in the gradient after acclimation to temperature cycling. While adult fish appear to function well across a wide range of temperatures, there may be negative effects on reproductive output. Temperature cycling reduced reproductive output under laboratory conditions through reduced egg production and poor rates of fertilization (Podrabsky et al., 2008).

10.2.2 Temperature Limits of Embryos

Developing embryos of annual killifishes appear to have upper thermal tolerances that are similar to other tropical fishes. For example, for early embryos of *Austrofundulus limnaeus* through diapause II (DII), a 2-h exposure to 42°C causes 100% mortality (Figure 10.1), and similar data are presented for DII embryos of *Nothobranchius guentheri* (Matias and Markofsky, 1978). In contrast, the lower thermal tolerance of *N. guentheri* embryos in DII is extended to very cold temperatures with about 67% of diapause I (DI) and 100% of DII embryos surviving for 2 weeks of exposure to 3.4°C (Matias and Markofsky, 1978). Thus, we conclude that annual killifish embryos have upper thermal tolerances that are very near the temperatures they likely experience in nature. In contrast, there appears to be an excess of tolerance to cold temperatures during DII in the tropical species, down to temperatures they are very unlikely to ever experience in their natural habitat. For those species living in southern Brazil, Uruguay, and Argentina, survival of extended times of cold temperature may be relevant, but at this point no laboratory data are available to assess their thermal tolerance limits.

10.2.3 Embryonic Incubation Temperature and the Regulation of Entry Into Diapause

There are two distinct developmental trajectories possible for embryos of annual killifishes, entrance into DII or "escape" from DII with direct development to diapause III (DIII) (Wourms, 1972a; Markofsky and Matias, 1977; Levels and Denuce, 1988; Podrabsky et al., 2010a). The timing

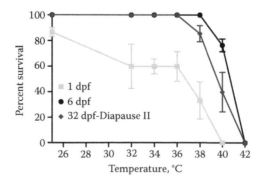

Figure 10.1 Survival of *Austrofundulus limnaeus* embryos exposed for 2 h to a range of temperatures. Groups of 10 embryos were placed in a thermal gradient at the appropriate temperature in 15 ml polypropylene culture tubes. Survival of embryos was determined 48 h after the initial exposure. Symbols are means ± sem (*n* = 3–9).

of morphological and physiological development is distinct in these two trajectories (Podrabsky et al., 2010a), and both trajectories produce a unique physiological phenotype at the completion of embryonic development (Chennault and Podrabsky, 2010). The incubation temperature experienced by embryos during early development can alter both the rate and trajectory of development. As expected, higher temperatures support faster rates of development (Levels and Denuce, 1988; Anderson and Podrabsky, 2014). In general, lower temperatures favor entry of embryos into DII in both African and South American annual killifishes, while even short exposures to higher temperatures can lead to production of escape embryos that do not enter DII (Levels and Denuce, 1988; Podrabsky et al., 2010a). The exact temperatures required to alter developmental trajectory are species specific. In *Austrofundulus limnaeus*, an embryonic incubation temperature of 20°C yields 100% embryos that enter DII, while 30°C leads to 100% escape embryos (Chapter 4, Figure 4.4; Podrabsky et al., 2010a). In *Nothobranchius korthausae,* temperatures from 18°C to 25°C favor DII embryos, while even short exposures to 30°C favor escape embryos (Levels and Denuce, 1988). In *N. guentheri*, incubation temperatures of 19.4°C lead to a high proportion of embryos entering DII, while temperatures of 22.7°C and 26.5°C lead to a high proportion of escape embryos (Markofsky and Matias, 1977). In some species, incubation at temperatures below 20°C appears to also induce entrance into DI (Wourms, 1972a; Levels and Denuce, 1988). Thus, warm temperatures tend to support shorter development times due to both an increase in developmental rate as well as an escape from entrance into diapause. The environmental relevance of this temperature effect is difficult to interpret at this point without detailed information about the thermal environment experienced by embryos under natural conditions.

10.2.4 Spawning Temperature and Regulation of Entry Into Diapause

Only one study has been published on the effects of adult temperature conditions on the length and occurrence of diapause. It appears that the incubation temperature of the embryo is more critical than adult spawning temperature in determining whether an embryo enters or escapes from DII. In *Nothobranchius guentheri*, incubation at 19.4°C favored entrance into DII regardless of the adult incubation temperature (Markofsky and Matias, 1977). However, it is important to note that significant intra and interindividual variation exists for the occurrence and duration of DII, and the reporting of general trends may overlook important relationships between temperature and the induction of diapause.

10.2.5 Temperature and Breakage of Diapause

Embryos that enter DII can be induced to break diapause (resume development) by increasing their incubation temperature. This is true for embryos of both South American and African annual killifishes. In *Nothobranchius korthausae*, a significant proportion of DII embryos broke diapause and resumed development when the incubation temperature was increased from 25°C to 30°C (Levels and Denuce, 1988). This effect was apparent in response to both short as well as chronic increases in temperature. Similar results have been observed for DII embryos of *Austrofundulus limnaeus* (J.E. Podrabsky, personal observations). Interestingly, it appears that DIII embryos may respond in the opposite manner to temperature. Levels and Denuce (1988) report that lowering the temperature from 25°C to 18°C leads to an increase in the proportion of DIII embryos successfully hatching, while increased temperature had the opposite effect. It is interesting that these two stages of diapause respond in an opposite manner to changes in temperature. Lower temperatures might be associated with new rain events as the soil becomes inundated with water, as previously reported for the habitat of *Nothobranchius* in eastern Africa (Tozzini et al., 2013). However, the reason that increased temperatures might break DII remains more speculative. Warmer temperatures are often associated with the rainy season in tropical habitats, and thus increased temperature might favor embryos that either do not enter DII (see above) or break DII early. This could lead to the ability to support multiple generations of fish within a single rainy season—either in a single inundation event or after multiple inundation events. Field studies will be required to evaluate the environmental importance of this well-established response under laboratory conditions.

10.3 OXYGEN

Dissolved oxygen (DO) content has been reported as extremely hypoxic (low oxygen content) and possibly anoxic (no oxygen) to hyperoxic (supersaturated with oxygen) in ponds inhabited by annual killifishes (Table 10.1). This high degree of variation can occur on many timescales, with large-scale fluctuations in DO (2%–256% saturation) recorded for many ponds on a circadian basis (Podrabsky et al., 1998). However, there also appears to be a great deal of spatial variation between ponds that is likely due to microhabitat differences such as types and amounts of vegetation. For instance, some ponds in the Maracaibo basin of Venezuela remain extremely hypoxic throughout the day, while others become hyperoxic in response to the intense midday sun (Podrabsky et al., 1998). Overall, it appears that ponds in grassland regions tend to have DO contents that are below saturation and sometimes extremely hypoxic (Nico and Taphorn, 1984; Podrabsky et al., 1998; Errea and Danulat, 2001; Volcan et al., 2011a,b; Volcan et al., 2014; Lanés et al., 2014a,b).

The DO environment of the pond sediments is likely very different from the overlying water. Pond sediments from a number of pools have been reported to be extremely hypoxic or anoxic (Podrabsky et al., 1998). It is highly likely that this is the case in the sediments of most if not all pools inhabited by annual killifishes due to the reduced diffusion of water through the sediments and the typically high rates of microbial respiration. Once the ponds dry, the partial pressure of oxygen (PO_2) in the soil is expected to increase, but due to the properties of the vertisol clay soils that comprise the substrate of most annual killifish ponds, the soils may remain hypoxic even after significant drying has occurred. In fact, PO_2 levels in dried gleyic vertisol soils drop from ~21 kPa (atmospheric levels of oxygen) to hypoxic levels that vary from 0–14 kPa within the first 5 mm of the soil edge (Zausig and Horn, 1992). Thus, annual killifish embryos likely never experience atmospheric levels of oxygen, but are rather probably always exposed to some level of hypoxia or anoxia.

10.3.1 Hypoxia Tolerance

Laboratory studies of the effects of chronic hypoxia during embryonic development of *Austrofundulus limnaeus* suggest that embryos of annual killifishes are not only highly tolerant of hypoxia but are adapted to develop normally even under extreme hypoxia (Anderson and Podrabsky, 2014). In this study, embryos were able to complete development at PO_2 levels as low as 2.2 kPa at temperatures of 25°C without any apparent teratogenic effects (Figure 10.2). This is in stark contrast to the embryos of other tropical fish such as zebrafish. However, when incubation temperatures were increased to 30°C, embryos were not able to complete development at PO_2 levels below 6.1 kPa. Thus, development under hypoxic conditions is likely only possible at moderate temperatures, and this could become an issue during the dry season, depending on how deep the embryos are buried in the soil. Hypoxia, of any level and at both temperatures, did not delay developmental progression until embryos reached stages of early organogenesis or after breakage of DII. This suggests that the rate of early development in embryos of *A. limnaeus* is rather insensitive to availability of oxygen in the environment at levels of 2.2 kPa and above.

10.3.2 Hypoxia and Entrance into Diapause II

As discussed above (see Section 10.2, "Temperature"), embryos of *Austrofundulus limnaeus* incubated at 30°C do not enter DII. This pattern was not altered under conditions of oxygen

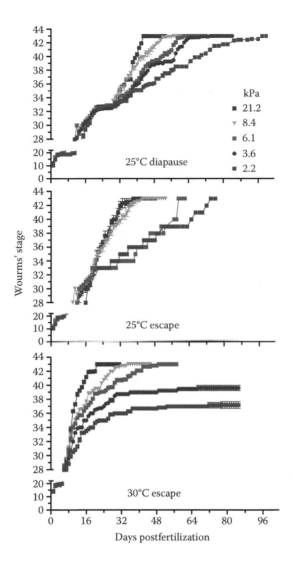

Figure 10.2 The effect of chronic hypoxia on the development of *Austrofundulus limnaeus* embryos at 25°C and 30°C. kPa = kilopascals. (Data are from Anderson, S. N. and J. E. Podrabsky. 2014. *Journal of Comparative Physiology B* 184:355–370.)

limitation, even by the most extreme levels of hypoxia tested. Moreover, neither embryos incubated at 25°C or 30°C arrested development in DI. It is important to note that embryos incubated at 30°C were not able to complete development, so they would have possibly benefited from entering DII under these conditions. Conversely, embryos that entered DII under low PO$_2$ conditions were not inhibited from breaking diapause by increasing their incubation temperature and exposing them to light (Anderson and Podrabsky, 2014). Thus, we conclude that entrance into DII is not altered by the amount of oxygen in the environment, and survival levels of hypoxia below 2.2 kPa may be dependent on embryos entering into a state of anoxia-induced quiescence (see Section 10.3.3).

Development under chronic hypoxia did lead to changes in the physiology and biochemistry of the embryos. Surprisingly, it led to embryos with a reduced total DNA content at the completion of development (WS 43; WS = Wourms' stage; Wourms, 1972c). Moreover, it appears that embryos do not increase their anaerobic or aerobic metabolic capacities to compensate for reduced oxygen levels, even at PO_2 levels that are known to affect rates of oxygen consumption. Rather, hypoxia leads to a reduction in metabolic activity as evidenced by decreased heart rates and the activity of key metabolic enzymes (Anderson and Podrabsky, 2014). Thus, it appears that embryos of *A. limnaeus* act as oxygen conformers during most of development and attempt to match metabolic capacity with available oxygen supplies. This leads to a slowing of development, as observed in this study, but allows for normal development across a wide range of oxygen partial pressures.

10.3.3 Anoxia Tolerance

Embryos of *A. limnaeus* are the most anoxia-tolerant vertebrate yet studied (Podrabsky et al., 2012b). When faced with anoxia, embryos of this species stop development and enter into a state of anoxia-induced dormancy termed *quiescence* (Podrabsky et al., 2007). Quiescence is distinct from diapause, because it is directly induced by environmental stress and is readily reversible. Thus, when oxygen is returned to the environment, the embryos will resume development. Anoxia tolerance increases during early development, peaking during DII, when some embryos survive more than 100 days of anoxia at 25°C (Podrabsky et al., 2007). This high level of tolerance is retained for the first 4–6 days of post-DII development, but then declines to a maximum survival time of about 24 h as embryos approach completion of development and entrance into DIII (Figure 10.3). This remarkable anoxia tolerance is unique to the embryos; the adult killifishes are not poised to survive prolonged anoxic exposure, enduring only a few hours without oxygen (Podrabsky et al., 2012b). The limited adult anoxia tolerance is not surprising in light of their ecology and the ability of adults to behaviorally avoid anoxia.

There are several apparently unique physiological and metabolic characters of *Austrofundulus limnaeus* embryos that appear to support long-term survival under anoxic conditions. Heart rate declines drastically over the first 24 h of anoxia and stops altogether in stages with extreme tolerance of anoxia (Fergusson-Kolmes and Podrabsky, 2007). Metabolic rate as estimated by heat dissipation is drastically reduced in response to anoxia in embryos that are actively developing, and there may be a slight decrease in metabolic rate for embryos already dormant in DII (Podrabsky et al., 2012a). Anoxic metabolism is supported by accumulation of millimolar quantities of lactate and γ-aminobutyric acid, with smaller amounts of alanine and succinate also accumulating (Podrabsky et al., 2007). Free amino acid levels shift during anoxia in a manner that suggests amino acid metabolism is critical for the support of metabolism during anoxia (Podrabsky et al., 2007). Importantly, embryos in DII and DIII have significantly reduced activity of mitochondrial respiratory complexes, and DII embryos have very low rates of ATP synthase activity (Duerr and Podrabsky, 2010). This suggests mitochondria poised not for ATP production but rather for support of intermediary metabolism.

When faced with anoxia, cells of *Austrofundulus limnaeus* embryos appear to arrest the cell cycle in the G_1 phase (Meller et al., 2012). These cells also appear to be resistant to the induction of apoptosis (programmed cell death) in response to anoxia (Meller and Podrabsky, 2013), despite an 80% decline in whole-embryo ATP levels (Podrabsky et al., 2012a) that would almost certainly lead to cell death in most mammalian cells. The tagging of proteins for degradation via ubiquitylation appears to be arrested during exposure to anoxia, although increased levels were observed during aerobic recovery (Meller et al., 2014). Moreover, patterns of protein modification by SUMOylation appear to be unique in embryos that can survive long-term anoxia (Meller et al., 2014). The exact importance of these posttranslational modifications is not yet clear but it promises to be a rich area for future exploration.

While almost everything that is known about anoxia tolerance in annual killifish embryos comes from studies of *A. limnaeus*, one comparative study has recently been conducted with DII embryos of *Nothobranchius furzeri*. Preliminary studies indicate that *N. furzeri* embryos may survive for up to 2 weeks of anoxia at 25°C, but they do not share the extreme tolerance of anoxia exhibited by embryos of *Austrofundulus limnaeus* (M. Polacik et al., personal observations). Additional comparative studies will be needed to assess the prevalence of anoxia tolerance across annual killifishes.

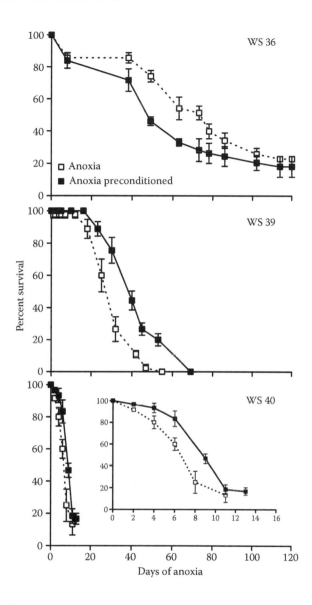

Figure 10.3   Survival of anoxia and the effect of anoxic preconditioning on anoxic survival during post-DII development in *Austrofundulus limnaeus*. Embryos with extreme tolerance of anoxia (WS 36, 4 days post-DII) can survive for more than 100 days of anoxia at 25°C but do not respond to anoxic preconditioning. Embryos at WS 39 and 40 (~9 and 12 days post-DII) have reduced survival of anoxia, but they do respond to anoxic preconditioning. Inset: Expanded axis for the WS 40 data. WS = Wourms' stage (Wourms, 1972c). (Data are from Podrabsky, J. E., C. L. Riggs, and J. M. Duerr. 2012b. *Anoxia*, 3–24. InTech.)

10.3.4 Anoxic Preconditioning

Anoxic preconditioning is a phenomenon in which an initial brief but sublethal exposure to anoxia actually prolongs the organism's survival of subsequent extended anoxic exposure. Ischemia (occlusion of blood flow, which may lead to hypoxia or anoxia) is known to cause major damage to tissues and organs in anoxia-sensitive mammalian models, and ischemic preconditioning in these models reduces cellular and tissue damage as a result of ischemia (Gidday, 2006). This same

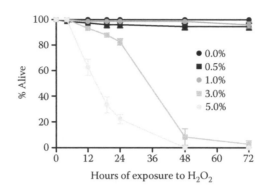

Figure 10.4 Survival of DII embryos of *Austrofundulus limnaeus* exposed to hydrogen peroxide (H$_2$O$_2$). Groups of 20 embryos were exposed in 100 × 20-mm plastic Petri dishes at 25°C.

phenomenon holds true for certain stages of post-DII embryos of *Austrofundulus limnaeus* (Figure 10.3). Importantly, DII embryos and those that are 4 days post-DII do not respond to anoxic preconditioning, which suggests they may already be poised for maximal anoxic survival. However, as embryos develop and lose extreme anoxia tolerance, the ability to increase survival time in anoxia through preconditioning develops. After 8–12 days of post-DII development, embryos exposed to an anoxic preconditioning regime of 24–48 h of anoxia followed by 24 h of aerobic recovery experience an increase in their survival time when exposed to a subsequent long-term anoxic exposure (Meller et al., 2012; Podrabsky et al., 2012b; Meller and Podrabsky, 2013; Meller et al., 2014). While anoxic preconditioning does increase time of survival, it does not induce the extreme tolerance of anoxia exhibited in earlier-stage embryos.

10.3.5 Tolerance of Oxidative Damage

Embryos of *Austrofundulus limnaeus* exhibit an amazing ability to survive and develop normally in the presence of high levels of oxidative damage. Diapause II embryos can survive for a seemingly indefinite amount of time in 1% hydrogen peroxide (H$_2$O$_2$) and more than 24 h in 5% H$_2$O$_2$ (Figure 10.4). It is unlikely that annual killifish embryos would ever be exposed to this level of oxidative damage as a result of their natural environment. However, it is possible that the embryos have a high antioxidant capacity to help protect them from production of reactive oxygen species during transitions into and out of anoxia or hypoxia. Alternatively, it could be an example of excessive tolerance that is often associated with organisms that can enter into states of metabolic dormancy (Jonsson, 2003). Future studies on the antioxidant capacity of these embryos may help shed light on the nature of this amazing capability.

10.4 DEHYDRATION

As discussed above, ephemeral ponds inhabited by annual killifishes may remain "dry" for months or perhaps even years at a time depending on local weather conditions. Survival of the dry season is invested exclusively in the drought-tolerant embryos encased in the drying pond sediments. There is no doubt that the embryos encased in the mud experience dehydration pressures, but the specifics of this environment remain to be characterized in detail for any annual killifish habitat. However, it is possible to make predictions on the soil conditions that embryos of annual killifishes must endure based on the properties of vertisol or vertisol-like soils that are typical of annual killifish habitats.

Vertisol clay soils dry more slowly than other soils, shrink when they lose water, and support rather low infiltration rates of water and air when wet or moist. Despite the ability to hold a significant amount of water even when seemingly dry, most of the interstitial water is likely not biologically available, and there are many examples of these soils with moisture contents of 20%–30% water (by weight) that are still dry enough to cause plants to wilt. Due to the nature of vertisol soils, it is highly unlikely that embryos experience direct exposure to the air phase for long periods of time when encased in the mud; furthermore, the relative humidity immediately around an embryo may remain high enough to support survival for long periods of time. Thus, the water balance between the embryos and the surrounding soil is likely a delicate equilibrium, and field studies will be needed to better understand water relations in embryos encased in the mud.

10.4.1 Survival of Dehydration by Embryos

The ability of embryos to develop in an aerial environment appears to be shared by many members of the Order Cyprinodontiformes as long as the relative humidity remains near saturation (Podrabsky et al., 2010b; J. E. Podrabsky personal observations). However, survival of water stress even slightly lower than 100% saturation appears to be more restricted, even within a single species. For example, embryos of *Austrofundulus limnaeus* can survive for weeks at nearly any developmental stage at relative humidity values of 85% or greater (Podrabsky et al., 2001). Lower levels of relative humidity can only be tolerated for prolonged periods of time by DII embryos (Figure 10.5d; Podrabsky et al., 2001). Since we do not currently have a good model of the soil environment that annual killifish embryos must endure, it is difficult to predict the ability of various developmental stages to survive in the soil. However, it is clear at this time that prolonged survival of even modest dehydration pressures will likely only be possible in DII embryos. Embryos of *Austrofundulus limnaeus* survive aerial dehydration by drastically reducing evaporative water loss and retaining water (Figure 10.5c). It is clear that a slow transition, on the order of 4–6 days, from "wet" to "dry" conditions is necessary for the embryos to survive (Podrabsky et al., 2001). During the initial drying event, embryos lose a significant amount of water (about 50% of the total embryonic water). This water can be completely accounted for by volume changes in the perivitelline space, while the embryo and yolk compartments appear to remain essentially fully hydrated (Figure 10.5a and b; Podrabsky et al., 2001). Thus, resistance to dehydration appears to be invested largely in the structure of the enveloping cell layer of the embryo and not in the egg envelope (chorion). However, an intact chorion is essential to survival during aerial exposure, and those embryos with defects or tears in the chorion quickly lose their embryonic water and die (J. E. Podrabsky, personal observation). Thus, we hypothesize that the mechanism that allows such surprising resistance to evaporative water loss depends on three essential components: (1) an enveloping cell layer with exceptionally low rates of water and ion exchange (Machado and Podrabsky, 2007), (2) secretion of large molecules into the perivitelline fluid that act as ion exchangers and have the ability to form a biological glass when dehydrated, and (3) a thick chorion that can maintain its integrity following dehydration and still provide physical protection and support for the embryo and retention of the large molecules in the perivitelline fluid (Podrabsky et al., 2001). To date, there is some evidence for these characteristics as described above, and we are currently working to provide evidence for the formation of a glass in the perivitelline fluid following the initial stages of dehydration.

10.4.2 Aerial Survival by Adults

A number of killifish species, especially those in the Family Rivulidae are able to survive for prolonged periods of emersion from water. Most of the work on this phenomenon has been done on *Kryptolebias marmoratus* from mangrove habitats (Taylor, 2012). However, despite this apparently

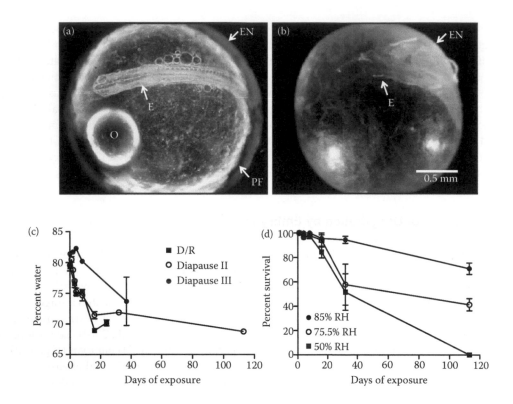

Figure 10.5 Embryos of *Austrofundulus limnaeus* survive exposure to dehydrating conditions by resisting evaporative water loss in the embryonic compartments. (a) A hydrated DII embryo. (b) A DII embryo exposed to 7 days of 75.5% relative humidity (RH). (c) Percent water for embryos exposed to 85% RH at three different developmental stages: dispersion/reaggregation (D/R), DII, and DIII. (d) Survival of DII embryos exposed to three levels of dehydration. PF = perivitelline fluid, O = oil droplet, E = embryo, EN = egg envelope. (Data are from Podrabsky, J. E., J. F. Carpenter, and S. C. Hand. 2001. *American Journal of Physiology* 280:R123–R131.)

widespread ability within the family, there is very little evidence that adult annual killifishes are able to survive for periods of more than a few hours of emersion. For *Austrofundulus limnaeus*, preliminary studies suggest survival times of a few hours under moist conditions (J. E. Podrabsky, personal observation). Thus, it is highly possible that annual killifishes have invested their ability to survive in ephemeral habitats exclusively in embryonic rather than adult adaptations. More systematic studies will be needed to resolve this apparent and interesting trade-off in strategies for dealing with unsuitable or temporary habitats within the Rivulidae.

10.5 SALINITY

Annual killifishes are found almost exclusively in freshwater habitats. While no experiments on the salinity tolerance of adult fish have been published, preliminary evidence on *Austrofundulus limnaeus* suggests that this species is an obligate freshwater fish. Under laboratory conditions, adult fish can tolerate salinities of up to about 10 parts per thousand (ppt, seawater is 32–35 ppt), which is just above the isosmotic point for the tissues of the fish. Salinities higher than this quickly lead to death (J. E. Podrabsky, personal observation). Salt levels in most killifish ponds are quite dilute, as illustrated by total conductivity of the water (Table 10.1). While salt concentrations do seem to increase as ponds begin to dry, they do not appear to reach extremely high levels even just prior to

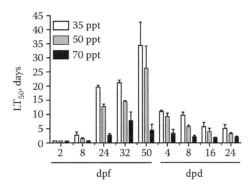

Figure 10.6 Lethal time to 50% mortality (LT₅₀) for *Austrofundulus limnaeus* embryos exposed to three levels of salinity at 25°C. Embryos enter DII at 24 days postfertilization (dpf). Data for embryos at 32 and 50 dpf are also in DII. Diapause II was experimentally broken at 50 dpf. dpf = days postfertilization; dpd = days post-DII; ppt = parts per thousand. (Data are from Machado, B. E. and J. E. Podrabsky. 2007. *Journal of Comparative Physiology B: Biochemical, Systemic, and Environmental Physiology* 177:809–820.)

the ponds drying, likely due to the high ion-exchange capacity of the clay soils. However, salts are abundant in the water and soils that comprise annual killifish habitats, with some soils becoming saturated with calcium, magnesium, and other salts upon drying.

Diapause II embryos of *Austrofundulus limnaeus* can survive for over a month in salinities of 35 or 50 ppt, and over a week in 70 ppt seawater (Figure 10.6; Machado and Podrabsky, 2007). Salt concentrations of this level have never been reported for any annual killifish habitat, and thus the ability to survive in a high-salt environment is likely only relevant during the dry season, when embryos may be encased in mud that is saturated with salts. Survival of high salt is not due to a high capacity for ion pumping but is rather due to the extremely low permeability of the embryos to salt and water. When placed in such hyperosmotic salt conditions, the embryos do not shrink. In fact, even after 4 days of incubation, they do not accumulate salts or organic osmolytes to offset the osmotic imbalance (Machado and Podrabsky, 2007). In addition, they do not up-regulate ion pumps that are associated with osmoregulation, but instead we observe a decrease in ion-pumping activity, just the opposite of what would be expected in an actively osmoregulating organism. The water permeability of intact embryos of *A. limnaeus* is quite remarkably low and is likely a major adaptation for surviving both salt stress and dehydration stress (see discussion above).

10.6 RADIATION

The most abundant source of shortwave radiation is from ozone-penetrating sunlight. Because annual killifishes generally bury eggs immediately after fertilization, the amount of solar radiation they may experience on a daily or seasonal basis is unknown. Additionally, naturally occurring ionizing radiation levels in annual killifish habitats have yet to be measured, but they likely experience lower than 400 mGy per year (Radiation, 2000). Radiation tolerance in embryos of the annual killifish *Austrofundulus limnaeus* likely far exceeds conditions they would normally experience and may represent yet another example of excessive tolerance to environmental stress associated with metabolic dormancy (Jonsson, 2003). However, studies of this tolerance are useful for testing life history hypotheses and determining how embryos may maintain genomic stability during dormancy.

10.6.1 Visible Light

Photoperiod is perhaps the most reliable predictor of seasonal change available, and organisms are exquisitely adapted to sense and respond to day/night length as well as changes in photoperiod. Both adults and embryos of annual killifishes are sensitive to photoperiod. Adult *Nothobranchius guentheri* produce embryos that enter DII when exposed to short-day photoperiods, while long-day photoperiods favored the production of escape embryos (Markofsky and Matias, 1977; Markofsky et al., 1979). Response to changes in photoperiod took several weeks in some cases. Moreover, a great deal of intraindividual variation was observed, and thus the response to photoperiod may be complicated in this species.

When DII embryos are exposed to constant light or to a photoperiod of any duration, the percentage of embryos breaking diapause increases (Levels and Denuce, 1988; Podrabsky and Hand, 1999). Exposure to a 12:12 photoperiod also increased the rate of hatching in DIII embryos of *N. korthausae* (Levels and Denuce, 1988). The environmental relevance of light detection in annual killifish embryos has yet to be explored.

10.6.2 UV Radiation

For most aquatic organisms, exposure to ultraviolet radiation (UV-R) from sunlight is inevitable. Exposure to UV-R can lead to oxidative damage to intercellular structures and, especially at lower wavelengths (100–315 nm), directly induce cytotoxic DNA lesions (Murphy, 1983; Rosenstein and Mitchell, 1987; Zhang et al., 1997). The most common DNA lesions are known to be *cis*-strand products in the form of cyclobutane pyrimidine dimers (CPDs) and 6–4 pyrimidine-pyrimidone photoproducts (6–4 PPs) (Sinha and Häder, 2002; Batista et al., 2009). If left unrepaired, these DNA lesions can lead to abnormal development or death in embryos, likely as a consequence of apoptotic mechanisms (Lesser et al., 2001; Yabu et al., 2001; Wiegand et al., 2004; Dong et al., 2007; Lima-Bessa et al., 2008; Batista et al., 2009; Nahon et al., 2009). Aquatic embryos generally rely on two pathways of enzymatic DNA repair following UV-C irradiation: (1) light-independent repair via nucleotide excision repair (NER), which requires the excision of approximately 27 nucleotide residues around the lesion before replacing the missing nucleotides using a DNA polymerase (Sinha and Häder, 2002) and (2) photorepair (PR) that requires activation of CPD or 6–4 PP-specific photolyases with visible light to reverse lesions (Sancar, 1994). The extent to which reliance on NER or PR to repair DNA lesions in embryos varies by species, and at least with PR, may be partially driven by life history (Akimoto and Shiroya, 1987; Hays et al., 1996; Langhelle et al., 1999; Häkkinen et al., 2001).

Although annual killifish eggs are generally thought to be deposited in sediment immediately after spawning, the depth at which they remain buried has yet to be determined (Wourms, 1972a). Nevertheless, embryos of *Austrofundulus limnaeus* have exceptional tolerance to DNA damage induced by UV-R (Wagner and Podrabsky, 2015a). Overall, *A. limnaeus* embryos are more tolerant to UV-C (254 nm) radiation than other teleost embryos irradiated with similar doses, with DII embryos being the most tolerant (Figure 10.7). Abnormal development is most pronounced after doses of 500 J m^{-2} in embryos irradiated during early somitogenesis (10 dpf), while those irradiated during dispersed phases (4 dpf) or DII are more frequently able to develop normally after the same dose of UV-C (Figure 10.8). Survival is significantly increased following UV-C irradiation for these three stages if embryos are allowed to recover under full-spectrum light. Coinciding with this increase in survival, assays using anti-CPD antibodies reveal that embryos are able to efficiently remove CPDs when recovered in visible light and suggest a competent photorepair system that involves a CPD-specific photolyase. Interestingly, DII embryos irradiated with 4860 J m^{-2} of UV-C accumulate approximately 5.6 CPDs per 10 kb of DNA and are capable of repairing over 90% of these CPDs following 48 h of light recovery, but no change in CPD frequency is observed

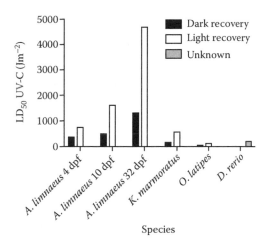

Figure 10.7 Lethal dose to produce 50% mortality (LD$_{50}$) for *Austrofundulus limnaeus* (4, 10, and 32 dpf), *Kryptolebias marmoratus* (blastula), *Oryzias latipes* (morula), and *Danio rerio* (1000-cell stage) embryos exposed to UV-C (254 nm) and recovered in either darkness or light. dpf = days post-fertilization. For *A. limnaeus*: 4 dpf = dispersed cell phase, 10 dpf = solid neural keel, 32 dpf = DII. (Data are from Wagner, J. T. and J. E. Podrabsky. 2015a. *Journal of Experimental Zoology* 323A:10–30.)

if embryos are recovered in 48 h of darkness. This result suggests that DII embryos do not have an active dark-repair system during dormancy and are instead reliant on PR. DII embryos also do not appear to have substantial levels of apoptotic cells while maintaining these high DNA lesion loads, agreeing with previous observations that diapausing embryos suppress apoptotic signaling even after experiencing conditions that should promote cell death (Meller and Podrabsky, 2013). Previous work on DNA content in DII embryos has suggested a lack of change in DNA content during dormancy, thus implying that DNA replication is arrested during dormancy (Podrabsky and Hand, 1999). The fact that both apoptotic signaling and dark repair following UV-induced DNA damage may require recognition of damaged sites during replication may explain how DII embryos can maintain high lesion loads without experiencing apoptotic events (Batista et al., 2009).

While *Austrofundulus limnaeus* embryos clearly have mechanisms to deal with UV-induced CPDs, it is not yet known how the embryos may deal with oxidative DNA damage. Even under

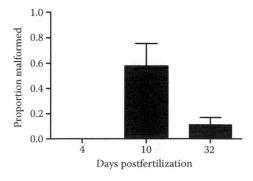

Figure 10.8 Proportion of *Austrofundulus limnaeus* embryos developing abnormally after a 500 J m^{-2} exposure to UV-C (254 nm) light. Embryos were exposed at 4 dpf (dispersed-cell), 10 dpf (solid neural keel), or 32 dpf (DII). Bars are means ± sem (*n* = 3). (Adapted from Wagner, J. T. and J. E. Podrabsky. 2015a. *Journal of Experimental Zoology* 323A:10–30.)

stable conditions, DNA is prone to significant levels of oxidative damage from endogenous cellular sources (Marnett, 2000). In particular, annual killifish embryos likely experience bouts of anoxia whereby entry back into normoxic conditions would cause oxidative DNA damage via interactions with released free radicals (McCord, 1985; Cui et al., 2000). Determining whether diapausing embryos continue to actively repair oxidative DNA damage or damage accumulates until exit of diapause will provide important insight into how genomic stability is maintained during periods of dormancy.

10.6.3 Ionizing Radiation

In living cells, exposure to ionizing radiation causes hydrolysis of water, leading to production of reactive-oxygen species (ROS) that can interact negatively with cellular structures. Additionally, DNA damage caused directly or indirectly by ionizing radiation exposure can include double/single-stranded breaks, damaged bases, abasic sites, sugar damage, and DNA-protein cross-links (Lehnert, 2007). The tolerance of *Austrofundulus limnaeus* embryos to ionizing radiation is similar to that described in zebrafish (Yabu et al., 2003), with nearly complete embryonic mortality occurring after exposures greater than 25 gray using a cesium-137 source (Figure 10.9). At this dose, there is also a sharp increase in embryos developing abnormally if exposure occurred during early somitogenesis (12 dpf) or DII (Figure 10.10) but not during dispersed phases (4 dpf). For embryos exposed to ionizing radiation at 12 dpf or DII, abnormalities include irregular somite formation and/or swollen otic vesicles (Figure 10.10). In zebrafish embryos, rates of apoptosis have been shown to increase mostly linearly with increasing radiation dose, likely through a mechanism involving caspase activation (Yabu et al., 2001; Bladen et al., 2007). The degree to which apoptosis may be a factor in the survival of *A. limnaeus* embryos exposed to ionizing radiation has yet to be determined, although DII embryos exposed to ionizing radiation would be predicted to behave like those exposed to UV-C or anoxia and exhibit limited capacity for apoptosis (Meller and Podrabsky, 2013; Wagner and Podrabsky, 2015a).

The observation that DII embryos are not any more resistant to ionizing radiation than other vertebrate embryos is initially surprising, because desiccation-tolerant organisms tend to also possess resistance to ionizing radiation. Perhaps the best-known example, the bacterium *Deinococcus radiodurans*, utilizes DNA repair pathways important for desiccation tolerance to survive doses of ionizing radiation that far exceed naturally occurring amounts (Cox and Battista, 2005). Since double-stranded DNA (dsDNA) breaks caused by ionizing radiation and desiccation are structurally similar, the capacity for *D. radiodurans* to maintain genome integrity after ionizing

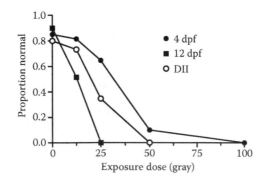

Figure 10.9 Survival and normal development of *Austrofundulus limnaeus* embryos following ionizing radiation exposure at three different developmental stages. Data are from 48 days postexposure. $n = 20$ for controls, $n = 60$ for exposures. dpf = days postfertilization, DII = DII.

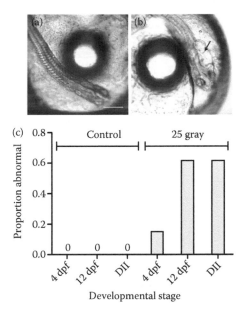

Figure 10.10 Representative photographs of embryos exposed to 25 gray of ionizing radiation at (a) 4 dpf or (b) DII (DII) at 58 days postexposure. While most embryos irradiated at 4 dpf with 25 gray or less developed normally (a), many embryos irradiated at DII with a similar dose developed abnormally (b). A swollen otic vesicle that has formed in the DII embryo is indicated with an arrow. Scale bar: 200 μm. (c): Proportion of embryos irradiated at 4 dpf, 12 dpf, or DII that developed abnormally after 25 gray of ionizing radiation exposure. Data are from 48 days postexposure. $n = 20$ for controls, $n = 60$ for exposures.

radiation is likely an incidental response to its ability to survive bouts of desiccation. When faced with dehydrating conditions *A. limnaeus* embryos resist desiccation (see Section 10.4). Therefore, embryonic cells would not be subject to the damaging effects of intracellular water loss and would presumably not be under selective pressure to maintain a high capacity for dsDNA break repair, an essential requirement for ionizing radiation tolerance.

10.7 pH

The pH of annual killifish habitats has been described to range from 5.48 to 9.07 and may fluctuate over 1 pH unit or greater on a daily basis (Table 10.1). Although the pH of pond sediments in annual killifish habitats has not yet been measured, laboratory studies suggest that embryos may be able to tolerate a wide range of pH environments. Tolerance of fish embryos to suboptimal pHs varies by species and by developmental stage, with lower limits previously described from pH 4.75–5.2 in embryos of the common carp (*Cyprinus carpio*) to 3.0–3.6 in embryos of Atlantic salmon (*Salmo salar*) (Daye and Garside, 1977; Oyen et al., 1991). Similar to the studies of Atlantic salmon embryos by Daye and Garside (1977, 1979), embryos of *Austrofundulus limnaeus* are generally more sensitive to acid pH at earlier stages. In embryos transferred into pH 4 immediately after spawning, embryo mortality is rapid, and nearly all embryos died within the first day (Figure 10.11). Embryos transferred at 5 dpf were much more resilient to low pH, agreeing with previous reports that embryonic resistance to denaturing conditions (i.e., bleach) is higher after the completion of epiboly (Podrabsky, 1999). DII embryos are quite resistant to acid pH, and we have observed an LT_{50} of 6.7 days in pH 2, with some individuals surviving for about a month (Figure 10.12).

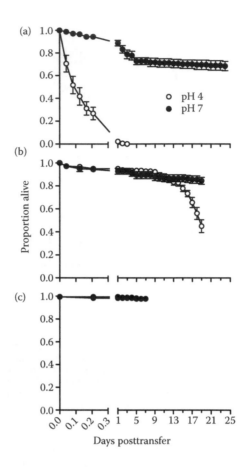

Figure 10.11 Survival of *Austrofundulus limnaeus* embryos transferred into low-pH media either immediately after spawning (a), after the completion of epiboly at 5 dpf (b), or during DII (c). In (c), the open and closed circles are overlapping. For pH transfers, embryos remained in regular embryo media as described by Podrabsky (1999) until time of transfer. pH media was made using regular embryo media supplemented with 10 mM 4-(2-hydroxyethyl)-1-piperazineethanesulfonic acid (HEPES) and then pH adjusted. When compared with a neutral pH, embryos transferred immediately after spawning (a) showed significant mortality after 1 h of transfer (2-way ANOVA with Bonferroni posttest, $p < 0.05$), while embryos transferred at 5 dpf did not have significant mortality until 16 days posttransfer (2-way ANOVA with Bonferroni posttest, $p < 0.05$). No significant death was seen after 7 days in pH 4 for the DII embryos. Symbols are means ± sem ($n = 12$ replicates of 15 embryos each).

10.8 GASEOUS AMMONIA

One study has been conducted on the effects of gaseous ammonia on the occurrence and duration of DII in embryos of *Nothobranchius guentheri* (Matias, 1983). When early embryos (epiboly stages) were exposed to the gaseous ammonia produced from 15 µl of a 1% solution of ammonium hydroxide for 30 min in a sealed 8-ml glass vial, the percentage of embryos entering DII was reduced from about 80 to 40%. However, when embryos in DII were exposed using the same techniques to a range of ammonium hydroxide concentrations ranging from 0 to 10%, an increase in the duration of DII was observed as the amount of ammonia increased. Increased levels of gaseous ammonia caused high rates of mortality in both epiboly- and DII-stage embryos. Exposure to gaseous 10% ammonium hydroxide led to 83% mortality in epiboly-stage embryos and only 21% in DII embryos. Thus, tolerance to exposure to gaseous ammonia is highest in DII embryos.

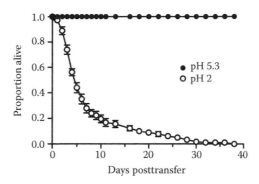

Figure 10.12 Survival of *Austrofundulus limnaeus* embryos when transferred into low pH media during DII. Regular embryo medium supplemented with 10 mM HEPES and pH adjusted. Symbols are means ± sem (*n* = 12 replicates of 10–15 embryos each).

The importance of ammonia as a potential stress during prolonged dehydration is a topic worthy of additional investigation.

10.9 GENERAL DISCUSSION AND CONCLUSIONS

Annual killifishes have evolved a suite of adaptations that allow them to thrive where no other fishes survive. These adaptations include the ability to arrest development in diapause but also to tolerate rather extreme environmental conditions across most of embryonic development. One of the most interesting outcomes of the studies on radiation exposure in *Austrofundulus limnaeus* embryos is the resistance of dispersed-phase embryos to abnormal development following UV-C or ionizing radiation. A similar observation has been seen in zebrafish embryos exposed to UV-B radiation, wherein severe malformations were observed when embryos were irradiated after an embryonic axis had formed but not during the late blastomere (~1000 cell) stage prior to axis formation (Dong et al., 2007). First suggested by Wourms (1972b) and later by Epel et al. (1999), dispersed-phase embryos that experience stress-induced cell death may be able to replace lost cells by mitotic events of surrounding pluripotent stem cells. This process of "developmental buffering" may benefit embryos by allowing the replacement of lost or damaged cells before the onset of sensitive morphogenic processes, such as gastrulation. Although *A. limnaeus* does not regularly arrest during dispersed phases under laboratory conditions, anecdotal evidence suggests that *A. limnaeus* and other species of annual killifishes may frequently arrest during dispersed phases in DI in response to stressful environmental conditions such as low temperatures or habitat desiccation (Wourms, 1972a; Markofsky and Matias, 1977; Genade et al., 2005). It is therefore likely that the dispersed-cell phases consist of embryonic cells that have yet to undergo differentiation (Wagner and Podrabsky, 2015b), thus allowing an opportunity to arrest before embryonic patterning begins.

Annual killifishes exhibit a wide range of environmental stress tolerances across their life history, as might be expected from their ability to thrive in small, isolated, temporary aquatic habitats. Most of their stress tolerance has been invested in embryonic rather than adult stages, as the adult fishes are not remarkable in terms of hypoxia, anoxia, temperature, or salinity tolerance compared with other tropical or temperate fishes. However, embryos of annual killifishes push the limits of vertebrate tolerance to temperature, dehydration, salt, anoxia, oxidative stress, hypoxia, UV radiation, gamma radiation, and pH. While extreme environmental tolerances are associated with entrance into DII, other diapause stages and even the developing embryos exhibit rather impressive tolerance to these environmental stresses. Despite these high levels of tolerance to environmental stress, annual killifishes are still threatened with extinction in many parts of their range due to human disturbance, as well as patterns of global climate change. For example, tropical annual killifishes appear to live very close to their thermal limits, and thus even small increases in air temperatures or incident solar radiation could tip the balance against survival of the dry season. Similarly, the strategy of avoiding evaporative water loss during the dry season places limits on the amount of time that an embryo can survive encased in the mud, as water loss cannot be completely stopped. Thus, changes in the timing of rainy seasons or increases in the variability in rain events could have dramatic effects on local populations and perhaps entire species of annual killifishes. In light of these limits to their physiology, it is especially alarming that we know so little about the actual microclimate conditions that embryos of annual killifishes experience in the field. Field studies must be conducted in order to better understand how these embryos survive such extreme conditions, how much plasticity in their physiology can be induced under natural conditions, and how close they actually are to the limits of their environmental tolerances. Continued studies of annual killifish embryos may lead to advances in biomedicine and bioengineering that allow vertebrate tissues to be engineered to survive in extreme environments or enter into metabolic dormancy.

ACKNOWLEDGMENTS

Data presented in Figure 10.1 on temperature tolerance of embryos were collected and analyzed by U. Olemgbe and K.M. Culpepper. Data on hydrogen peroxide tolerance presented in Figure 10.4 were collected by M.J. Knapp. Data presented in Figures 10.9 and 10.10 on tolerance of ionizing radiation were collected by S. Vege. K. Hanslits assisted in the collection of data on pH tolerance that are presented in Figures 10.11 and 10.12.

This work was supported by NIH grant HL095454 and NSF grant IOS 1354549 to JEP.

REFERENCES

Akimoto, Y. and T. Shiroya. 1987. Photoreversibility of UV-induced thymine dimers and abnormal morphogenesis in sea urchin embryos. *Photochemistry and Photobiology* 45:403–406. doi: 10.1111/j.1751-1097.1987.tb05393.x

Anderson, S. N. and J. E. Podrabsky. 2014. The effects of hypoxia and temperature on metabolic aspects of embryonic development in the annual killifish *Austrofundulus limnaeus*. *Journal of Comparative Physiology B* 184:355–370. doi: 10.1007/s00360-014-0803-6

Bailey, R. G. 1972. Observations on the biology of *Nothobranchius guentheri* (Pfeffer) (Cyprinodontidae), an annual fish from the coastal region of east Africa. *The African Journal of Tropical Hydrobiology and Fisheries* 2:33–43.

Batista, L. F., B. Kaina, R. Meneghini, and C. F. Menck. 2009. How DNA lesions are turned into powerful killing structures: Insights from UV-induced apoptosis. *Mutation Research/Reviews in Mutation Research* 681:197–208. doi: 10.1016/j.mrrev.2008.09.001

Bladen, C. L., M. A. Flowers, K. Miyake et al. 2007. Quantification of ionizing radiation-induced cell death *in situ* in a vertebrate embryo. *Radiation Research* 168:149–157. doi: 10.1667/RR0803.1

Chennault, T. and J. E. Podrabsky. 2010. Aerobic and anaerobic capacities differ in embryos of the annual killifish *Austrofundulus limnaeus* that develop on alternate developmental trajectories. *Journal of Experimental Zoology Part A: Ecological Genetics and Physiology* 313A:587–596. doi: 10.1002/jez.632

Cox, M. M. and J. R. Battista. 2005. *Deinococcus radiodurans*—The consummate survivor. *Nature Reviews Microbiology* 3:882–892. doi: 10.1038/nrmicro1264

Cui, J., E. H. Holmes, T. G. Greene, and P. K. Liu. 2000. Oxidative DNA damage precedes DNA fragmentation after experimental stroke in rat brain. *The FASEB Journal* 14:955–967. doi: 10.1096/fj.1530-6860

Daye, P. G. and E. T. Garside. 1977. Lower lethal levels of pH for embryos and alevins of Atlantic salmon, *Salmo salar* L. *Canadian Journal of Zoology* 55:1504–1508. doi: 10.1139/z77-194

Daye, P. G. and E. T. Garside. 1979. Development and survival of embryos and alevins of the Atlantic salmon, *Salmo salar* continuously exposed to acidic levels of pH, from fertilization. *Canadian Journal of Zoology* 57:1713–1718. doi: 10.1139/z79-222

Dong, Q., K. Svoboda, T. R. Tiersch, and W. T. Monroe. 2007. Photobiological effects of UVA and UVB light in zebrafish embryos: Evidence for a competent photorepair system. *Journal of Photochemistry and Photobiology B: Biology* 88:137–146. doi: 10.1016/j.jphotobiol.2007.07.002

Duerr, J. M. and J. E. Podrabsky. 2010. Mitochondrial physiology of diapausing and developing embryos of the annual killifish *Austrofundulus limnaeus*: Implications for extreme anoxia tolerance. *Journal of Comparative Physiology B: Biochemical, Systemic, and Environmental Physiology* 180:991–1003. doi: 10.1007/s00360-010-0478-6

Epel, D., K. Hemela, M. Shick, and C. Patton. 1999. Development in the floating world: Defenses of eggs and embryos against damage from UV radiation. *American Zoologist* 39:271–278. doi: 10.1093/icb/39.2.271

Errea, A. and E. Danulat. 2001. Growth of the annual fish, *Cynolebias viarius* (Cyprinodontiformes), in the natural habitat compared to laboratory conditions. *Environmental Biology of Fishes* 61:261–268. doi: 10.1023/A:1011047611009

Fergusson-Kolmes, L. and J. E. Podrabsky. 2007. Differential effects of anoxia on heart rate in developmental stages of the annual killifish *Austrofundulus limnaeus* that differ in their tolerance of anoxia. *Journal of Experimental Zoology* 307A:419–423. doi: 10.1002/jez.395

Genade, T., M. Benedetti, E. Terzibasi et al. 2005. Annual fishes of the genus *Nothobranchius* as a model system for aging research. *Aging Cell* 4:223–233. doi: 10.1111/j.1474-9726.2005.00165.x

Gidday, J. M. 2006. Cerebral preconditioning and ischaemic tolerance. *Nature reviews. Neuroscience* 7:437–448. doi:10.1038/nrn1927

Häkkinen, J., S. Pasanen, and J. V. Kukkonen. 2001. The effects of solar UV-B radiation on embryonic mortality and development in three boreal anurans (*Rana temporaria, Rana arvalis* and *Bufo bufo*). *Chemosphere* 44:441–446. doi: 10.1016/S0045-6535(00)00295-2

Hays, J. B., A. R. Blaustein, J. M. Kiesecker et al. 1996. Developmental responses of amphibians to solar and artificial UVB sources: A comparative study. *Photochemistry and Photobiology* 64:449–456. doi: 10.1111/j.1751-1097.1996.tb03090.x

Jonsson, I. K. 2003. Causes and consequences of excess resistance in cryptobiotic metazoans. *Physiological and Biochemical Zoology* 76:429–435.

Lanés, L. E. K., Â. C. Gonçalves and M. V. Volcan. 2014a. Discovery of endangered annual killifish *Austrolebias cheradophilus* (Aplocheiloidei: Rivulidae) in Brazil, with comments on habitat, population structure and conservation status. *Neotropical Ichthyology* 12:117–124. doi: 10.1590/S1679-62252014000100012

Lanés, L. E. K., F. W. Keppeler, and L. Maltchik. 2014b. Abundance variations and life history traits of two sympatric species of Neotropical annual fish (Cyprinodontiformes: Rivulidae) in temporary ponds of southern Brazil. *Journal of Natural History* 1–18. doi: 10.1080/00222933.2013.862577

Langhelle, A., M. J. Lindell, and P. Nyström. 1999. Effects of ultraviolet radiation on amphibian embryonic and larval development. *Journal of Herpetology* 33:449–456.

Lehnert, S. 2007. *Biomolecular Action of Ionizing Radiation*. Taylor & Francis, Boca Raton, FL. doi:10.1201/9781420011920

Lesser, M. P., J. H. Farrell, and C. W. Walker. 2001. Oxidative stress, DNA damage and p53 expression in the larvae of Atlantic cod (*Gadus morhua*) exposed to ultraviolet (290–400 nm) radiation. *Journal of Experimental Biology* 204:157–164.

Levels, P. J. and J. M. Denuce. 1988. Intrinsic variability in the frequency of embryonic diapauses of the annual fish *Nothobranchius korthausae*, regulated by light: Dark cycle and temperature. *Environmental Biology of Fishes* 22:211–223. doi: 10.1007/BF00005382

Lima-Bessa, K. M., M. G. Armelini, V. Chigancas et al. 2008. CPDs and 6-4PPs play different roles in UV-induced cell death in normal and NER-deficient human cells. *DNA Repair* 7:303–312. doi: 10.1016/j.dnarep.2007.11.003

Machado, B. E. and J. E. Podrabsky. 2007. Salinity tolerance in diapausing embryos of the annual killifish *Austrofundulus limnaeus* is supported by exceptionally low water and ion permeability. *Journal of Comparative Physiology B: Biochemical, Systemic, and Environmental Physiology* 177:809–820. doi: 10.1007/s00360-007-0177-0

Markofsky, J. and J. R. Matias. 1977. The effects of temperature and season of collection on the onset and duration of diapause in embryos of the annual fish *Nothobranchius guentheri*. *Journal of Experimental Zoology* 202:49–56. doi: 10.1002/jez.1402020107

Markofsky, J., J. R. Matias, K. Inglima, J. H. Vogelman, and N. Orentreich. 1979. The variable effects of ambient and artificial light: Dark cycles on embryonic diapause in a laboratory population of the annual fish *Nothobranchius guentheri*. *Journal of Experimental Biology* 83:203–215.

Marnett, L. J. 2000. Oxyradicals and DNA damage. *Carcinogenesis* 21:361–370. doi: 10.1093/carcin/21.3.361

Matias, J. R. 1983. The effect of exposure to gaseous ammonia on the duration of diapause II in the embryos of the annual fish, *Nothobranchius guentheri*. *Experientia* 39:1148–1150. doi: 10.1007/BF01943154

Matias, J. R. and J. Markofsky. 1978. The survival of embryos of the annual fish *Nothobranchius guentheri* exposed to temperature extremes and the subsequent effects on embryonic diapause. *Journal of Experimental Zoology* 204:219–228. doi: 10.1002/jez.1402040209

McCord, J. M. 1985. Oxygen-derived free radicals in postischemic tissue injury. *The New England Journal of Medicine* 312:159–163. doi: 10.1056/NEJM198501173120305

Meller, C. L., R. Meller, R. P. Simon, K. M. Culpepper, and J. E. Podrabsky. 2012. Cell cycle arrest associated with anoxia-induced quiescence, anoxic preconditioning, and embryonic diapause in embryos of the annual killifish *Austrofundulus limnaeus*. *Journal of Comparative Physiology B* 182:909–920. doi: 10.1007/s00360-012-0672-9

Meller, C. L., R. Meller, R. P. Simons, and J. E. Podrabsky. 2014. Patterns of ubiquitylation and SUMOylation associated with exposure to anoxia in embryos of the annual killifish *Austrofundulus limnaeus*. *Journal of Comparative Physiology B* 184:235–247. doi: 10.1007/s00360-013-0791-y

Meller, C. L. and J. E. Podrabsky. 2013. Avoidance of apoptosis in embryonic cells of the annual killifish *Austrofundulus limnaeus* exposed to anoxia. *PLoS ONE* 8:e75837. doi: 10.1371/journal.pone.0075837

Murphy, T. M. 1983. Membranes as targets of ultraviolet radiation. *Physiologia Plantarum* 58:381–388. doi: 10.1111/j.1399-3054.1983.tb04198.x

Nahon, S., V. A. Castro Porras, A. M. Pruski, and F. Charles. 2009. Sensitivity to UV radiation in early life stages of the Mediterranean sea urchin *Sphaerechinus granularis* (Lamarck). *Science of the Total Environment* 407:1892–900. doi: 10.1016/j.scitotenv.2008.11.044

Nascimento, W. S., M. E. Yamamoto, and S. Chellappa. 2012. Proporção sexual e relação peso-comprimento do peixe anual *Hypsolebias antenori* (Cyprinodontiformes: Rivulidae) de poças temporárias da região semiárida do Brasil. *Biota Amazônia* 2:37–44.

Ng'oma, E., S. Valdesalici, K. Reichwald, and A. Cellerino. 2013. Genetic and morphological studies of *Nothobranchius* (Cyprinodontiformes) from Malawi with description of *Nothobranchius wattersi* sp. nov. *Journal of Fish Biology* 82:165–188. doi: 10.1111/jfb.12001

Nico, L. G. and D. C. Taphorn. 1984. Limnology of Orinoco basin annual killifish pools. *Journal of the American Killifish Association Technical Paper* 24:3–16.

Nico, L.G., D.C. Taphorn, and J. E. Thomerson. 1987. Datos limnologicos sobre el habitat de los peces anuales (Cyprinodontidae) de los Llanos Venezolanos con una clave para su identificacion. *Biollania* 1987:129–144.

Oyen, F. G. F., L. E. C. M. M. Camps, and S. E. Wendelaar Bonga. 1991. Effect of acid stress on the embryonic development of the common carp (*Cyprinus carpio*). *Aquatic Toxicology* 19:1–12. doi:10.1016/0166-445X(91)90024-4

Podrabsky, J. E. 1999. Husbandry of the annual killifish *Austrofundulus limnaeus* with special emphasis on the collection and rearing of embryos. *Environmental Biology of Fishes* 54:421–431. doi: 10.1023/A:1007598320759

Podrabsky, J. E., J. F. Carpenter, and S. C. Hand. 2001. Survival of water stress in annual fish embryos: Dehydration avoidance and egg envelope amyloid fibers. *American Journal of Physiology* 280:R123–R131.

Podrabsky, J. E., D. Clelen, and L. I. Crawshaw. 2008. Temperature preference and reproductive fitness of the annual killifish *Austrofundulus limnaeus* exposed to constant and fluctuating temperatures. *Journal of Comparative Physiology A: Neuroethology, Sensory, Neural, and Behavioral Physiology* 194:385–393. doi: 10.1007/s00359-008-0313-7

Podrabsky, J. E. and S. C. Hand. 1999. The bioenergetics of embryonic diapause in an annual killifish, *Austrofundulus limnaeus*. *Journal of Experimental Biology* 202:2567–2580.

Podrabsky, J. E., I. D. F. Garrett, and Z. F. Kohl. 2010a. Alternative developmental pathways associated with diapause regulated by temperature and maternal influences in embryos of the annual killifish *Austrofundulus limnaeus*. *Journal of Experimental Biology* 213:3280–3288. doi: 10.1242/jeb.045906

Podrabsky, J. E., T. Hrbek, and S. C. Hand. 1998. Physical and chemical characteristics of ephemeral pond habitats in the Maracaibo basin and Llanos region of Venezuela. *Hydrobiologia* 362:67–78. doi: 10.1023/A:1003168704178

Podrabsky, J. E., J. P. Lopez, T. W. M. Fan, R. Higashi, and G. N. Somero. 2007. Extreme anoxia tolerance in embryos of the annual killifish *Austrofundulus limnaeus*: Insights from a metabolomics analysis. *Journal of Experimental Biology* 210:2253–2266. doi: 10.1242/jeb.005116

Podrabsky, J. E., M. A. Menze, and S. C. Hand. 2012a. Rapid communication: Long-term survival of anoxia despite rapid ATP decline in embryos of the annual killifish *Austrofundulus limnaeus*. *Journal of Experimental Zoology A: Ecological Genetics and Physiology* 317:524–532. doi: 10.1002/jez.1744

Podrabsky, J. E., C. L. Riggs, and J. M. Duerr. 2012b. Anoxia tolerance during vertebrate development— Insights from studies on the annual killifish *Austrofundulus limnaeus*. In *Anoxia*, ed. P. Padilla, 3–24. InTech. doi: 10.5772/39079

Podrabsky, J. E. and G. N. Somero. 2003. Changes in gene expression associated with acclimation to constant temperatures and fluctuating daily temperatures in an annual killifish *Austrofundulus limnaeus*. *Journal of Experimental Biology* 207:2237–2254. doi: 10.1242/jeb.01016

Podrabsky, J. E., A. Tingaud-Sequeira, and J. Cerdà. 2010b. Metabolic dormancy and responses to environmental desiccation in fish embryos. In *Dormancy and Resistance in Harsh Environments*, eds. E. Lubzens, J. Cerda and M. S. Clark, pp. 203–226. Springer, Heidelberg, Germany. doi: 10.1007/978-3-642-12422-8_12

Reichard, M. 2010. *Nothobranchius kadleci* (Cyprinodontiformes: Nothobranchiidae), a new species of annual killifish from central Mozambique. *Zootaxa* 2332:49–60.

Reichard, M., M. Polacik, and O. Sedlacek. 2009. Distribution, colour polymorphism and habitat use of the African killifish *Nothobranchius furzeri*, the vertebrate with the shortest life span. *Journal of Fish Biology* 74:198–212. doi: 10.1111/j.1095-8649.2008.02129.x

Rosenstein, B. S. and D. L. Mitchell. 1987. Action spectra for the induction of pyrimidine (6–4) pyrimidone photoproducts and cyclobutane pyrimidine dimers in normal human skin fibroblasts. *Photochemistry and Photobiology* 45:775–780. doi: 10.1111/j.1751-1097.1987.tb07881.x

Sancar, A. 1994. Structure and function of DNA photolyase. *Biochemistry* 33:2–9. doi: 10.1021/bi00167a001

Sinha, R. P. and D. P. Häder. 2002. UV-induced DNA damage and repair: A review. *Photochemical and Photobiological Sciences* 1:225–36. doi: 10.1039/B201230H

Taylor, D. S. 2012. Twenty-four years in the mud: What have we learned about the natural history and ecology of the mangrove rivulus, *Kryptolebias marmoratus*? *Integrative and Comparative Biology* 52:724–36. doi: 10.1093/icb/ics062

Tozzini, E. T., A. Dorn, E. Ng'oma et al. 2013. Parallel evolution of senescence in annual fishes in response to extrinsic mortality. *BMC Evolutionary Biology* 13:1–12. doi: 10.1186/1471-2148-13-77

United Nations. Scientific Committee on the Effects of Atomic Radiation. 2000. *Sources and Effects of Ionizing Radiation: Sources*. United Nations Publications, New York, USA.

Valdesalici, S. and G. Amato. 2011. *Nothobranchius oestergaardi* (Cyprinodontiformes: Nothobranchiidae), a new annual killifish from Mweru Wantipa Lake drainage basin, northern Zambia. *Aqua* 17:111–199.

Volcan, M. V., A. P. Fonseca, and R. B. Robaldo. 2011a. Reproduction of the threatened annual killifish *Austolebias nigrofasciatus* (Cyprinodontiformes: Rivulidae), confined to a natural environment. *Journal of Threatened Taxa* 3:1864–1867.

Volcan, M. V., A. C. Goncalves, and L. E. K. Lanes. 2014. *Austrolebias quirogai* (Actinopterygii: Cyprinodontiformes: Rivulidae) in Brazil: Occurrence, population parameters, habitat characteristics, and conservation status. *Acta Ichthyologica et Piscatoria* 44: doi: 10.3750/AIP2014.44.1.05

Volcan, M. V., L. E. K. Lanés, A. C. Gonçalves, and M. M. Cheffe. 2011b. First record of annual killifish *Austrolebias melanoorus* (Amato, 1986) (Cyprinodontiformes: Rivulidae) from Brazil, with data on habitat and conservation. *Journal of Applied Ichthyology* 27:1120–1122. doi: 10.1111/j.1439-0426.2010.01626.x

Wagner, J. T. and J. E. Podrabsky. 2015a. Extreme tolerance and developmental buffering of UV-C induced DNA damage in embryos of the annual killifish *Austrofundulus limnaeus*. *Journal of Experimental Zoology* 323A:10–30. doi: 10.1002/jez.1890

Wagner, J. T. and J. E. Podrabsky. 2015b. Gene expression patterns that support novel developmental stress buffering in embryos of the annual killifish *Austrofundulus limnaeus*. *EvoDevo* 6:2. doi:10.1186/2041-9139-6-2

Watters, B. 2009. The ecology and distribution of *Nothobranchius* fishes. *Journal of the American Killifish Association* 42:37–76.

Wiegand, M. D., D. L. W. Young, B. M. Gajda et al. 2004. Ultraviolet light-induced impairment of goldfish embryo development and evidence for photorepair mechanisms. *Journal of Fish Biology* 64:1242–1256. doi: 10.1111/j.0022-1112.2004.00388.x

Wourms, J. P. 1972a. The developmental biology of annual fishes III. Pre-embryonic and embryonic diapause of variable duration in the eggs of annual fishes. *Journal of Experimental Zoology* 182:389–414. doi: 10.1002/jez.1401820310

Wourms, J. P. 1972b. The developmental biology of annual fish II. Naturally occurring dispersion and reaggregation of blastomeres during the development of annual fish eggs. *Journal of Experimental Zoology* 182:169–200. doi: 10.1002/jez.1401820203

Wourms, J. P. 1972c. Developmental biology of annual fishes I. Stages in the normal development of *Austrofundulus myersi* Dahl. *Journal of Experimental Zoology* 182:143–168. doi: 10.1002/jez.1401820202

Yabu, T., Y. Ishibashi, and Y. Michiaki. 2003. Stress-induced apoptosis in larval embryos of Japanese flounder. *Fisheries Science* 69:1218–1223. doi: 10.1111/j.0919-9268.2003.00748.x

Yabu, T., S. Todoriki and M. Yamashita. 2001. Stress-induced apoptosis by heat shock, UV and γ-ray irradiation in zebrafish embryos detected by increased caspase activity and whole-mount TUNEL staining. *Fisheries Science* 67:333–340. doi: 10.1046/j.1444-2906.2001.00233.x

Zausig, J. and R. Horn. 1992. Soil water relations and aeration status of single soil aggregates, taken from a gleyic vertisol. *Zeitschrift für Pflanzenernährung und Bodenkunde* 155:237–245. doi: 10.1002/jpln.19921550314

Zhang, X., B. S. Rosenstein, Y. Wang, M. Lebwohl, D. M. Mitchell, and H. Wei. 1997. Induction of 8-oxo-7, 8-dihydro-2'-deoxyguanosine by ultraviolet radiation in calf thymus DNA and HeLa cells. *Photochemistry and Photobiology* 65:119–124. doi: 10.1111/j.1751-1097.1997.tb01886.x

CHAPTER 11

Annual Fishes (Rivulidae) from Southern Brazil
A Broad-Scale Assessment of Their Diversity and Conservation

Matheus Vieira Volcan, Ândrio Cardozo Gonçalves, Luis Esteban Krause Lanés, and Demetrio Luis Guadagnin

CONTENTS

11.1 INTRODUCTION

Human activities have changed most ecosystems through deforestation, land fragmentation, pollution, overexploitation of resources, and introduction of exotic species, resulting in habitat loss and a reduction in biodiversity (Groom et al., 2006). Among these ecosystems, aquatic environments and particularly freshwater systems are among the most affected because of their increasing reduction in recent decades and unprecedented levels of human disturbance (Santamaría and Klaassen, 2002; Saunders et al., 2002). Wetlands are considered among the most threatened environments, with estimates that 87% of the planet's natural wetlands have been lost since the early eighteenth century (Davison, 2014). Brazil has the largest area of wetlands in South America (Naranjo, 1995), and its current state of conservation is similar to the rest of the world.

Among wetlands, ephemeral wetlands (EW), that is, wetlands that undergo periods of drought, are the most affected. Usually, EW are relatively small and shallow bodies of water with an annual dry phase of variable duration that support unique communities of aquatic organisms (Collinson et al., 1995; Maltchik et al., 2003; Lanés et al., 2014b). Furthermore, EW usually occurs in patches, scattered in the landscape, and associated with rivers, streams, and permanent lagoons (Sanderson et al., 2005). The populations of these unique communities are small and isolated and are consequently more vulnerable to extinction (Gibbs, 2000). EW are highly vulnerable to human activities and are threatened in many regions of the planet; however, their biodiversity value is often overlooked, contributing to negligence and making their management inadequate (Semlitsch and Bodie, 1998; Snodgrass et al., 2000). Generally, these ecosystems are poorly known and most studies are limited to specific groups of organisms (e.g., Volcan et al., 2010, 2011a,b; Lanés et al., 2014a). Studies about the structure and function of these ecosystems, which are needed to understand and manage their sustainable use, are scarce (Burger, 2000; Lanés, 2009; Lanés et al., 2014a).

Neotropical annual fishes of the family Rivulidae are particularly affected by the loss and fragmentation of EW because they exclusively inhabit ephemeral environments (Costa, 2002a). The low ecological plasticity, the restricted area of distribution, and the low vagility of most species, coupled with the fact that the wetlands are being lost at an increasing pace, make annual fishes one of the most threatened vertebrate groups (Costa, 2002a; Rosa and Lima, 2008; ICMBio, 2013).

Rivulidae is among the most diversified families of Neotropical freshwater fishes, with about 350 described species; Brazil has the highest diversity, with about one-third of the species (Costa, 2008). Thirty species of Rivulidae are known from the state of Rio Grande do Sul, Brazil, of which 24 belong to the genus *Austrolebias*, five to *Cynopoecilus*, and one to the *Atlantirivulus* (Costa, 2002b, 2006; Costa and Lanés, 2009; Lanés et al., 2014a,b; Volcan et al., 2014a,b). Among these, only *Atlantirivulus riograndensis* is known to have a nonannual life cycle (Lanés et al., 2012); the conservation status of these species is similar to that of other annual fishes in Brazil. About 70% of species of threatened freshwater fishes in Rio Grande do Sul are Rivulidae (SEMA, 2014) and only *Cynopoecilus melanotaenia* is not considered threatened under IUCN criteria (SEMA, 2014).

The extinction of endemic fish species is increasingly directly proportional to the destruction of aquatic ecosystems. This also applies to Rivulidae, with several species on the edge of extinction (Rosa and Lima, 2008; Costa, 2009, 2012; ICMBio, 2013; SEMA, 2014). There are increasing efforts in Brazil to identify the types of impacts and ways to mitigate them and define actions and priority areas for biodiversity conservation (MMA/SBF, 2000; ICMBio, 2013). However, while conservation priorities are being developed, many species are lost (Costa, 2012).

In this chapter, we summarize current data about the diversity, distribution, risk of extinction, and conservation status of annual fish species collected between 2011 and 2014 in Rio Grande do Sul State, southern Brazil. The results are discussed in their relation to land use, climate change, management policies, actions, and conservation strategies.

11.2 MATERIAL AND METHODS

Study area. Rio Grande do Sul is the largest state in southernmost part of Brazil with a territory of 281,730 km² and about 3% of the country's area. Its economy is based mainly on the production of grain and livestock (SEPLAG, 2014). Currently, it is the fourth-largest economy of Brazil by gross domestic product (GDP), with 6.7% of the national GDP. The per capita GDP also has an advantaged position with values above the national average (SEPLAG, 2014).

Rio Grande do Sul has a subtropical climate with relatively well-distributed annual rainfall ranging from 1300 to 1800 mm (Köppen, 1931) and average temperatures between 15°C and 18°C with a minimum of −10°C in winter and a maximum of 40°C in summer (Radambrasil, 1986).

Rio Grande do Sul is part of two larger biomes, the northern half of the state consists of Atlantic Forest, which extends for approximately 37% of the territory, and the Pampa biome on the southern half of the state, which extends for approximately 63% of the territory. The Pampa is restricted to the Rio Grande do Sul in Brazil and is defined by a set of lowland vegetation that extends to Uruguay and Argentina (IBGE, 2004). The relief of Rio Grande do Sul has different geomorphological units that vary in altitudes and geology. The predominant formations are (1) the Escudo Cristalino (Crystalline Shield), consisting of granitic rocks that form rounded hills, with altitudes exceeding 300 m; (2) the Planalto Meridional, which consists of high altitudes reaching over 1000 m in some places; (3) the Depressão Central (Central Depression), which is a range of relatively low and flat or gently rolling land resembling a plain and is crossed from east to west by the state's most important river, the Jacuí River; and (4) the Planície Costeira (Coastal Plains), consisting of Atlantic Ocean coastal lowlands characterized by dry and seasonally and/or permanently flooded fields with the presence of scattered and numerous lagoons (SEPLAG, 2014).

11.2.1 Sampling

We summarize data collected in 50 municipalities across Rio Grande do Sul between 2011 and 2014. Each municipality was sampled for periods ranging from 1 to 3 days. We also use data from museums and studies in Rio Grande do Sul (e.g., Costa, 2002b, 2006; Ferrer et al., 2008; Volcan et al., 2010, 2011a,b) to locate areas with occurrence of annual fishes, as well as to analyze the presence of species in protected areas.

In order to define the sampling strategy, we analyzed satellite images from Google Earth (earth.google.com). Fieldwork was carried out by traveling the main highways to access previously selected areas; at every site we performed an active search for fish. We collected fish using hand nets (60×40 cm and 2-mm mesh), and to estimate the density of fish in each pond, we carried out 25 sweeps of 0.6 m², corresponding to approximately 15 m² of sampled area per studied pond. We used the catch per unit area (CPUA, N° Ind./m²) as a measure of fish density in each wetland.

Collected fish were euthanized in a eugenol (70 mg/L) bath and were subsequently fixed in 10% commercial formalin and transferred to alcohol 70%. Voucher specimens are deposited at the Museu de Ciência da Pontifícia Universidade Católica do Rio Grande do Sul (MCP—PUCRS) and at the Museu da Universidade Federal do Rio Grande do Sul (MUFRGS).

The area of each pond was estimated using a measuring tape for smaller ponds less than a hectare and applying the Google Earth program for larger areas. We visually assessed the types of impacts that affected every pond in relation to the occurrence of annual fishes. The following impacts were considered: cultivation of rice, soy, forestry, roads and highways, pollution (garbage disposal, discharge of domestic effluents, etc.), channeling and drainage, dams, wind farms, livestock, meadows, and the presence of exotic fishes.

The extent of occurrence (EOO) for each species was calculated by the distance between the limits of its distribution within the territory of Rio Grande do Sul using satellite imagery available in Google Earth. The EOO (in km²) is defined as the area contained within the shortest continuous imaginary boundary that can be drawn to encompass all the known, inferred, or projected points for the actual presence of a taxon (IUCN, 2014). To calculate the area of occupancy (AOO), we added the area of each pond to the occurrence of each species. The AOO is defined as the area that is occupied by a taxon within its extent of occurrence. This measure reflects the fact that a taxon does not usually occur throughout its extent of occurrence, which may contain unsuitable or unoccupied portions of habitats (IUCN, 2014). Currently this approach is applied in the revaluation of the red lists of threatened fauna for Rio Grande do Sul State (SEMA, 2014) and for Brazil (ICMBio, 2013). The categories of threat presented herein followed SEMA (2014) that used the data presented in this study for the evaluation of Rivulidae species in Rio Grande do Sul, with the exception of species recorded and/or described after the conclusion of the evaluation (Volcan et al., 2014a,b).

11.3 RESULTS

A total of 238 annual fish populations were sampled across 179 ponds (Figure 11.1). The sampled populations belong to 37 species, of which 31 belong to *Austrolebias* and six to *Cynopoecilus* (Table 11.1, Figure 11.2). Among them, five species are new to science and are currently under description and an additional three need additional analysis to confirm their taxonomic status.

Most recorded species are endemic to Rio Grande do Sul (22 species), 14 species also occur in Uruguay, and *A. alexandri* occurs in Rio Grande do Sul, Uruguay, and Argentina. Of the species previously evaluated for Rio Grande do Sul (SEMA, 2014), 66% are considered "Critically Endangered," 17% "Endangered," and 14% "Vulnerable." Only *C. melanotaenia* is not considered threatened and is listed as "Least Concern" (LC). The still undescribed species are likely to be listed/considered as "Critically Endangered," given that they have restricted distribution, few populations (Table 11.1), and a reduced quality of their environment.

Among the three hydrographic systems of Rio Grande do Sul, the Patos-Mirim lagoon system was the richest, with 86.5% of the recorded species, while the Uruguay River basin has 13.5% of the species, and the Tramandaí-Mampituba rivers system recorded only one species (2.7%; *C. multipapillatus*). Only *A. melanoorus* presented a disrupted distribution and was recorded in two hydrographic systems (Uruguay River and Patos-Mirim lagoon system).

Among the geomorphological formations sampled, the Planície Costeira and the Depressão Central showed the highest diversity, with 22 and 10 species, respectively; whereas the Planalto Merdional and the Escudo Cristalino have three and two species each, respectively.

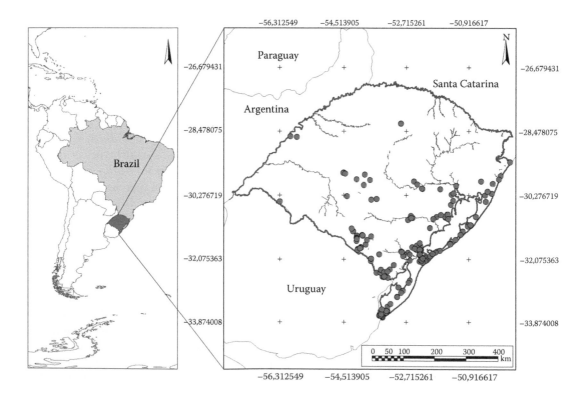

Figure 11.1 Map showing the 179 biotopes sampled between 2011 and 2014, distributed in 50 municipalities of Rio Grande do Sul, southern Brazil.

Table 11.1 List and Status of Threatened Species Recorded in Rio Grande do Sul, Southern Brazil, during the Samplings Carried Out between 2011 and 2014 and the Respective Numbers of Populations Registered, CPUA, Area of Occupancy (AOO in m²), and Extent of Occurrence (AEE in km²)

Species	N° pop.	CPUA	AOO	AEE	Status
Austrolebias adloffi (Ahl, 1922)	4	0,48	55,000	1,232	CR
Austrolebias aff. *gymnoventris* (Amato, 1986)	3	0,89	13,350	2,080	–
Austrolebias aff. *reicherti* (Loureiro and García, 2004)	1	1,32	4,000	12	–
Austrolebias alexandri (Castello and López, 1974)	2	1,12	10,800	78	CR
Austrolebias arachan (Loureiro, Azpelicueta and García, 2004)	2	0,58	1,150	6	CR
Austrolebias bagual (Volcan, Lanés and Gonçalves, 2014)	1	0,40	10,450	1	CR
Austrolebias charrua (Costa and Cheffe, 2001)	24	1,69	53,150	6,510	EN
Austrolebias cheradophilus (Vaz-Ferreira, Sierra de Soriano and Scaglia de Paulete, 1965)	2	0,20	7,500	2	CR
Austrolebias cyaneus (Amato, 1987)	4	1,43	3,750	780	CR
Austrolebias ibicuiensis (Costa, 1999)	3	1,33	3,550	9	CR
Austrolebias jaegari (Costa and Cheffe, 2002)	2	0,10	10,250	120	CR
Austrolebias juanlangi (Costa, Cheffe, Salvia and Litz, 2006)	19	0,61	74,120	6,900	CR
Austrolebias litzi (Costa, 2006)	6	0,73	6,128	1,800	CR
Austrolebias luteoflammulatus (Vaz-Ferreira, Sierra de Soriano and Scaglia de Paulete, 1965)	8	0,79	5,434	1,053	CR
Austrolebias melanoorus (Amato, 1986)	5	0,30	15,100	3,010	EN
Austrolebias minuano (Costa and Cheffe, 2001)	14	0,65	91,650	2,700	EN
Austrolebias nachtigalli (Costa and Cheffe, 2006)	9	2,41	30,470	3,400	CR
Austrolebias nigrofasciatus (Costa and Cheffe, 2001)	9	1,88	49,565	142	EN
Austrolebias paucisquama (Ferrer, Malabarba and Costa, 2008)	2	0,82	4,100	1	CR
Austrolebias periodicus (Costa, 1999)	5	0,92	17,100	16,296	EN
Austrolebias prognathus (Amato, 1986)	6	0,01	13,900	10	CR
Austrolebias quirogai (Loureiro, Duarte and Zarucki, 2011)	3	0,15	955	100	CR
Austrolebias sp. A	4	1,25	9,450	198	–
Austrolebias sp. B	4	0,85	250	4	–
Austrolebias sp. C	3	1,09	238	100	–
Austrolebias sp. D	2	3,70	6,450	9	–
Austrolebias sp. E	2	0,66	9,400	1	–
Austrolebias univentripinnis (Costa and Cheffe, 2005)	5	2,15	20,970	2,812	CR
Austrolebias varzeae (Costa, Reis and Behr, 2004)	2	0,58	10,500	2	CR
Austrolebias vazferreirai (Berkenkamp, Etzel, Reichert and Salvia, 1994)	2	0,16	1,600	60	CR
Austrolebias wolterstorffi (Ahl, 1924)	9	0,14	56,800	29,700	CR
Cynopoecilus aff. *melanotaenia* (Regan, 1912)	10	0,24	25,945	4,752	–
Cynopoecilus fulgens (Costa, 2002)	13	0,81	69,950	3,585	VU
Cynopoecilus intimus (Costa, 2002)	3	0,24	600	1,504	VU
Cynopoecilus melanotaenia (Regan, 1912)	32	0,58	171,203	18,100	LC
Cynopoecilus multipapillatus (Costa, 2002)	7	0,73	36,315	3,360	VU
Cynopoecilus nigrovittatus (Costa, 2002)	4	0,31	11,140	1900	VU

Note: The threat status is presented based on the assessment of fish species threatened with extinction in Rio Grande do Sul (SEMA, 2014), with the exception of *A. quirogai*, valued in Volcan et al. (2014a) and *A. bagual*, valued in Volcan et al. (2014b).
LC: least concern; VU: vulnerable; EN: endangered; CR: critically endangered.

Figure 11.2   Image of representative *Austrolebias* and *Cynopoecilus* species recorded in Rio Grande do Sul.
(a) *A. vazferreirai*, (b) *A. univentripinnis*, (c) *A. wolterstorffi*, (d) *A. jaegari*, (e) *A. nigrofasciatus*,
(f) *A. bagual*, (g) *A. ibicuiensis*, (h) *C. melanotaenia*.

Among the 50 sampled municipalities, those with the highest species richness were Jaguarão
(six), followed by Pelotas (five), Chui, Pedro Osório, and Rio Grande (four). Nine municipalities have
three species, four municipalities two species, and 24 others only one species (Table 11.2). Regarding
the occurrence of species by municipality, *C. melanotaenia* was recorded in eight municipalities, *C.
multipapillatus* in seven, *A. wolterstorffi* in six, and *A. juanlangi* and *A. periodicus* in five; 10 other
species were recorded in two and 11 species in a single municipality (Table 11. 2).

The species generally showed very restricted distribution, with EOO that ranged from about
1 km² for *A. bagual*, *Austrolebias* sp. E, and *A. paucisquama* to about 29,700 km² for *A. wolterstorffi*. However, most species (64%) showed areas less than 2000 km² of EOO. As for AOO, all
species showed an area less than 1 km², with areas ranging from 238 to 250 m² for *Austrolebias* sp.
C and *Austrolebias* sp. B, respectively, to 17,1203 m² for *C. melanotaenia*.

The areas of the ponds inhabited by rivulids ranged from 12,000 to about 15,000 m². Among
the 238 populations sampled, 59% were recorded in ponds with less than 2000 m² area and only
19% of them in areas more than 10,000 m² (Table 11.1). All sampled biotopes had some type of
environmental impact. About 36% of biotopes showed one type of impact, 41% two types, 16%
three types, 4.5% four types, and in 2.5% we observed up to five different types of impacts affecting a single pond. The most frequent impact was the presence of cattle (74.7%), followed by the
road effects/impact (49.4%), rice fields (16.7%), urbanization (13.6%), and exotic forestation (11.7%)

Table 11.2 Number of Municipalities of Occurrence and Distribution of Species Recorded in the 50 Municipalities of Rio Grande do Sul during Field Sampling Carried Out between 2011 and 2014

Species	Number	Municipalities of Occurrence
A. adloffi	3	Eldourado do Sul, Gravataí, Porto Alegre
A. aff. gymnoventris	3	Jaguarão, Pedro Osório, São Lourenço do Sul
A. aff. reicherti	2	Jaguarão, Piratini
A. alexandri	1	São Borja
A. arachan	1	Pedras Altas
A. bagual	1	Encruzilhada do Sul
A. charrua	3	Santa Vitoria do Palmar, Chuí, Rio Grande
A. cheradophilus	1	Jaguarão
A. cyaneus	2	Rio Pardo, Minas do Leão
A. ibicuiensis	2	São Vicente do Sul, São Pedro do Sul
A. jaegari	2	Pelotas, Capão do Leão
A. juanlangi	5	Aceguá, Bagé, Hulha Negra, Herval, Jaguarão
A. litzi	4	Dilermano de Aguiar, Santa Maria, São Gabriel, São Sepé
A. luteoflammulatus	2	Santa Vitoria do Palmar, Chuí
A. melanoorus	4	Bagé, Candiota, Herval, Hulha Negra
A. minuano	4	Mostardas, Rio Grande, São José do Norte, Tavares
A. nachtigalli	3	Jaguarão, Pedro Osório, Arroio Grande
A. nigrofasciatus	2	Pelotas, Capão do Leão
A. paucisquama	1	Caçapava do Sul
A. periodicus	5	Quaraí, Santana do Livramento, Dom Pedrito, Alegrete, Rosário do Sul
A. prognathus	1	Chuí
A. quirogai	1	Aceguá
Austrolebias sp. A	1	Encruzilhada do Sul
Austrolebias sp. B	1	Pedro Osório
Austrolebias sp. C	1	Pelotas
Austrolebias sp. D	1	Rio Grande
Austrolebias sp. E	1	São Lourenço
A. univentripinnis	1	Jaguarão
A. varzeae	1	Carazinho
A. vazferreirai	2	Aceguá, Bagé
A. wolterstorffi	6	Eldorado do Sul, Pelotas, Rio Grande, São José do Norte, Tavares, Gravataí
C. aff. melanotaenia	4	Arambaré, Barra do Ribeiro, Camaquã e Tapes
C. fulgens	3	Mostardas, São José do Norte, Tavares
C. intimus	3	Caçapava do Sul, São Gabriel e Santa Maria
C. melanotaenia	5	Arroio Grande, Camaquã, Capão do Leão, Chuí, Jaguarão, Pelotas, Rio Grande, Santa Vitoria do Palmar
C. multipapillatus	7	Capão da Canoa, Osório, Palmares do Sul, Pinhal, Santo Antônio da Patrulha, Torres, Barão do Triunfo
C. nigrovittatus	4	Guaíba, Eldorado do Sul, Porto Alegre, Gravataí

(Figure 11.3). Soybean cultivation, canals, dams, wind farms, pastures, and the presence of the exotic fish *Cyprinus carpio* represented less than 10% of the impacts observed in rivulid environments.

11.4 DISCUSSION

Our study recorded all species previously described or cited for Rio Grande do Sul. Additionally, we recorded six new species (one, *A. bagual*, was recently described, Volcan et al., 2014b, the

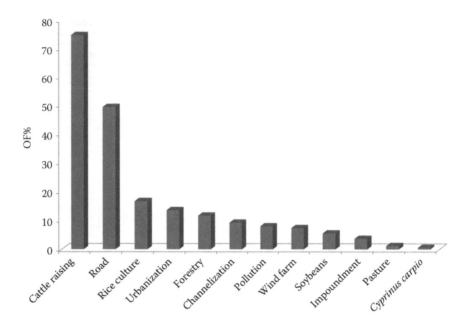

Figure 11.3 Frequency of occurrence (%) of different types of environmental impact present in biotopes with the presence of annual fishes in the Rio Grande do Sul, Brazil.

other five are under description), and three others may represent new taxa. Also, *A. cheradophilus*, *A. quirogai*, and *A. arachan* represent the first report for Brazil (Lanés et al., 2013, 2014a; Volcan et al., 2014a). Furthermore, the study expands the distribution for all species previously recorded for Rio Grande do Sul.

These results are extremely important for conservation strategies, especially for those species with few populations and under greater decline and/or extinction threat. The data obtained over a relatively short period of time represent a significant increase in the number of species and populations recorded for Rio Grande do Sul.

We observed that most species of annual fishes in Rio Grande do Sul live in small, isolated, and restricted populations usually in areas with high anthropogenic use. This threat most commonly relates to the expansion of agricultural frontiers and growth of urban areas, a pattern that has already been highlighted in various species in Rio Grande do Sul (Costa, 2002a; Reis et al., 2003; Ferrer et al., 2008; Volcan et al., 2011a,b, 2014a,b; Lanés et al., 2013, 2014a).

The high degree of threat and loss of EW and annual fish populations is not exclusive to Rio Grande do Sul. A previous study in Rio de Janeiro found that 90%–99.6% of the biotopes of five species of annual fishes was lost over the previous 17 years of monitoring and left the species on the limit of extinction (Costa, 2012). Similar problems were reported for other Brazilian regions and biomes and numerous other species (Costa, 2002a, 2009; Rosa and Lima, 2008; ICMBio, 2013). In the Atlantic Forest, the species occur in forest areas and the problems arise from deforestation. In the Amazon, programs of energy generation (hydropower plants) and building of large dams threaten several endemic species (ICMBio, 2013). Similar conservation problems also affect neighboring countries such as Uruguay, where annual fishes are considered the most threatened fish group in the country (Loureiro et al., 2013). However, regardless of the type of impact, biome, and country, the main threat to annual killifishes is the loss of their habitat.

Below we discuss the main land uses and major threats that affect and/or threaten the rivulid populations in southern Brazil (Rio Grande do Sul state), as well as actions, policies, and strategies for the conservation of the annual fish species.

11.4.1 The Effect of Cattle

We observed that, although livestock are found in most ponds as a result of cattle farming, which is usually performed extensively in Rio Grande do Sul, they seem to have no significant effect on rivulid populations and environmental quality. The ecological processes that maintain biodiversity in grasslands are very different from those in forest ecosystems (Pilar and Velez, 2010). In ecosystems where the climate favors woody vegetation, ecological disturbances such as grazing are important for the maintenance of the temporal and spatial of physiognomy of grasses (Bond and Parr, 2010; Pillar and Velez, 2010). Livestock are not necessarily detrimental to the communities, and in some cases they contribute to the maintenance of native plants and aquatic diversity in temporary ponds (Marthy, 2005). Therefore, cattle may play an important role in maintaining the biodiversity of EW (Collins et al., 1998; Harrison, 1999; Maestas et al., 2003), mainly in the southern fields, which is an area that historically has been a habitat for now extinct large herbivores (Lopes, 2005); cattle may be the functional equivalent of these extinct species. However, the high densities of cattle can result in deleterious effects on the aquatic community, particularly in small EW. Unfortunately, there is no data available that correlates the presence and density of cattle with the occurrence and abundance of annual fishes.

Our observations at the type locality of *A. jaegari* suggest that the absence of cattle is apparently affecting the biotope and its population, since the site is dominated by dense vegetation and reduced water surface (Figure 11.4). The removal of cattle has altered the ecological succession due to the absence of grazing and caused the disappearance of the population. Over the past four years, we recorded a few individuals of *A. jaegari* restricted to a small man-made canal, next to a pond by the access road (Figure 11.4), and that the species was no longer present at the original pond. However, some studies suggest a relationship between grazing and the increasing diversity in EW (Paine, 1966), positing that predators prevent the monopolization of resources by dominant species, causing grazed areas generally to exhibit greater diversity than areas not grazed (e.g., Marty, 2005).

Figure 11.4 Images from the type locality of *A. jaegari* in Pelotas obtained over 12 years. The 2002 photo was obtained when the biotope still had grazing cattle and water lamina can be verified. Over the following years, the advancement of grassy and shrubby vegetation and the absence of water lamina were observed. In the image at the bottom right the small canal beside the road, where the occurrence of the species was restricted, can be seen.

It is likely that the management of this locality with the presence of cattle (at appropriate densities) could reduce the density of vegetation and assist in the recovery of environmental conditions for the occurrence of *A. jaegari*. Grazing can prevent shrub/woody invasion and increase herbaceous diversity (Bond and Parr, 2010). Moreover, the invasion of woody plants and forest expansion are promoted when the disturbance ends (Duarte et al., 2006), which is consistent with our observations over 12 years of monitoring the pond (M.V. Volcan et al., unpublished data).

11.4.2 Occurrence Near Roads and Highways

We observed that a considerable number of ponds are located marginal to roads and highways. Probably these areas are remnants of historically larger natural areas. Furthermore, areas near roads are under government administration, and theoretically they cannot be used for agriculture or urbanization. Although these ponds are "protected," they are subject to impacts from expansion, duplication and paving of roads, and accidents with toxic cargos. All known populations of *A. ibicuiensis*, *A. bagual*, *A. arachan*, and *Austrolebias* sp. E. were found by roads. Roads also increase isolation of populations, preventing or hindering gene flow. In order to prevent the destruction of ponds and to reduce the impact caused by roads, environmental licensing should assess the occurrence of annual fishes to reduce their impact.

11.4.3 The Impact of Rice Cultivation

The presence of only 17% of biotopes under the direct influence of rice cultivation disguises the alarming fact that most EW has already been lost through the grain cultivation. Rio Grande do Sul is the largest rice producer in Brazil (representing about two-thirds of the national production). The prediction for 2014–2015 is that 1,116,458 hectares will be seeded for grain (IRGA, 2014). A conservative estimate points out that less than 10% of the original wetlands areas remain in the state (Guadagnin and Maltchik, 2007; Stenert and Maltchik, 2007; Maltchik et al., 2014), most of it lost to grain cultivation. Rice cultivation has a greater impact on the Planície Costeira, Depressão Central, and western part of the Planalto Meridional, which are lower and flatter areas and thus more favorable for cultivation. These areas coincide with areas of higher grain production in the state and also with areas of higher population density for annual fishes. Most populations of known species have been impacted by grain cultivation, which occurs without management or correct environmental licensing. Besides the degradation caused directly by cultivation, draining of ponds, leveling of soil, and use of fertilizers and pesticides, rice cultivation in Rio Grande do Sul is carried out during the period when ponds should be dry, that is, it has an inverse hydrological cycle to EW characteristic of annual fishes; this reality enhances the harmful effects of grain production on native species.

In the past, wetlands environments were considered economically unproductive areas, unhealthy and required to be "sanitized." "Sanitation" meant conversion into a different environment by draining, grounding, or any other means of elimination. Thus, from the 1960s, much of the wetlands was drained for agricultural use through the encouragement of the Pró-Várzea program of the federal government (Burger, 2000). According to FZB (2002), the state originally had 5.3 million hectares of wetlands. In 1986, a flight inspection over the wetlands of the state noticed that, except in specific localities, most environments were highly altered and degraded, mainly due to rice cultivation (FZB, 2002). These environments are still suffering from many impacts and their total area continues to be drastically reduced.

11.4.4 Urbanization

Urbanization has an important impact on the environment, especially in larger centers and development areas, where processes are further accelerated. In an annual monitoring between 2011 and 2014 of 59 EW in the municipality of Pelotas and surroundings, we recorded a loss of about 8% of EW due to the growth of the urban area. This data is extremely worrying for *Austrolebias* sp. C., which was recorded in four ponds, of which two were lost during this period and the long-term viability of other two is critical. Similarly, in the Pedro Osório municipality over the same period, 25% of ponds occupied by *Austrolebias* sp. B were lost, and the remaining species are threatened by further urbanization. Other species directly impacted by the growth of the urban areas in the state are: *A. luteoflammulatus*, *A. prognathus*, and *A. charrua* in Chuí; *A. minuano* in Rio Grande and São José do Norte; *A. nigrofasciatus* in Pelotas; *A. wolterstorffi* in Porto Alegre, Gravataí, Pelotas, Rio Grande, and São José do Norte; *A. cheradophilus*, *A. nachtigalli*, and *A. univentripinnis* in Jaguarão; as well as all species of *Cynopoecilus* across the state. The intensity of the threat to the species differs in each region and is usually related to local environmental characteristics and economic development.

Urbanization is a problem for the overall conservation of Rivulidae, affecting species in different biomes and groups (e.g., Costa, 2002a, 2009, 2012; Rosa and Lima, 2008; Costa et al., 2012). The allocation of protected areas on coastal regions and within urban centers, such as ecological parks, would be a high priority for conservation of annual fishes; municipal master plans should include wetlands areas. The allocation of resources resulting from environmental compensation for the acquisition and protection of sites of occurrence of Rivulidae is another important action.

11.4.5 Climatic Changes

Global climate change may alter various hydrological parameters on which the wetlands and the species that inhabit them depend (IPCC, 2007). Rising temperatures caused by climate change will result in increasingly frequent prolonged droughts and more restricted and shorter rains in the Pampa biome (MMA, 2011), thus increasing the vulnerability of temporary aquatic ecosystems. The EW are usually small and shallow, respond quickly to environmental changes, and are considered particularly vulnerable to climate change because of their fragility and limited adaptability. Changes in the geographic range of the species are also expected, causing part of the current distribution of many species to become inadequate. This is more critical, since rivulid species have low vagility and no ability to migrate to more favorable environments. The periods of prolonged drought and increased temperatures may affect the entire life cycle of the species. Many biological processes are strictly controlled by temperature. Therefore, a change in thermal regime (e.g., extreme temperatures, duration, and seasonal rates of change) can directly regulate physiological characteristics such as growth rates, behavioral performance, and habitat preference. In annual fishes, temperature affects most life history traits, such as embryonic development (Arenzon et al., 2002; Arezo et al., 2007), growth (Liu and Walford, 1970; Errea and Danulat, 2001; Fonseca et al., 2013), reproduction (Arenzon et al., 1999; Volcan et al., 2013b), and lifespan (Liu and Walford, 1966, 1970).

Effects caused by global warming, such as rising sea levels and consequently more frequent floods and rising tides may have significant effects on species inhabiting coastal areas, for example, *C. melanotaenia*, *C. fulgens*, *C. multipapilatus*, *A. wolterstorffi*, and *A. minuano*. Changes in rainfall patterns may also affect the population biology of annual fishes as well as the hydrological cycles of the ponds. The increase in the intensity and frequency of El Niño events already has caused droughts and/or floods in many parts of the Southern Hemisphere (Ramsar, 2002). Thus, the more frequent occurrence of these events may significantly affect populations of annual fishes. During two El Niño events in Rio Grande do Sul (2010 and 2014, V. Matheus, personal observation), we observed that many ponds with confirmed previous occurrence of Rivulidae lacked annual fishes or they were present at low density. Monitoring of some areas in the southern coastal region of the state indicates that during El Niño events, the EW flooded early or simply did not dry up during the summer and had high abundance of nonannual fishes. However, the factors that led to absence or low abundance of annual fishes during these events need further investigation.

11.4.6 Assessment According to the Criteria of the IUCN and Endangered Species

Several methods have been proposed to assess the conservation status of plants and animals. The largest classification system used internationally was adopted and developed by IUCN and offers categories to indicate the degree of threat for the species in their habitats.

The results of AOO and the number of registered populations for some species, for example, *A. charrua*, *A. minuano*, and *Cynopoecilus* spp., may be underestimated, since we did not sample every single pond within the limits of species distribution. Nevertheless, we believe that even if all ponds were sampled, this would hardly change the threat status of the species, since, under IUCN criteria and as noted in the samplings, no species has AOO greater than 2000 km² or AEE higher than 20,000 km² (except *A. wolterstorffi*). Moreover, all species have severely fragmented environments and continued decline in area of occupancy, quality of habitat, and number of locations or subpopulations, and may be included in both criteria B and D.

As annual fishes occur in small and isolated patches, the measurement of the area of each pond coupled with the sum of their areas provide an estimate of AOO for each species. As most species occur in degraded and fragmented landscapes, with continued loss and reduced quality of habitat, the use of the criterion B2 is the most suitable for the evaluations. Furthermore for more widely distributed species with a possibility of additional unknown populations, the D2 criterion is more applicable, since they could have a larger area than estimated but never with an AOO exceeding 20 km².

Although most of the species recorded in Rio Grande do Sul are endemic, a considerable proportion is also shared with neighboring countries such as Uruguay and Argentina. This situation requires collective international efforts in order to assess the conservation status of species globally, as well as the requirement of proposed actions and international cooperation to assist in the conservation of these species. Until now, no recorded species in Rio Grande do Sul has had global assessment (IUCN, 2014). Despite national and regional listings, which indicate that the majority of the Rivulidae species are threatened with extinction (e.g., ICMBio, 2013; Loureiro et al., 2013; SEMA, 2014), only 18 species are included in any category of threat of the IUCN Red List of Threatened Species (IUCN, 2014).

Independent of the type of threat, size of ponds, number of recorded populations, density of populations, and other factors, we consider *A. jaegari*, *A. bagual*, *A. ibicuiensis*, *A. varzeae*, *Austrolebias* sp. B, *Austrolebias* sp. C, *A. prognathus*, *A. cheradophilus*, *A. quirogai*, and *A. arachan* to be among the most critically threatened species in Rio Grande do Sul. Such species need urgent conservation efforts, due to extremely restricted geographical areas, low-density populations, and imminent threat. Conservation efforts and the allocation of areas for conservation in the Rio Grande do Sul should primarily consider biotopes of these species to ensure their conservation.

11.4.7 Presence in Protected Areas (PAs)

Records of annual fishes in protected areas of Rio Grande do Sul are extremely scarce and scattered throughout the literature (Buckup and Malabarba, 1983; Koch et al., 2000; Costa, 2002a,b; Reis et al., 2003; Dufech and Fialho, 2009; Costa and Lanés, 2009; Lanés and Maltchik, 2010; Lanés, 2011; Volcan et al., 2013a). In the 104 registered PAs in the state, only 15 populations of annual fishes are recorded, covering only 10 species (Lanés, 2011; Table 11.3). Despite the fact that Rivulidae are the most threatened freshwater fish in the state (SEMA, 2014), no PA was created with the purpose of protecting some species. The low representation of annual fishes in protected areas (PAs; only 24% of the species are recorded in PAs) is extremely worrying, since all species are threatened with extinction (except *C. melanotaenia*) and current population declines due to loss and fragmentation of their biotopes. Moreover, the records of Rivulidae in these PAs are uncommon,

Table 11.3   List of Annual Fish Species Recorded in Protected Areas in the State of Rio Grande do Sul, Southern Brazil

Species	Conservation Units
A. adloffi	Parque Estadual do Delta do Jacuí (Reis et al., 2003).
A. alexandri	Reserva Biológica do Banhado São Donato (Reis et al., 2003).
A. charrua	Estação Ecológica do Taim (unpublished data).
A. minuano	Parque Nacional da Lagoa do Peixe (Lanés, 2011; Lanés et al., 2014a).
A. nigrofasciatus	RPPN do Pontal da Barra (Reis et al., 2003; Volcan et al., 2013a,b).
A. periodicus	APA do Ibirapuitã (Bertaco and Azevedo, 2013).
A. wolterstorffi	Parque Estadual do Delta do Jacuí (Reis et al., 2003), Parque Nacional da Lagoa do Peixe (Lanés and Maltchik, 2010), Estação Ecológica do Taim (unpublished data), RPPN do Pontal da Barra (Reis et al., 2003; Volcan et al., 2013a,b).
C. melanotaenia	RPPN do Pontal da Barra (Volcan et al., 2013a,b), Estação Ecológica do Taim (Costa, 2002a,b; unpublished data).
C. fulgens	Parque Nacional da Lagoa do Peixe (Lanés, 2011; Lanés et al., 2014a).
C. nigrovittatus	Parque Estadual do Delta do Jacuí (Koch et al., 2000), Parque Estadual de Itapuã (Dufech and Fialho, 2009).

and little is known about population parameters, conditions, and representation of species in these areas and how they are effective for the conservation of annual fishes. Therefore, it is essential to conduct studies aimed at mapping and monitoring populations to verify the presence of the species in protected areas and to undertake searches in PAs where specific inventories of annual fishes have not yet been conducted.

11.4.8 Priority Areas for Conservation of Annual Fishes in Rio Grande do Sul

Despite the anthropogenic change scenario, areas with great potential for the conservation of populations of annual fishes were also identified. This is especially important in the Pampa biome, which has the lowest representation in the National System of Protected Areas (NSPA) and is considered the second most devastated biome in the country after the Atlantic Forest. Beyond the protected areas already established in Rio Grande do Sul, 107 other areas were identified as priorities for biodiversity conservation of the Pampa biome in the country (MMA, 2007); among these areas are important and significant remaining wetlands, which harbor unique communities and species, including annual fishes.

We highlight the following floodplain areas as a priority for the conservation of annual fishes: the Canal São Gonçalo in Pelotas and Rio Grande (*A. nigrofasciatus, A. wolterstorffi,* and *C. melanotaenia*), the Bolaxa stream in Rio Grande (*A. minuano, A. wolterstorffi,* and *C. melanotaenia*), the Chuí stream in the homonym municipality (*A. charrua, A. luteoflammulatus, A. prognathous,* and *C. melanotaenia*), the Del Rey stream in Santa Vitória do Palmar (*A. charrua* and *C. melanotaenia*), the low course of Jaguarão River in the homonym municipality (*A. cheradophilus, A. nachtigalli, A. univentripinnis, A. juanlangi,* and *C. melanotaenia*), the Várzea River in Carazinho (*A. varzeae*), the Toropi River in São Pedro do Sul (*A. ibicuiensis*), the Jacuí and Guaíba rivers (*A. adloffi, A. wolterstorffi,* and *C. nigrovittatus*), the Jaguarão-Chico River in the municipalities of Aceguá and Pedras Altas (*A. juanlangi, A. quirogai, A. arachan,* and *A. vazferreirai*), the Francisquinho and Lajeadinho streams in Minas do Leão (*A. cyaneus*), the low course of the Piratini River in Pedro Osório (*Austrolebias* sp. B), the Abranjo stream in Encruzilhada do Sul (*A. bagual*), the Vacacaí River in Santa Maria (*A. litzi* and *C. intimus*), the Candiota and Seival streams in Candiota (*A. melanoorus*), and the Santa Maria River in Dom Pedrito (*A. periodicus*). The creation of PAs in these locations could ensure conservation of most species registered in the territory of Rio Grande do Sul.

11.4.9 Conservation Actions and Environmental Policies

Until now few efforts have been made to assist the conservation of rivulids. Government actions such as the construction of lists of threatened fauna has already been performed for more than 20 years in Rio Grande do Sul (Marques et al., 2002; Fontana et al., 2003; SEMA, 2014) and are part of the strategy adopted in Brazil and worldwide to support policies and planning projects, environmental management, and biodiversity monitoring. These listings recognize the vulnerability of species and provide measures for their conservation.

Although listings have considered most rivulid species as threatened (Reis et al., 2003; Rosa and Lima, 2008; SEMA, 2014) and conservation efforts have provided evidence for many species, this has not been enough to prevent the loss of many ponds and populations. That is, although the species and the environments in which they live are protected by national and state laws, it seems that even with all efforts to disseminate information about endangered species and environments in Rio Grande do Sul, there is still negligence on the part of the licensing bodies about the importance of conserving wetlands for biodiversity.

Because of the high degree of threat to most species, the National Action Plan for the Conservation of Rivulid Fishes (PAN; ICMBio, 2013) was recently launched in Brazil, with its main goal "to establish mechanisms for protecting annual fishes and void the loss of habitat of the focal species in the next five years." In this plan, focal species were defined as those with greater urgency for conservation efforts in Brazil and those with greater genetic diversity (to protect more basal phylogenetic groups), giving priority to sympatric species. In total, 64 species were included in this plan, of which 20 were focal species. Of the species recorded in Rio Grande do Sul, 12 were included in the plan, six being considered focal (ICMBio, 2013).

The inclusion of a species in an action plan does not ensure its conservation and does not guarantee that the goals proposed will be achieved, but the PAN serves as a guidance for companies, institutions, environmental compensation, and public agencies that want to finance projects focused on biodiversity conservation. In this sense, several funding sources are already giving higher priority to projects that address focal species to meet the specific demands of national action plans. Thus, although the PAN does not have specific resources to achieve its goals, it has been an important tool to raise funds and awareness for conservation projects in Brazil.

Another effort to protect rivulid fishes in Rio Grande do Sul was the publication of Environmental Zoning for Activity of Silviculture (ZAS, 2008), which considered wetlands as priority environments and species of annual fishes as one of the target groups for management and activity monitoring. Furthermore, it suggests the mapping of wetlands to prevent their loss and suppression. Currently, a similar zoning exercise is in preparation for the wind power projects in the state. However, the actions that were most effective in ensuring the viability of any given population or species of rivulids, such as the allocation of protected areas and specific management plans for EW, are unknown. In the wind farms, for example, despite several ponds being lost or degraded by their creation and despite the existence of several documents protecting species of rivulids, no area was acquired with the intention of conserving annual fishes as a form of environmental compensation.

Thus, despite relatively recent conservation efforts, the advance of the loss of habitat of the species has been faster than the actions to conserve them. The lack of oversight, billing, and technical knowledge exhibited by environmental agencies and the existence of government pressure to release the projects and not prevent "economic development" of a region has been the main argument used by decision makers to justify the impacts caused to natural environments. Thus, it has become necessary to implement the creation of policies to encourage research in these environments addressing monitoring, dynamics, and evaluation of key impacts; the development of environmental education programs; development and implementation of national policies and strategies for conservation;

development of sustainable use programmes; and creation of conservation units (Burger, 2002; Lanés, 2011; ICMBio, 2013; Lanés et al., 2014a,b; Volcan et al., 2014a,b).

As in other grasslands of South America, the conservation of the Pampa biome has been neglected, since it has historically received less attention in the conservation agenda of Brazil (Overbeck et al., 2007). In Rio Grande do Sul, only 0.15% of the total area covered by pasture is included in strictly protected areas (Devely et al., 2008). As most species of rivulids occur in the grasslands environment, the creation of new protected areas in the Pampa could ensure the viability and conservation of various species.

Currently, the wetlands and its species have become more valued, but unfortunately, the new concepts still do not seem to have reached the decision makers, politicians, media, and the vast majority of the population (Burger, 2000; Junk et al., 2014). According to Loyola (2014), Brazil is an international environmental leader that has conducted important negotiations to define sustainable development goals over the last decade. However, in terms of national policy, Brazil is always making decisions that go against the global policies that it ratifies, such as the revision of Brazil's Forest Code, which is the main environmental legislation in the country, and that represented a setback for conservation and protection of natural environments, reducing the area covered by some CUs and permanent preservation areas. Among other consequences, these controversial actions taken by national leaders could increase CO_2 emissions and boost a massive loss of species, with irreversible consequences for ecosystems (Loyola, 2014). The problem is even greater because there is not enough political pressure or environmental awareness on the part of a considerable portion of the population to demand an effective implementation of a system of PAs, as well as changes in environmental policy in Brazil, in order to assist in the conservation of species and their native environments.

11.5 CONCLUSION

Rio Grande do Sul has a great diversity of annual fishes, but the majority of the species are endangered. Conservation of annual fishes is an issue that transcends the boundaries of Rio Grande do Sul, because beyond the state, some species are shared with Uruguay and Argentina, so the impacts resulting from habitat loss of the species are reflected not only in the state but in all regions where human beings interact with EW. Conservation strategies for the species, in turn, depend on a number of factors and initiatives, but in general are related to the incentive to research, search for and map species, define taxonomic problems, correct environmental licensing, implement environmental education programs and an incentive plan, and encourage the creation of CUs. Only through collective efforts can the reality of imminent extinction for most species be altered, mostly because legal competence and issues related to the preservation and management of wetlands and endangered species belong to local, state, and federal levels, but above all, because it is the duty of the population as a whole. Finally, we expect our data to draw the attention of the scientific community, politicians, decision makers, and the general public to the importance of conserving annual fishes and their habitats. Thus, the support of the government and society is an urgent requirement to create protected areas for the protection of annual fishes in Rio Grande do Sul and in southern Brazil.

ACKNOWLEDGMENTS

We thank Michel Corrêa for providing the map of the study area. This study was funded by Fundação Grupo Boticário de Proteção à Natureza as part of the project "Peixes Anuais do Pampa." We thank ICMBio by issuing a collecting license (#18334). M.V.V. thanks CAPES for the PhD Grant. CNPq provides research fellowship to D.L.G. (309298/2009-1).

REFERENCES

Arenzon, A., C.A. Lemos, and M.B.C. Bohrer. 2002. The influence of temperature on the embryonic development of the annual fish *Cynopoecilus melanotaenia* (Cyprinodontiformes: Rivulidae). *Brazilian Journal of Biology* 62: 743–747.

Arenzon, A., A.C. Peret, and M.B.C. Bohrer. 1999. Reproduction of the annual fish *Cynopoecilus maelanotaenia* (Regan 1912) based on a temporary water body population in Rio Grande do Sul State, Brazil. *Hydrobiologia* 411: 65–70. doi: 10.1023/A:1003868711295

Arezo, M.J., S. D'Alessandro, N.G. Papa, R.O. de Sá, and N. Berois. 2007. Sex differentiation pattern in the annual fish *Austrolebias charrua* (Cyprinodontiformes: Rivulidae). *Tissue and Cell* 39: 89–98. doi: 10.1016/j.tice.2007.01.004

Bertaco, V.A., and M.A. Azevedo. 2013. Fishes from Rio Ibirapuitã basin, Environmental Protection Area of Ibirapuitã, Pampa Biome. *Check List* 9: 966–972.

Bond, W.J., and C.L. Parr. 2010. Beyond the forest edge: Ecology, diversity and conservation of the grassy biomes. *Biological Conservation* 143: 2395–2404. doi: 10.1016/j.biocon.2009.12.012

Buckup, P.A., and L.R. Malabarba. 1983. A list of the fishes of the Taim Ecological Station, Rio Grande do Sul, Brazil. *Iheringia. Série Zoologia* 63: 103–114.

Burger, M.I. 2000. *Situação e ações prioritárias para conservação de banhados e áreas úmidas da Zona Costeira.* http://www.anp.gov.br/brnd/round6/guias/perfuracao/perfuracao_r6/refere/banhados.pdf. Accessed on 20 September 2014.

Collins, S.L., A.K. Knapp, J.M. Briggs, J.M. Blair, and E.M. Steinauer. 1998. Modulation of diversity by grazing and mowing in native tallgrass prairie. *Science* 280: 745–747. doi: 10.1126/science.280.5364.745

Collinson, N.H., J. Biggs, A. Corfield et al. 1995. Temporary and permanent ponds: An assessment of the effects of drying out on the conservation value of aquatic macroinvertebrate communities. *Biological Conservation* 74: 125–133. doi: 10.1016/0006-3207(95)00021-U

Costa, W.J.E.M. 2002a. *Peixes Anuais Brasileiros: Diversidade e Conservação.* Curitiba, ed. UFPR. 238p.

Costa, W.J.E.M. 2002b. The annual fish genus *Cynopoecilus* (Cyprinodontiformes, Rivulidae): Taxonomic revision, with descriptions of four new species. *Ichthyological Exploration of Freshwaters* 13: 11–24.

Costa, W.J.E.M. 2006. The South American annual killifish genus *Austrolebias* (Teleostei: Cyprinodontiformes: Rivulidae): Phylogenetic relationships, descriptive morphology and taxonomic revision. *Zootaxa* 1213: 1–162.

Costa, W.J.E.M. 2008. *Catalog of Aplocheiloid Killifishes of the World.* 1st edn. Rio de Janeiro: Reproarte. 127pp.

Costa, W.J.E.M. 2009. *Peixes Aplocheiloideos da Mata Atlântica: História, diversidade e conservação.* Rio de Janeiro, ed. UFRJ. 172p.

Costa, W.J.E.M. 2012. Delimiting priorities while biodiversity is lost: Rio's seasonal killifishes on the edge of survival. *Biodiversity and Conservation* 21: 2443–2452. doi: 10.1007/s10531-012-0301-7

Costa, W.J.E.M., P.F. Amorim, and L.O. Mattos. 2012. Species delimitation in annual killifishes from the Brazilian Caatinga, the *Hypsolebias flavicaudatus* complex (Cyprinodontiformes: Rivulidae): Implications for taxonomy and conservation. *Systematics and Biodiversity* 10: 71–91. doi: 10.1080/14772000.2012.664177

Costa, W.J.E.M., and L.E.K. Lanés. 2009. *Rivulus riograndensis*, a new aplocheiloid killifish from southern Brazil (Cyprinodontiformes: Rivulidae). *Ichthyological Exploration of Freshwaters* 20: 91–95.

Davison, N. 2014. How much wetland has the world lost? Long-term and recent trends in global wetland area. *Marine and Freshwater Research* 65: 934–941. doi: 10.1071/MF14173

Devely, P.F., R.B. Setubal, R.A. Dias, and G. Bencke. 2008. Grasslands bird and biodiversity conservation aligned with livestock production. *Revista Brasileira de Ornitologia* 16: 308–315.

Duarte, L.D.S., M.M.G. Dos-Santos, S.M. Hartz, and V.D. Pillar. 2006. Role of nurse plants in Araucaria Forest expansion over grassland in south Brazil. *Austral Ecology* 31: 520–528. doi: 10.1111/j.1442-9993.2006.01602.x

Dufech, A.P.S., and C.B. Fialho. 2009. Estudo comparado da taxocenose de peixes em dois ambientes aquáticos do Parque Estadual de Itapuã, sul do Brasil. *Iheringia. Série Zoologia* 99: 177–188.

Errea, A. and E. Danulat. 2001. Growth of the annual fish, *Cynolebias viarius* (Cyprinodontiformes), in the natural habitat compared to laboratory conditions. *Environmental Biology of Fishes* 61: 261–268. doi: 10.1023/A:1011047611009

Ferrer, J., L.R. Malabarba, and W.J.E.M. Costa. 2008. *Austrolebias paucisquama* (Cyprinodontiformes: Rivulidae), a new species of annual killifish from southern Brazil. *Neotropical Ichthyology* 6: 175–180.

Fonseca, A.P., M.V. Volcan, L.A. Sampaio, L.A. Romano, and R.B. Robaldo. 2013. Growth of critically endangered annual fish *Austrolebias wolterstorffi* (Cyprinodontiformes: Rivulidae) at different temperatures. *Neotropical Ichthyology* 11: 837–844.

Fontana, C.S., G.A. Bencke, and R.E. Reis. 2003. *Livro vermelho da fauna ameaçada de extinção no Rio Grande do Sul*. Porto Alegre: Edipucrs. 632p. ISSN 0100-5363.

FZB. 2002. Fundação Zoobotânica do Rio Grande do Sul. *Mapeamento, Diagnóstico e Gerenciamento de Ambientes de Áreas Úmidas na Bacia do Guaíba, tendo em vista sua Preservação ou Conservação. Pró-Guaíba—Subprograma Parques e Reservas—Projeto II*. Estudos para consolidação do Sistema de Parques e Reservas Naturais no Bacia do Guaíba. Porto Alegre. 189p.

Gibbs, J.P. 2000. Wetland loss and biodiversity conservation. *Conservation Biology* 14: 314–317. doi: 10.1046/j.1523-1739.2000.98608.x

Groom, M.J., G.K. Meffe, and C.R. Carroll. 2006. *Principles of Conservation Biology*, 3rd edn. Sinnauer Associates, Sunderland.

Guadagnin, D.L., and L. Maltchik. 2007. Habitat and landscape factors associated with neotropical waterbird occurrence and richness in wetland fragments. *Biodiversity and Conservation* 16: 1231–1244. doi: 10.1007/s10531-006-9127-5

Harrison, S. 1999. Native and alien species diversity at the local and regional scales in a grazed California grassland. *Oecologia* 121: 99–106. doi: 10.1007/s004420050910

IBGE. 2004. Instituto Brasileiro de Geografia e Estatística. Mapa da vegetação do Brasil e Mapa de Biomas do Brasil. www.ibge.gov.br. Accessed on December 2013.

ICMBio. 2013. Instituto Chico Mendes de Conservação da Biodiversidade. Sumário Executivo do Plano de Ação Nacional para a Conservação dos Peixes Rivulídeos Ameaçados de Extinção. Brasília. http://www.icmbio.gov.br/portal/images/stories/docs-plano-de-acao/pan-rivulideos/sumario-executivo-rivulideos.pdf. Accessed on April 2013.

IPCC. 2007. *Climate Change 2007: Synthesis Report*. Contribution of Working Groups I, II and III to the Fourth Assessment Report of the Intergovernmental Panel on Climate.

IRGA. 2014. Instituto Rio Grandense do Arroz. Secretaria de Agricultura, Pecuária e Agronegócio. http://www.irga.rs.gov.br. Accessed on October 2014.

IUCN. Standards and Petitions Subcommittee. 2014. Guidelines for Using the IUCN Red List Categories and Criteria. Version 11. Prepared by the Standards and Petitions Subcommittee. http://www.iucnredlist.org/documents/RedListGuidelines. Accessed on March 2014.

Junk, W.J., M.T.F. Piedade, R. Lourival et al. 2014. Brazilian wetlands: Their definition, delineation, and classification for research, sustainable management, and protection. *Aquatic Conservation: Marine and Freshwater Ecosystems* 24: 5–22. doi: 10.1002/aqc.2386

Koch, W.R., P.R. Milani, and K.M. Grosser. 2000. *Guia Ilustrado: Peixes Parque Delta doJacuí*. Porto Alegre, Fundação Zoobotânica do Rio Grande do Sul, 91p.

Köppen, W. 1931. *Climatologia*. Fundo de Cultura Econômica, México.

Lanés, L.E.K. 2011. *Dinâmica e conservação de peixes anuais (Cyprinodontiformes: Rivulidae) no Parque Nacional da Lagoa do Peixe*. MSc Dissertation UNISINOS.

Lanés, L.E.K., Â.C. Gonçalves, and M.V. Volcan. 2013. *Austrolebias arachan* Loureiro, Azpelicueta and García 2004 (Cyprinodonfiformes: Rivulidae) in Rio Grande do Sul, Brazil: Occurrence, length-weight relationships and condition factor. *Journal of Applied Ichthyology* 29: 252–256. doi: 10.1111/j.1439-0426.2012.02052.x

Lanés, L.E.K., A.C. Gonçalves, and M.V. Volcan. 2014a. Discovery of endangered annual killifish *Austrolebias cheradophilus* (Aplocheiloidei: Rivulidae) in Brazil, with comments on habitat, population structure and conservation status. *Neotropical Ichthyology* 12: 117–124.

Lanés, L.E.K., F.W. Keppeler, and L. Maltchik. 2012. Abundance, sex-ratio, length-weight relation, and condition factor of non-annual killifish *Atlantirivulus riograndensis* (Actinopterygii: Cyprinodontiformes: Rivulidae) in Lagoa do Peixe National Park, a Ramsar Site of Southern Brazil. *Acta Ichthyologica et Piscatoria* 42: 247–252. doi: 10.3750/AIP2011.42.3.09

Lanés, L.E.K., F.W. Keppeler, and L. Maltchik. 2014b. Abundance variations and life history traits of two sympatric species of Neotropical annual fish (Cyprinodontiformes: Rivulidae) in temporary ponds of southern Brazil. *Journal of Natural History* 48: 1971–1988. doi: 10.1080/00222933.2013.862577

Lanés, L.E.K., and L. Maltchik. 2010. Discovery of the critically endangered annual killifish, *Austrolebias wolterstorffi* (Ahl, 1924) (Rivulidae: Cyprinodontiformes) in Lagoa do Peixe National Park, Rio Grande do Sul, southern Brazil. *Journal of Threatened Taxa* 2: 1282–1285.

Liu, R.K., and R.L. Walford. 1966. Increased growth and life-span with lowered ambient temperature in the annual fish *Cynolebias adloffi*. *Nature* 212: 1277–1278. doi: 10.1038/2121277a0

Liu, R.K., and R.L. Walford. 1970. Observations on the lifespans of several species of annual fishes and of the world's smallest fishes. *Experimental Gerontology* 5: 241–246. doi: 10.1016/0531-5565(70)90044-6

Loyola, R. 2014. Brazil cannot risk its environmental leadership. *Diversity and Distributions* 20(12): 1365–1367. doi: 10.1111/ddi.12252

Lopes, R.P., F.S.C. Buchmann, F. Caron, and M.E.G.S. Itusarry. 2005. Barrancas Fossilíferas do Arroio Chuí, RS: Importante megafauna pleistocênica no extremo sul do Brasil. In: *Sítios geológicos e paleontológicos do Brasil*, eds. M. Winge, C. Schobbenhaus, M. Berbert-Born et al. Brasília: CPRM. 515p. http://www.unb.br/ig/sigep/sitio119 /sitio119.pdf. Accessed on August 2013.

Loureiro, M., M. Zarucki, I. González, N. Vidal, and G. Fabiano. 2013. Peces continentales. Pp. 91–112, In: *Especies prioritarias para la conservación en Uruguay. Vertebrados, moluscos continentales y plantas vasculares*, eds. A. Soutullo, C. Clavijo, and J.A. Martínez-Lanfranco. snap/dinama/mvotma y dicyt/mec, Montevideo. 222pp.

Maestas, J.D., R.L. Knight, and W.C. Gilgert. 2003. Biodiversity across a rural land-use gradient. *Conservation Biology* 17: 1425–1434. doi: 10.1046/j.1523-1739.2003.02371.x

Maltchik, L., L.E.K. Lanés, F.W. Keppeler, A.S. Rolon, and C. Stenert. 2014. Landscape and habitat characteristics associated with fish occurrence and richness in southern Brazil palustrine wetland systems. *Environmental Biology of Fishes* 97: 297–308. doi: 10.1007/s10641-013-0152-4

Maltchik, L., E. Schneider, G. Becker, and A. Escobar. 2003. Inventory of wetlands of Rio Grande do Sul (Brazil). *Pesquisas Botânica* 53: 89–100.

Marques, A.A.B., C.S. Fontana, E. Vélez, G.A. Bencke, M. Schneider, and R.E. Reis. 2002. *Lista das espécies da fauna ameaçadas de extinção no Rio Grande do Sul*. FZB-MCT-PUCRS-Pangea, Porto Alegre. (Publicações Avulsas FZB).

Marthy, J.T. 2005. Effects of cattle grazing on diversity in ephemeral wetlands. *Conservation Biology* 19: 1626–1632. doi: 10.1111/j.1523-1739.2005.00198.x

MMA. 2007. *Áreas Prioritárias para Conservação, Uso Sustentável e Repartição dos Benefícios da Biodiversidade Brasileira—Bioma Pampa*. Ministério do Meio Ambiente, Secretaria de Biodiversidade e Florestas. Brasília: Ministério do Meio Ambiente. 105p.

MMA. 2011. *Primeiro Inventário Nacional de Emissões Atmosféricas por Veículos Automotores Rodoviários. Ministério do Meio Ambiente (MMA), Secretaria de Mudanças Climáticas e Qualidade Ambiental*, Departamento de Mudanças Climáticas, Gerência de Qualidade do Ar. Brasília, DF.

MMA/SBF. 2000. *Avaliação e ações prioritárias para a conservação da biodiversidade da Mata Atlântica e Campos Sulinos/ in: Conservation International do Brasil, Fundação SOS Mata Atlântica, Fundação Biodiversitas, Instituto de Pesquisas Ecológicas, Secretaria do Meio Ambiente do Estado de São Paulo*, SEMAD/Instituto Estadual de Florestas-MG. Brasília, 2000. 40p.

Naranjo, L.G. 1995. An evaluation of the first inventory of South American wetlands. In: *Classification and Inventory of the World's Wetlands, Advances in Vegetation Science* 16, Kluwer Academic Publishers, Dordrecht, The Netherlands, 125–129.

Overbeck, G.E., S.C. Müller, A. Fidelis et al. 2007. Brazil's neglected biome: The South Brazilian Campos. *Perspectives in Plant Ecology, Evolution and Systematics* 9: 101–116. doi: 10.1016/j.ppees.2007.07.005

Paine, R.T. 1966. Food web complexity and species diversity. *American Naturalist*. 100:65-75.

Pillar, V.D.P., and E. Velez. 2010. Extinction of the southern plains in conservation areas: A natural phenomenon or an ethical problem? *Natureza e Conservação* 8:84–86. doi: 10.4322/natcon.00801014

Radambrasil. 1986. Levantamento de recursos naturais. IBGE, Rio de Janeiro.

Ramsar. 2002. The convention on wetlands. Resolution VIII.33: Identification of Temporary pools. http://www.ramsar.org/res/key_res_viii_33_e.htm. Accessed on December 2013.

Reis, R.E., Z.M.S. Lucena, C.A.S. Lucena, and L.R. Malabarba. 2003. Peixes. In: *Livro vermelho da fauna ameaçada de extinção no Rio Grande do Sul*, eds. C.S. Fontana, G.A., Bencke, and R.E. Reis. Porto Alegre: Edipucrs, 632 pp.

Rosa, R.S., and F.C.T. Lima. 2008. Peixes. In: *Livro Vermelho da Fauna Brasileira Ameaçada de Extinçã*, eds. A.B.M. Machado, G.M. Drummond, and A.P. Paglia. Brasília, Ministério do Meio Ambiente. pp. 9–285.

Sanderson, R.A., M.D. Eyre, and S.P. Rushton. 2005. Distribution of selected macroinvertebrates in a mosaic of temporary and permanent freshwater ponds as explained by autologistic models. *Ecography* 28: 355–362. doi: 10.1111/j.0906-7590.2005.04093.x

Santamaría, L., and M. Klaassen. 2002. Waterbird-mediated dispersal of aquatic organisms: An introduction. *Acta Oecologica* 23: 115–119. doi: 10.1016/S1146-609X(02)01144-X

Saunders, D.L., J.J. Meeuwig, and A.C.J. Vincent. 2002. Freshwater protected areas: Strategies for conservation. *Conservation Biology* 16: 30–41. doi: 10.1046/j.1523-1739.2002.99562.x

SEMA. 2014. Secretaria Estadual de Meio Ambiente. Reavaliação da Lista das Espécies da Fauna Silvestre Ameaçadas de Extinção no Rio Grande do Sul. http://www.liv.fzb.rs.gov.br/livcpl/?idmodulo=1andid_uf=23. Accessed on August 2014.

Semlitsch, R.S., and R. Bodie. 1998. Are small, isolated wetlands expendable? *Conservation Biology* 12: 1129–1133. doi: 10.1046/j.1523-1739.1998.98166.x

SEPLAG. 2014. Secretaria do Planejamento, Gestão e Particiáção Cidadã.www.seplag.rs.gov.br. Accessed on October 2014.

Snodgrass, J.W., M.J. Komoroski, A.L. Bryan, and J. Burger. 2000. Relationships among isolated wetland size, hydroperiod, and amphibian species richness: Implications for wetland regulations. *Conservation Biology* 14: 414–419. doi: 10.1046/j.1523-1739.2000.99161.x

Stenert, C., and L. Maltchik. 2007. Influence of area, altitude and hydroperiod on macroinvertebrate communities in southern Brazil wetlands. *Marine and Freshwater Research* 58: 993–1001.

Volcan M.V., Â.C. Gonçalves, and D.L. Guadagnin. 2013a. Length–weight relationship of three annual fishes (Rivulidac) from temporary freshwater wetlands of southern Brazil. *Journal of Applied Ichthyology* 29(3): 1188–1190. doi: 10.1111/jai.12214

Volcan, M.V., Â.C. Gonçalves, and L.E.K. Lanés. 2011a. Distribution, habitat and conservation status of two threatened annual fishes (Rivulidae) from southern Brazil. *Endangered Species Research* 13: 79–85. doi: 10.3354/esr00316

Volcan, M.V., Â.C. Gonçalves, and L.E.K. Lanés. 2014a. *Austrolebias quirogai* (Actinopterygii: Cyprinodontiformes: Rivulidae) in Brazil: Occurence, population parameters, habitat characteristics, and conservation status. *Acta Ichthyologica et Piscatoria* 44: 37–44. doi: 10.3750/AIP2014.44.1.05

Volcan, M.V., L.E.K. Lanés, and M.M. Cheffe. 2010. Distribuição e Conservação de Peixes Anuais (Cyprinodontiformes, Rivulidae) no município do Chuí, Brasil. *Biotemas* 23: 51–58. doi: 10.5007/2175-7925.2010v23n4p51

Volcan, M.V., L.E.K. Lanés, A.C. Gonçalves, and M.M. Cheffe. 2011b. First record of annual killifish *Austrolebias melanoorus* (Amato, 1986) (Cyprinodontiformes: Rivulidae) from Brazil, with data on habitat and conservation. *Journal of Applied Ichthyology* 27: 1120–1122. doi: 10.1111/j.1439-0426.2010.01626.x

Volcan, M.V., L.E.K. Lanés, and Â.C. Gonçalves. 2014b. *Austrolebias bagual,* a new species of annual fish (Cyprinodontiformes: Rivulidae) from Southern Brazil. *Aqua, International Journal of Ichthyology* 20: 3–14.

Volcan, M.V., L.A. Sampaio, D.C. Bongalhardo, and R.B. Robaldo. 2013b. Reproduction of the annual fish *Austrolebias nigrofasciatus* (Rivulidae) maintained at different temperatures. *Journal of Applied Ichthyology* 29(5): 648–652. doi: 10.1111/jai.12013

ZAS. 2008. *Zoneumento Ambiental para Atividade de Silvicultura.* Coordenação: FEPAM. Consultoria Técnica: Biolaw Consultoria Ambiental e Fundação Zoobotânica do Rio Grande do Sul. 305pp.

PART III

Evolution

CHAPTER 12

Reproductive Behavior and Sexual Selection in Annual Fishes

Carlos Passos, Bettina Tassino, Gil G. Rosenthal, and Martin Reichard

CONTENTS

12.1 INTRODUCTION

While the most noteworthy feature of annual fishes may be their unusual life history, these animals have long been appreciated by scientists and hobbyists for the striking sexual dimorphism and dichromatism they share with other cyprinodontiforms. Indeed, one of the first studies on sexual selection in fishes was on the African annual *Nothobranchius guentheri* (Haas, 1976a). Males of most annual fish species are larger than females, exhibit bright coloration in the body flanks and have strongly pigmented unpaired fins, perform elaborate courtship displays, and engage in ritualized aggressive contests with other males for access to mates. By contrast, females are small and cryptic with few distinguishing characteristics across species and express behavioral preferences for distinct male phenotypes (Figure 12.1). This pronounced sexual dimorphism suggests that sexual selection shapes the reproductive behavior in this clade (Box 12.1) and that direct and indirect competition for mates should be intense.

Furthermore, these fishes live in small, geographically isolated, and ephemeral pools. Thus, annual fishes provide a unique opportunity for empirical studies of the interaction between sexual selection and extrinsic variables like sex ratio and water quality, as well as of the role of sexual selection in the evolution of reproductive barriers among populations. Nonoverlapping generations and numerous disjunct populations make annuals an ideal system for integrating field data with theoretical predictions. Although courtship and mating behaviors have been described in detail in several species, it is only recently that we have begun studying sexual selection in this group of fishes systematically. Two very distinctive advantages are that we can rule out cross-generation learning (Verzijden et al., 2012) and that the short window for reproduction makes it ideal for understanding life history trade-off involving mating decisions. Moreover, from a

Figure 12.1   Male (top) and female (down) *Austrolebias charrua* (unscaled). (Photos by M. Casacuberta.)

BOX 12.1   SEXUAL SELECTION

Sexual selection is a powerful mechanism of evolutionary change proposed by Darwin in 1871 that results in differential reproductive success among individuals of the same sex and explains both the emergence and the evolution of sexually dimorphic traits. The extravagant male ornaments in many species are good indicators of the action of sexual selection. These ornaments are usually disadvantageous to a male's viability or survival, but they are advantageous in that they increase a male's mating success. Sexual selection is revealed through direct competition among members of the same sex (intrasexual selection) or by the choice of one sex over individuals of the other sex (intersexual selection); the relative importance of these two processes differs among species. In general, females invest a large amount of energy in egg formation, and their reproductive potential is often limited by resource availability. In contrast, the number of sperm that males can produce rarely limits their reproductive success. Consequently, females are most often selective in their mating decisions, whereas males usually actively compete for access to females or resources to attract them, and there is more variance in male than in female reproductive success (Andersson, 1994). Although the evolution of certain male traits can be promoted exclusively by one of the two components of sexual selection, they often act simultaneously (in concert or in opposition) because many traits are involved in both female mating decisions and male contest outcomes (Berglund et al., 1996; Candolin, 2004; Moore and Moore, 1999; Qvarnström and Forsgren, 1998).

In intersexual selection, individuals can base their choice on signals (often displayed during courtship behavior) that may or may not be associated with latent traits in potential mates (Andersson, 1994; Darwin, 1871). In some cases, the benefits of mate choice could be directly related to survival or fecundity. For instance, selection will favor the mating preferences for individuals with higher fertility, or who provide better quality or quantity of resources and parental care. More controversially, mate choice may be adaptive according to indirect benefits achieved by mating with certain individuals and thereby obtaining offspring with genotypes that increase the viability and survival ("good genes" model, Zahavi, 1975; but see Kirkpatrick, 1987). A more general mechanism for the elaboration of male traits and female preferences is Fisher's runaway model (Fisher, 1930), in which traits and preferences are favored merely as a consequence of the established genetic correlation between female preference and male trait (Andersson, 1994). Finally, mating preferences may have arisen independent of male traits, in response to a preexisting sensory bias that has evolved in a nonreproductive context (Endler and Basolo, 1998; Macías Garcia and Ramirez, 2005; Ryan, 1998). All of these mechanisms may act alone or, more likely, in concert to determine the evolution of mate choice and sexual characters (Kokko et al., 2003).

In a microevolutionary context, sexual selection affects population genetic structure, but also acts as a driving force of morphological and behavioral innovations linked to systems of mating signals. It can therefore have a leading role in the establishment and strengthening of barriers to gene exchange between populations, and therefore impacts speciation (Butlin and Ritchie, 1994) and hybridization (Rosenthal, 2012). Both theoretical and empirical studies (reviewed in Panhuis et al., 2001; Ritchie, 2007) have shown that sexual selection can drive speciation (due to reproductive isolation via the generation of assortative mating) by producing divergence between populations in sexual traits (female preferences and male signals). Sexual selection is considered one of the most crucial processes responsible for the generation and maintenance of biological diversity (Boake, 2002).

practical point of view, reproductive behavior can be observed easily, as both males and females perform reproductive activity daily.

Most research on sexual selection in annual fishes has focused on two systems: the *Austrolebias charrua—A. reicherti* system in Uruguay and *N. furzeri–N. orthonotus* system in Mozambique. Additional insight comes from work on captive populations of *N. korthausae*. Throughout the chapter, we will try to demonstrate the close relationship between reproductive behavior and particular life history of annual fishes, using examples from *Austrolebias* and *Nothobranchius*, the two most diverse genera of annual fishes. We will discuss the sexual signals involved in intrasexual competition, courtship, and mate choice. Finally, we will discuss the potential role of sexual selection on the divergence of this group of fishes.

12.2 INTRASEXUAL COMPETITION

In *Cynolebias* and *Simpsonichthys*, females aggressively suppress reproduction in other females; in an aquarium, one dominant female will exclude the others through chases and bites, and consequently only the dominant female spawns (Belote and Costa, 2002, 2003). In *Simpsonichthys,* females can even fight to the death (Belote and Costa, 2002). In contrast, there is no fight among females in *Austrolebias* and *Nothobranchius*, and two *Austrolebias* females may try oviposit with a single male at the same time (Belote and Costa, 2004; Haas, 1976b). In *Nothobranchius*, additional females may remain in close proximity to the spawning pair and spawn later with the male after ongoing oviposition is resumed, with no aggression among females observed (Haas, 1976b). This difference in female aggressive behavior among genera could be related to the fact that females of *Cynolebias* play an active role during the courtship behavior, while in *Austrolebias* and

Nothobranchius, females are limited to observing male courtship displays and to following males to spawn (see the following section).

Regarding male–male competition, *Austrolebias* males are very aggressive and quickly escalate to physical fights in aquaria. Experiments addressing social interactions in aquaria with restricted spawning areas have found that males of *A. reicherti* defend and exclude rivals from a spawning territory. In addition, both territorial residence and dominance are stable over time, although females move freely around the aquarium (Fabra, 2011). Social interactions likely establish dominance ranks that determine access to females and/or best areas to spawn in the heterogeneous conditions of temporary ponds they inhabit (Passos et al., 2013a). These findings also suggest that female choice plays an important role, since male coercion of females is limited. Future studies should address whether male–male aggression acts to physiologically suppress reproduction in losers as it does in other fishes (e.g., Desjardins and Fernald, 2008; Fernald, 2009).

Male–male competitive interactions in *A. reicherti* and *A. charrua* often begin with displays similar to courtship toward females: males exhibit lateral displays alternated with sigmoid displays and place their flanks together in parallel or antiparallel position while quivering (Passos et al., 2013a; Table 12.1). Conflicts are often resolved through mutual displays, with one fish retreating. If not, one fish usually charges and bites the other male, often targeting the fins. The bites can be brief or may last several seconds when the aggressor uses his jaws to seize his opponent, causing evident injuries to the fins. Sometimes the fish receiving the first attack retreats and the contest is resolved; at other times the fish under attack fights back and the contest escalates. The outcome of the contest is clearly indicated by morphological and behavioral differences between the contenders. The subordinate male "turns off" quickly, decreasing the intensity of coloration, reducing the contrast of flanking vertical bars, and folding the dorsal fin. A defeated male resembles a female, confusing even an experienced observer. Once dominance is established, the subordinate male remains mostly

Table 12.1 Behavioral Unit Description of *Austrolebias*

Behavioral Units	Context	Description
Males		
Lateral display	Courtship and agonistic interactions	Body still, dorsal and anal fins extended, with or without quivering
Sigmoid display	Courtship and agonistic interactions	Lateral display with body undulations
Invitation to follow	Courtship	Successive short swimming away and stops
Invitation to dive	Courtship	Body in vertical or oblique position, with head over the substrate, and quivering of extended dorsal and anal fins
Mating	Courtship	Body placed by the side of female, pressing its body to the substrate. Unit performed from near the surface to completely buried
Lateral display with contact	Agonistic interactions	Lateral display, above, with both actors touching in a parallel or antiparallel position
Attack	Agonistic interactions	Actor bites or attempts to bite recipient, often directed at the fins
Submission	Agonistic interactions	Rapid flight to sides or corner, accompanied by color loss and folding of the dorsal fin
Females		
Following acceptance	Courtship	Female follows male (displaying IF)
Diving acceptance	Courtship	Females placed by the side of male, touching its body
Mating	Courtship	Female buried in the substrate (partially or completely)

Source: Adapted from García, D., M. Loureiro, and B. Tassino. 2008. *Neotropical Ichthyology* 6: 243–248; Passos, C. et al. 2013a. *Behavioral Process* 96:20–26.

inactive and flees the dominant male when approached. In contrast, a dominant male has darker coloration, enhancing both body and fin pigmentation design, unfolds his dorsal fin, and moves freely around the aquarium, performing aggressive displays against the other males. Once the hierarchy is established, the overall level of aggression decreases, and the dominant male's lateral and sigmoid displays are the most frequently observed agonistic interactions (Passos et al., 2013a).

In the laboratory, the hierarchy between individuals is stable for several days, even though coloration, fin deployment, and behavior are dynamic and can change in few seconds. The "turn on" coloration likely signals social status and acts as a reliable indicator of the competitive abilities of a male, in the sense that "dishonest" individuals will be exposed to fight escalation (Maynard Smith and Harper, 1988). Sexually dimorphic pigment patterns in other fish species exhibit rapid changes with aggression and social status (Hurd, 1997; Morris et al., 1995; Nelissen, 1991). Chemical cues may also play a role in short- and long-term dominance interactions (Almeida et al., 2005; Barata et al., 2007).

In *Nothobranchius*, males also engage in fierce fights, with the level of male–male aggression varying across species (Polačik and Reichard, 2011; Reichard et al., 2014). It is not clear whether males patrol territories in the wild, but the dominant male usually guards the spawning site with a substrate in an aquarium and repels other males from its vicinity (Reichard and Polačik, 2010). The bottom of *Nothobranchius* pools is typically more uniform than those of *Austrolebias* (see Chapter 9); substrate suitable for spawning and egg development is distributed across the entire pool. At least under captive conditions, dominant males have the highest mating success due to exclusive access to prime spawning sites; in a study under seminatural conditions, spawning only took place in deeper parts of the pool (Haas, 1976b). Males actively explore their home range and search for females, with dominant males apparently controlling the largest areas. In the wild, dominant males likely become most conspicuous to females by moving freely around a pool and enjoy the most frequent female encounters and mating. Dominant males also actively disrupt the spawning of other males (Haas, 1976b).

Male–male combats are comparable to *Austrolebias*. Males rapidly approach each other and perform lateral displays using the spread of unpaired fins. Unlike *Austrolebias* males, male *Nothobranchius* prominently extend the lateral projections of their opercular and branchiostegal membranes (Haas, 1976b). The branchiostegal membrane is exceptionally well developed in males and conceivably has an important function in agonistic signaling similar to other fishes, for example, the Siamese fighting fish, *Betta splendens* (Rosenthal and Lobel, 2006). The lateral displays include tail beating and attempts to bite an opponent's fins and flanks, in addition to mutual threats performed with the branchiostegal membranes prominently displayed (Haas, 1976b). When a contest is not resolved with these displays, males may lock their jaws and remain in the locked position for several seconds to minutes. At least in aquaria, males rapidly establish linear hierarchies (Haas, 1976b), and male–male aggression subsides to a relatively low level. There are major interspecific differences in the intensity and frequency of male–male aggressive interactions, with large species such as *N. orthonotus* being more aggressive than smaller species, for example, *N. pienaari* and *N. korthausae* (Polačik and Reichard, 2009, 2011). A slight increase in male–male aggression associated with the peak of mating activity is observed during the midday hours, even in the established hierarchies (Haas, 1976b).

Dominant males have brighter coloration than subordinates. However, in contrast to *Austrolebias*, changes in male coloration are not abrupt and do not respond rapidly to the subtleties of social situations. Perhaps this is because the changes in coloration do not involve melanocyte-related and carotenoid patterns, which are known to be involved in quick responses to individual status (Kodric-Brown, 1998). Instead, *Nothobranchius* male coloration is dominated by structural colors (Ng'oma et al., 2014), and its intensity is likely mediated hormonally rather than neuronally.

As in other fishes (Beaugrand et al., 1996; Benson and Bassolo, 2006; Moretz, 2003), larger *Austrolebias* males are more aggressive and socially dominant (Passos et al., 2013a), though the association between the body size and dominance is much weaker in *Nothobranchius* (Poláčik and Reichard, 2009; Reichard and Poláčik, 2010). Fight duration between males of *A. charrua* decreases with increasing differences in the body size of opponents (Passos et al., 2013). When the size difference is large, dominance is established by lateral and sigmoid displays, without escalation. Lateral displays, in which one individual exposes his flank to the opponent with fin extension and oscillation, and likely provide the receiver with visual and mechanosensory information about the body size of the sender (Bleckmann, 1993). Sigmoid displays, in which an individual performs high-speed undulating body movements, may be predictors of the body size, motivation, and locomotor performance (Passos et al., 2013a).

Several researchers have suggested that a negative relationship between contest duration and body-size asymmetry indicates mutual assessment in animal contests (e.g., Bridge et al., 2000; Enquist et al., 1990; Hack, 1997). In mutual assessment models, individuals assess their own body size relative to their opponent, and contest duration is expected to be inversely correlated with the relative body size of contestants, because an animal can detect and act on a large difference quicker than on a small difference (Enquist and Leimar, 1983; Maynard Smith and Parker, 1976). However, recent studies have shown that this result is also consistent with a self-assessment model (Prenter et al., 2008; Taylor and Elwood, 2003), in which animals make decisions based on their own fighting skills. Fighting is costly in terms of time and energy, risk of physical injury, and predation. Furthermore, selection should favor any mechanism allowing accurate gathering of information on the likely costs and benefits of continuing a conflict and to enable appropriate tactical decisions (Maynard Smith and Parker, 1976; Parker, 1974).

In several taxa, including many fishes, males are polymorphic. Intraspecific polymorphism typically involves differences in coloration among the morphs (i.e., polychromatism; see Gray and McKinnin, 2007). Male coloration may be continuous, but overall distinct morphs are often clearly separated. In several *Nothobranchius* species, males have two discrete color morphs, while other species show a continuous gradation of male coloration across populations (Wildekamp, 2004). The genetic basis of discrete polymorphism for some species is understood and relatively simple. In *N. furzeri*, the allele for yellow color is dominant, whereas the red allele is recessive (Valenzano et al., 2009). The role of such dimorphism is important in the establishment of male hierarchies; in many systems, color phenotypes predict dominance in male–male contests (Pryke and Griffith, 2006), and dominant males are preferred by females or mate by coercion. In contrast, other color morphs are subordinate in hierarchies and mate largely by alternative tactics (Sinervo and Lively, 1996). Recent experiments with *N. furzeri* revealed that male coloration (red or yellow caudal fins) has no impact on the outcome of dominance hierarchies, at least in simple dyadic combats (Reichard et al., unpublished data).

12.3 COURTSHIP SEQUENCE AND SEXUAL SIGNALS

Courtship and mating behavior have been described in a number of species, notably in *Austrolebias* (*A. bellottii*: Vaz-Ferreira et al., 1964; Vaz-Ferreira and Sierra, 1972; *A. nigripinnis*: Belote and Costa, 2004; Vaz-Ferreira et al., 1964; Vaz-Ferreira and Sierra, 1972; *A. viarius*: Vaz-Ferreira et al., 1964; Vaz-Ferreira and Sierra, 1972; *A. luteoflammulatus*: Vaz-Ferreira et al., 1964; Vaz-Ferreira and Sierra, 1972; *A. cheradophilus*: Vaz-Ferreira et al., 1964; *A. cyaneus*: Belote and Costa, 2004; *A. charrua*: Belote and Costa, 2004; *A. reicherti*: García et al., 2008; and *A. affinis*: Libran, 2011) and *Nothobranchius* (*N. guentheri*: Haas, 1976a,b; *N. korthausae*: Poláčik and Reichard, 2009; Reichard and Poláčik, 2010; *N. furzeri*: Poláčik and Reichard, 2011; and *N. orthonotus*: Poláčik and Reichard, 2011). Reproductive behavior is highly conserved in the genus *Austrolebias*; all species exhibit similar courtship displays with addition or slight modification of the same behavioral units (Table 12.1). Other South American annuals also use the same basic behavioral patterns (e.g., *Simpsonichthys* and *Cynolebias*; Belote and Costa, 2002, 2003). In contrast, courtship displays in *Nothobranchius* are relatively simple (Haas, 1976a; Reichard and Polačik, 2010).

Courtship sequences for *A. reicherti* and *A. charrua* have been described in detail (Belote and Costa, 2004; García et al., 2008; Passos, 2013; Table 12.1; Figure 12.2). A courting male typically follows a female and begins to display when he comes within her field of view. Male courtship consists of lateral displays with rapid quivering and expanded fins, alternating with sigmoid displays with fast body undulations. Subsequently, the male darts a short distance away and makes a new sequence of lateral and sigmoid displays. If the female is responsive, she approaches and follows the male, who swims near the bottom searching for a suitable site for oviposition. Then, the male places his snout into the substrate in a 45° to 90° angle and then vigorously undulates his body. If the female does not follow the male, then the male quickly rises and restarts lateral and sigmoid displays. Alternatively, the female comes to the side of the male and touches his flank in a parallel position, in which case the pair digs under the substrate while quivering. Oviposition and fertilization occur in the substrate, with the male pressing the female against the substrate with his body lying over female (Table 12.1; Figure 12.2). After a few seconds, the couple emerges, either together or at different times, and a new courtship cycle starts. The duration of the courtship and the frequency of spawning depend mainly on the whether the female stays with the male and whether there is a presence of competitors. It is not unusual for a male to abandon courtship and switch to confront a rival male. However, if the couple is kept isolated from other individuals, the courtship continues until the female becomes unresponsive. Usually, females of most annual fishes do not show elaborate behaviors during mating; they mostly observe the courtship display of the males and follow a male to a spawning site. However, the female of *C. albipunctatus* touches its snout to the urogenital region of the male, sometimes exhibiting brief and subtle courtship-like waving movements (Belote and Costa, 2003).

In *Nothobranchius*, the spawning sequence is very similar across species. Males initiate mating, but females sometimes approach a male first, especially under experimental conditions of physical isolation behind a barrier (Reichard and Polačik, 2010). The male approaches a female and

Figure 12.2 Courtship sequence of *Austrolebias charrua*: male display (top left), invitation to dive (top right), diving acceptance (down left), and mating (down right) (see Table 12.1). (Photos by M. Casacuberta.)

Figure 12.3   Spawning behavior of *Nothobranchius eggersi*: male folds female using his dorsal and anal fins, and female stiffens her anal fin (left); the male then raises his head and pushes female toward substrate (right); this is followed by oviposition. (Photos by K. Zahrádka.)

frequently stops to engage in displays. At this stage, a female unwilling to spawn swims away, while a receptive female allows the male to approach. The male makes displays laterally to the female but without spreading his fins, as occurs during agonistic encounters (Haas, 1976b). Individual lateral displays are interrupted by sequential approaches, with the two behaviors being swapped in succession. In *N. korthausae*, the male's head points slightly upward at an angle of approximately 30° to 45° during displays (Reichard and Polačik, 2010). Alternation of displays and approaches results in a darting approach movement of the male toward the female, which is followed by physical contact. The male places his lower jaw on the upper part of the female's head or slightly behind, directing her toward a spawning site. Then, the male moves slowly alongside the female, keeping physical contact with her and clasping her from above using his dorsal fin. Maintaining physical contact, the male raises his head and pushes his caudal fin away from the female with rapid flicking of the pectoral fins. He folds the female using his anal fin (Figure 12.3), and the female stiffens the anterior part of her anal fin (possessing rigid prolonged rays) and inserts it into the substrate. The pair remains motionless for few seconds, followed by oviposition of a single egg on the substrate during a jerked downward movement. This final phase disturbs the substrate, and oviposition can thus be easily recognized. The pair may return to clasping, and the second part of spawning sequence is finalized by another oviposition (Haas, 1976b; Reichard and Polačik, 2010). Haas (1976b) observed two to seven successive ovipositions in *N. guentheri*, but the series can be longer, at least under experimental conditions, in *N. furzeri* (Blažek et al., 2013).

The sexual dimorphism of male form and color, coupled with intensified contrast of body markings during courtship, suggests that visual cues play an important role during reproductive interactions in annual fishes. There is, however, limited experimental evidence for the role of visual signals in mate choice (see the following section).

Males of *A. reicherti* produce chemical cues during courtship that cause behavioral changes in conspecific females (Passos et al., 2013b). Chemical cues were obtained by placing individual males next to individual females for 24 h to elicit courtship. Females preferred male-conditioned water when tested both against control water and female-conditioned water (Figure 12.4a). In fishes, semiochemicals have been shown to play an important role in mate choice (Fisher and Rosenthal, 2006; Milinski et al., 2005), intrasexual competition (Almeida et al., 2005; Barata et al., 2007), and speciation processes (Plenderleith et al., 2005). Olfactory cues may also play particularly important roles in recognition of conspecifics and hence act as a mechanism of reproductive isolation among closely related species (Kodric-Brown and Strecker, 2001; McLennan and Ryan, 1997, 1999; Rafferty and Boughman, 2006; Wong et al., 2005). This has been convincingly shown in swordtails (Poeciliidae), which have been extensively studied for mate choice and species recognition mechanisms (Crapon de Caprona and Ryan, 1990; Fisher et al., 2006, 2009; McLennan and Ryan, 1997, 1999; Wong et al., 2005). Sympatric *Xiphophorus* species use chemical signals for species recognition, and the

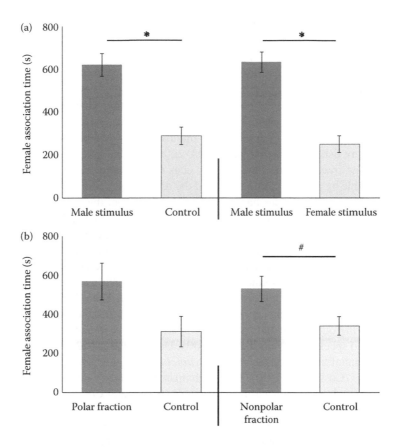

Figure 12.4 Association time (mean ± SE) of females *Austrolebias reicherti* with chemical cues. Male-conditioned water vs. control water, and male- vs. female-conditioned water (a); polar fraction of the male-conditioned water vs. control water, and nonpolar fraction of the male-conditioned water vs. control water (b) (*$p < 0.05$; #$0.05 < p < 0.1$). (Modified from Passos, C. et al. 2013b. *Ethology* 119: 891–897.)

disruption of chemical communication results in interbreeding of natural populations of swordtails (Fisher et al., 2006; Rosenthal et al., 2003). Chemical cues in *Austrolebias* could be important during the initial stages of mate location and recognition, when distance, occluding vegetation, and turbidity within the pond would limit visual detection. This is also pertinent to *Nothobranchius*, which often inhabit pools with very turbid water, but the role of chemical cues in *Nothobranchius* has not been investigated. Also, chemical cues could reinforce visual cues at close proximity, for example, during courtship.

Characterization of the chemical signals in annual fishes will likely be a challenging task. Specialized scent glands or overt marking behaviors are not common in fishes. Consequently, it is difficult to obtain clean samples for chemical analysis. Indeed, in some species, it has been shown that courting males release their pheromones in the urine, which increases the chemical complexity of the samples (Maruska and Fernald, 2012; Rosenthal et al., 2011). Most fish pheromones that have been chemically identified to date comprise prostaglandins and steroids, but there have been reports of a communicative role in bile acids and amino acids (reviewed in Sorensen and Hoye, 2010). These compounds differ in their polarity; therefore, a crude separation of polar and nonpolar compounds, combined with activity bioassays, would provide primary information as to the chemical nature of an olfactory cue. In *A. reicherti*, female preference for male cues was reduced or abolished when cues were separated into polar and nonpolar fractions. However, a tendency was observed for attraction toward the organic fraction that included compounds of medium polarity (Passos et al., 2013b; Figure 12.4b). Alternatively, polar and nonpolar compounds may combine to produce a multicomponent cue that is attractive to females (Passos et al., 2013b). Indeed, in goldfish, *Carassius auratus,* both polar and nonpolar fractions of pheromone blends were necessary to achieve biological activity (Levesque et al., 2011).

In addition, other modalities may play a role. The undulating movements and quivering of a courting male generate currents, and receivers could simultaneously attend to chemical, visual, and mechanosensory information. Vibratory and undulatory movements of males during courtship are common in many fishes (Rosenthal and Lobel, 2006). Several studies have reported that these displays produce low-frequency mechanosensory cues (Bleckmann, 1993; Nelissen, 1991; Sargent et al., 1998) detectable by the lateral line system of the receiver (Bleckmann, 1993). Although mechanosensory communication in fishes has received little attention, there is compelling evidence to suggest that this modality could play an important role in courtship behavior and mate choice (e.g., Satou et al., 1994a,b). Annual fishes have a highly developed laterosensory system (Belote and Costa, 2003; García et al., 2008), and the close-range vibrations and motor displays performed during courtship would generate hydrodynamic cues that could be detected by the females.

Finally, many other fishes use sound in the context of courtship and mating. While sound production is taxonomically widespread (reviewed in Bass and McKibben, 2003), the best evidence for the role of acoustic cues in mate choice comes from cichlids (Maruska et al., 2012; Verzijden et al., 2010). Differences in courtship sounds among sympatric cichlid species may contribute to reproductive isolation and speciation (Amorim et al., 2008; Danley et al., 2012; Lobel, 1998). At least two South American annual fishes produce sound: *Cynolebias albipunctatus* (Belote and Costa, 2003) and *C. parnaibensis* (Costa et al., 2010). The mechanism of sound production in these species is not known but may involve the pharyngeal jaw apparatus (Belote and Costa, 2003; Costa et al., 2010). The bony support of the gill arches in these fishes is highly modified (Costa, 2001) and skeletal traits may be related to sound production (e.g., friction of teeth, Belote and Costa, 2003).

Sound production is also plausible in *Nothobranchius.* During experimental work with *N. korthausae,* a female apparently responded to the initial stage of male courtship behavior (i.e., darting approach) with an approach, despite the male being visually and chemically isolated in an adjacent aquarium (Reichard and Polačik, 2010). Intriguingly, the sulcus (a specialized part of the saccular otolith in the inner ear of fishes) morphology is divergent in sympatric *Nothobranchius* species (Reichenbacher and Reichard, 2014). The sulcus is in contact with a sensory epithelium (sensory hair cells), and its species-specific morphology may indicate specialization in hearing abilities (Fekete, 2003). Species-specific characteristics of the sulcus were confirmed in both tested pairs of sympatric species, *N. rubripinnis* and *N. ruudwildekampi* from coastal Tanzania and *N. orthonotus* and *N. furzeri* from Mozambique (Reichenbacher and Reichard, 2014). Intraspecific auditory communication would have obvious advantages for annual fishes that often live in very turbid waters. Additional work may reveal a role for acoustic communication in courtship and mating.

12.4 MATE CHOICE

The sexual dimorphism and small size of annual fishes make them a good model for laboratory studies of sexual selection (Box 12.1). Surprisingly, apart from one early study on *N. guentheri* (Haas, 1976a), mate choice has not been addressed until recently. When given a choice between two males varying in size, females of both *A. charrua* and *A. reicherti* prefer to mate with larger males (Passos et al., 2013a, 2014). The preference for larger males is shared with many fishes (Andersson, 1994), including other cyprinodontiform fishes (e.g., MacLaren and Daniska, 2008; Ptacek and Travis, 1997; Rosenthal and Evans, 1998; Ryan et al., 1990). Several nonexclusive factors may favor the evolution or maintenance of preferences for large size (Box 12.1). First, larger males could generate great visual stimulation and therefore elicit sensory bias underlying female preferences (Rosenthal and Evans, 1998). Second, female preference could also evolve via indirect selection on male traits (Andersson, 1994; Fisher, 1930; Zahavi, 1975). Third, females may obtain direct benefits by mating with larger males, since body size is correlated with dominance status (see above). In South American annual fishes, direct benefits of mating with dominant males could include additional spawning sites and refuges, which may be limiting factors for reproduction and viability of eggs. In contrast, female choice in *Nothobranchius* is independent of male body size, except when the difference between males is relatively large (Reichard and Polačik, 2010), though the effect of male body size has not been tested directly.

Male–male interactions may be a source of information in mate choice (Candolin, 1999; Doutrelant and McGregor, 2000). Direct experimental evidence in fishes indicates that witnessing male–male interaction influences subsequent female choice (Cox and LeBoeuf, 1977; Montgomerie and Thornhill, 1989), although sometimes females prefer losing males (Wong and Candolin, 2005). Indeed, socially dominant males may often be worse at providing direct benefits like nest defense (e.g., Wong, 2004) and may behave aggressively toward females (Reichard et al., 2007). The interaction between male–male competition and female choice in *A. charrua* was tested experimentally (Passos et al., 2013a). Females were given a simultaneous choice between two males and consequently witnessed agonistic encounters between the males. Subsequent testing showed that females were consistent in their choice and selected the same male in both tests, independent of whether they had observed the male contest or not, suggesting that competition does not increase the attractiveness of the dominant male over the subordinate (Passos et al., 2013a). However, given the difficulty of disentangling the highly correlated effects of body size and social hierarchy (see above), it is not possible to conclude that female perception of male dominance in *A. charrua* is irrelevant in female choice. A possible approach to distinguish the effects of body size from the effects of social hierarchy would be to engineer dominance relationships among males, such that the smaller male is dominant. This could be achieved through the effect of prior social experience on agonistic behavior, in which individuals modify their behavior based on the outcome of previous contests. There is evidence that individuals who were winners in a fight increase their chances to win in the future (winner effect), while losing individuals decrease their chances to win in other contests (loser effect) (Hsu et al., 2006). Therefore, previous agonistic trials could reverse the outcome of the interaction between competitors, such that the smaller male wins. Moreover, field studies should address the relationship between male aggression and egg viability.

Females of *N. korthausae* did not lay more eggs with socially dominant males. In noncompetitive trials, in which males with previously ranked dominance were sequentially presented to females with no contact to any rivals during the observation, females did not discriminate among males on the basis of their dominance or body size. There were also no benefits to mating with large or dominant males in terms of increased fertilization or hatching success in aquaria (Reichard and Polačik, 2010). In choice trials with *N. korthausae*, females made their partner choices, but these were not related to male dominance or body size. In these trials, however, male dominance was tested after completion of female choice trials. Therefore, females were unable to make decisions on the basis of "eavesdropping" on male contests, and males were unable to modulate their display behavior according to winner/loser effect (Polačik and Reichard, 2009).

Several studies have documented mate choice based on visual displays of male coloration in cyprinodontiform fishes. In swordtails, females attend to vertical bars on the lateral surface of males, which intensify during courtship and aggressive interactions (e.g., Morris et al., 1995; Morris et al., 2003). *Austrolebias* males have similar vertical bars, but there is no detectable relationship between the number, width, or intensity of bars and female mating preference (Passos, 2013). Techniques like computer-animation playback (Culumber and Rosenthal, 2013; Rosenthal, 2000) could be useful in isolating preferences for signal components.

In *N. guentheri*, females prefer to associate with brighter males (Haas, 1976a). However, perhaps surprisingly, discrete male morph (red or yellow caudal fin) had no effect on female choice in *N. furzeri*. Given the potential for assortative mating and inability to phenotypically score female color alleles, females were tested twice, each time with a different pair comprising a red and a yellow male. Females were choosy and associated with one male preferentially, but their preference was not consistently related to male color morph. A female often preferred a male of one color in her first trial, but a male of the other color morph in the second trial. Red and yellow males in this experiment were wild captured and originated from a single population (Reichard et al., unpublished data). In *N. korthausae*, males of red and yellow morphs came from allopatric populations, and experimental fish were descendants of already captive fish. Virgin females mated indiscriminately, but nonvirgin females from both populations preferred red males (that were sympatric for females from one population but allopatric for females from a second population). Red males displayed to females at a disproportionally higher rate; it cannot be separated whether females responded to red coloration, cues from other sensory modalities, or higher courtship intensity (Reichard and Polačik, 2010). The role of discrete male color morphs on mate choice in *Nothobranchius* requires further investigation. Ideally, future studies should include sympatric and allopatric populations and control samples for a potential effect of inbreeding in captive populations, since female choice for optimal outbreeding (e.g., detected by chemosignals) may override preferences related to coloration (Agbali et al., 2010).

Female mate choice is a contextual phenomenon modulated by different factors (Jennions and Petrie, 1997). The fitness effects for females of choosing or of mating with certain males may vary in magnitude and direction as environmental conditions or individual phenotypes change (Borg et al., 2006; Forsgren et al., 2004; Jennions and Petrie, 1997; Qvarnström, 2001). Mate choice can be influenced by several factors, for example, availability of resources (Fisher and Rosenthal, 2006; Moskalik and Uetz, 2011), intensity of intrasexual competition (Lehtonen and Lindström, 2008), and individual condition (Amundsen and Forsgren, 2003; Burley and Foster, 2006; Hunt et al., 2005). One important variable is the operational sex ratio (OSR; Owens and Thompson, 1994), defined as the ratio of males and females available to mate at a given time. Given daily reproduction of male and female annual fishes, the OSR generally equals the adult population sex ratio. Natural populations of *A. charrua* and *A. reicherti* vary markedly in OSR over the course of one season. At the beginning of the reproductive season, the OSR did not deviate from parity, whereas later in the reproductive season the OSR became female biased (Passos, 2013; Passos et al., 2014). In captivity, we have not detected any sex differences in mortality rate or biases in the adult sex ratio, suggesting that increased extrinsic mortality in males could lead to the observed biases in the OSR at the end of the season in natural populations. *Austrolebias* species exhibit intense male–male competition, often involving serious injuries and elaborate male courtship displays during reproduction (see above). A plausible explanation is that high male mortality stems from physical exhaustion resulting from courtship displays and male–male competition (Passos et al., 2014). Furthermore, males exhibit brighter coloration and show sexual displays that may incur an increased predation risk. Female-biased sex ratio in wild populations is also widespread in *Nothobranchius*, in which predation from predatory birds and hemipterans were implicated as potential sources of male-biased mortality (Haas, 1976a; Reichard et al., 2014). Mortality directly related to costly male–male interactions is also plausible in *Nothobranchius* (Reichard et al., 2014). Intriguingly, in both African and South American annual fishes, adult sex ratio often stabilizes at a ratio of 1 male to 2 females (Passos et al., 2014; Reichard et al., 2014), and the same adult sex ratio is reported for other small fishes with striking sexual dimorphism (Arendt et al., 2014).

In concert with the change in the OSR, female choosiness changes during the reproductive season in *A. reicherti* (Passos et al., 2014). At the start of the reproductive season, females strongly preferred larger males in simultaneous choice trials. In addition, the strength of female preference, measured as the difference in association time between stimuli, was positively related to the

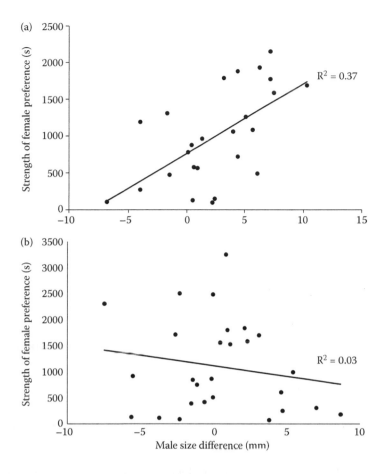

Figure 12.5 Female mate choice and male body size. Relationship between male size difference (standard length of preferred male—standard length of nonpreferred male) and strength of female preference (difference in time spent with preferred and nonpreferred males) in the annual killifish *Austrolebias reicherti* at the beginning (a) and end (b) of the reproductive season. Negative scores indicate more time with smaller males, positive scores indicate more time with larger males. (Modified from Passos, C. et al. 2014. *PLoS ONE* 9: e101648.)

asymmetry in male body size. However, female preference for larger males vanished entirely at the end of the reproductive season. Females were equally likely to choose small or large males later in the season, and female preference was not affected by the size difference between stimulus males (Figure 12.5). Different alternatives, not mutually exclusive, could explain this change (Passos et al., 2014). First, the reduction in choosiness could arise from diminished reproductive prospects due to a decline in male availability (Borg et al., 2006; Kokko and Johnstone, 2002). As the sex ratio becomes female biased late in the season, the opportunity for female choice may be reduced by a decline in rate of encounters with males (Kokko and Mappes, 2005). Second, the distinctive life history of annual fishes, where both life expectancy and mating opportunities decline rapidly over time, could account for the seasonal change in female choice. At the end of the reproductive season, all adults die due to extreme abiotic conditions. As this critical time approaches, a female may be faced with a trade-off in which she will have to decrease choosiness or risk losing the opportunity to fertilize her eggs. Finally, while female preference for body size is reduced over the course of the season, it is not necessarily the case that females become entirely nonselective. Females may assign varying weights to different male traits over the course of the season (Borg et al., 2006;

Forsgren et al., 2004). In *N. korthausae*, virgin females mated indiscriminately, while nonvirgin females of the same age were more selective. While this change in female choosiness runs counter the pattern in *Austrolebias*, indiscriminate mating of virgin females may have been due to a decrease in choosiness arising from their elevated motivation to spawn (Moore and Moore, 2001).

The fitness consequences for females of mating with certain males may vary throughout the season; females might then be expected to change how they respond to multiple male cues as environmental conditions or individual phenotypes change (Candolin, 2003; Jennions and Petrie, 1997; Qvarnström et al., 2000). Larger males have a competitive advantage in male–male competition (Passos et al., 2013a), and females could obtain direct benefits from mating with them early in the reproductive season to gain access to preferred areas for oviposition. However, the advantages of male size and the benefits to females of choosing large males may be reduced late in the reproductive season, because lower male density may decrease the intensity of male–male competition. The reduced direct benefits are a potential driver of the lack of preference for male size at the end of the reproductive season. Therefore, female choice could be based on other cues if traits other than male size become relatively more important for female reproductive success later in the reproductive season (Passos et al., 2014). Moreover, the extremely variable habitat provided by temporary ponds exposes fishes to drastic changes in water quality (particularly turbidity), and females may use different sensory modalities in mate choice across the life cycle. Future work will need to address whether females attend to different male traits over the course of a season and to disentangle the mechanisms leading to female behavioral changes, notably direct responses to demographic factors, environmental cues, or intrinsic changes during senescence. The general characteristics of *Austrolebias*, especially those related to their life cycle under great environmental and demographic changes, make annual fishes an exceptional and innovative model to contribute to the overall understanding of seasonal variation in mate choice.

There is also the potential for mate selection in males. Though traditionally it is argued that males are not selective and maximize reproductive success by increasing the number of copulations versus quality of partners (Bateman, 1948), theoretical approaches also suggest the occurrence of mutual choice even in species where each sex conforms to conventional roles (Bergstrom and Real, 2000; Edward and Chapman, 2011; Kokko and Johnstone, 2002). Male guppies (Herdman et al., 2004) and swordtails (Wong et al., 2005) show mating preferences despite females being the limiting sex. In fishes, bigger is often better from the point of view of males choosing females (e.g., Côté and Hunte, 1989; Pelabon et al., 2003). The reason is that female fecundity is typically an increasing function of body size (Charnov, 1993); this is indeed the case in *A. vandenbergi* (Schalk et al., 2014), *Cynopoecilus melanotaenia* (Gonçalves et al., 2011), and *N. furzeri* (Vrtílek and Reichard, 2015). There may therefore be a benefit to males of mating with larger females, although male preference remains to be evaluated in this system.

12.5 REPRODUCTIVE BEHAVIOR AND SPECIATION

The understanding of how new species arise, how they remain separate from other species, and how distinct species exchange genes are all major goals of current research in evolutionary biology. According to the biological species concept (Mayr, 1963), species are considered to constitute pools of interbreeding individuals that are substantially reproductively isolated from other such pools. Reproductive isolation is manifested through reproductive isolating barriers that decrease the probability of offspring production between individuals from different pools. The central problem of speciation is therefore the study of how reproductive isolation evolves to prevent actual or potential gene flow among populations. Its solution requires identification of the traits involved in the reduction of gene flow among populations and understanding of the evolutionary forces that have acted on these traits (Coyne and Orr, 2004). One of the most significant recent developments in speciation theory has been the increased attention given to sexual selection as an evolutionary force capable of rapidly inducing reproductive isolation among populations (Box 12.1). Although long recognized as important in intrapopulation evolution, sexual selection has more recently been invoked as a key promoter of both the origin and maintenance of new biological species (Kraaijeveld et al., 2011; Maan and Seehausen, 2011; Panhuis et al., 2001; Reding et al., 2013; Ritchie, 2007). Sexual selection can cause the rapid divergence of sexually dimorphic traits associated with mate acquisition (Andersson, 1994; Fisher, 1930). Many of these traits are involved in mate recognition and result in assortative mating within and among species (Coyne and Orr, 2004). Therefore, it has been proposed that sexual selection can facilitate divergence of mate-recognition systems among populations and thus, incidentally, lead to speciation (review in Panhuis et al., 2001; Ritchie, 2007).

Both theoretical (Higashi et al., 1999; Lande, 1981; Pomiankowski and Iwasa, 1998; Schluter and Price, 1993; Turelli et al., 2001; West-Eberhard, 1983) and empirical studies (e.g., Alexander and Breden, 2004; Boul et al., 2007; Seehausen and van Alphen, 1998; Seehausen et al., 1997) have shown the potential of sexual selection to contribute to the formation of reproductive barriers. Comparative studies have also suggested that sexual selection can contribute to and accelerate speciation (reviewed in Ritchie, 2007). Sexual selection is thought to have played an important role in major adaptive radiations, for example, Hawaiian *Drosophila* (Boake, 2005) and East African cichlid fishes (Seehausen et al., 1997).

Annual fishes constitute a useful model system for studying the role of sexual selection in reproductive isolation and speciation. The natural habitats of these fishes are temporary ponds that can be isolated from one another fairly easily, so interruption of gene flow among populations occurs quite often (Bartáková et al., 2013), though ponds may also be periodically connected during major floods. The potential repercussions of sexual selection for divergence and speciation are especially evident in geographically isolated populations, since sexual selection can drive the evolution of signaling and preference traits in divergent directions and thus result in prezygotic reproductive isolation among populations. On the other hand, those cases in which sexual selection has led to speciation should be characterized by significant differentiation in male sexual signals (Ritchie, 2007). In the "*A. adloffi*" species group, species differ almost exclusively in male coloration, while the general morphology and female coloration are quite uniform (Costa, 2006; Loureiro, 2004). Until now, there have been no indications of a notable ecological differentiation within of the "*A. adloffi*" species group, and molecular phylogenetic studies of this group showed that species divergence likely occurred recently through a burst process of multiple simultaneous speciation (García, 2006). This pattern is consistent with arbitrary divergence of male traits and female preferences via a Fisherian runaway process (Lande, 1981).

Within this group, *A. reicherti* and *A. charrua* in Uruguay have parapatric distributions along the southwestern lowlands of Laguna Merín, with a contact zone in the lower basin of the Río Cebollati (García et al., 2009), and hybrids are found wherever these species come into contact (Passos et al., unpublished data). Like other species in the group, *A. reicherti* and *A. charrua* are similar in morphology and behavior as well as in female coloration; however, they are strikingly different in male coloration, exhibiting divergent patterns of pigmentation of unpaired fins and in the design of dark vertical bands on body flanks (Loureiro and García, 2008). In *A. reicherti* males, the dark bands are narrower or equal in width to the lighter spaces between bands, and the unpaired fins are uniformly pigmented with a vertical black line on the posterior edge of the dorsal and anal fins. In *A. charrua*, the dark bands are equal or wider than the lighter spaces between bands, and the design of the unpaired fins consist of light spots over a dark background (Loureiro and García, 2008). These species are a very promising model for investigating the role of sexual selection in reproductive isolation and elucidating mechanisms underlying divergence and speciation in this lineage. We used choice experiments to investigate reproductive isolation between *A. reicherti* and *A. charrua*. In both species, females preferred to mate with their conspecific males, while males did not discriminate between females. Moreover, heterospecific discrimination by females was stronger in areas of sympatry than allopatry

(C. Passos et al., unpublished data). Currently, we are investigating the dimensionality of mate choice underlying reproductive isolation by identifying the cues that mediate assorted mating among species. Hence, determining the traits that underlie assortative mating will allow us to understand how reproductive isolation emerges during speciation and how it is maintained.

Nothobranchius furzeri and *N. orthonotus* are closely related species (Dorn et al., 2014) that regularly occur syntopically. Given the relatively smaller role that female choice plays in the reproductive system of *Nothobranchius* compared to *Austrolebias*, females may be forced to heterospecific matings. It is notable that heterospecific mating may have been driven by male coercion, that is, large *N. orthonotus* males were able to coerce mating from *N. furzeri* females, while spawning in the opposite combination was difficult to induce experimentally (Polačik and Reichard, 2011). The cost to females from a heterospecific mating has been suggested to be lower than the potential cost from male aggression when a female refuses to spawn (Polačik and Reichard, 2011). However, in wild populations females may have more opportunities to escape male attention than under laboratory conditions. Despite that, heterospecific mating was observed in the wild, and in the laboratory the F1, F2, and backcross progeny between *N. furzeri* and *N. orthonotus* is at least partly viable (Polačik and Reichard, 2011). Several *Nothobranchius* individuals of potentially hybrid origin (based on phenotypic traits) were also collected in the wild and await confirmation by genotyping. The sister pair of species, *N. furzeri* and *N. kadleci*, readily hybridize in the laboratory (Ng'oma et al., 2014). The two species have allopatric distributions but would likely naturally hybridize to an extent similar to *A. reicherti* and *A. charrua* should they come into secondary contact. In contradiction to previous assumptions (Dorn et al., 2011), a recent advanced calibration of its phylogenetic tree proposes that most *Nothobranchius* species have a Quaternary origin (Dorn et al., 2014). If this is true, range shifts and speciation in *Nothobranchius* are very dynamic, and prezygotic barriers to heterospecific mating in areas of secondary sympatry are crucial for reproductive isolation (or the lack of it). Annual fishes can therefore become a very useful group for studies on the role of sexually selected reproductive isolation in secondary sympatry, as opposed to the sympatric divergence implicated in radiation of lacustrine fish species (Seehausen et al., 2008).

ACKNOWLEDGMENTS

CP and BT were supported by the Sistema Nacional de Investigadores (ANII), Uruguay. MR acknowledges financial support from the Czech Science Foundation, GACR P505/12/G112.

REFERENCES

Agbali, M., M. Reichard, A. Bryjová, J. Bryja, and C. Smith. 2010. Mate choice for non-additive genetic benefits correlate with MHC dissimilarity in the rose bitterling (*Rhodeus ocellatus*). *Evolution* 64: 1683–1696. doi: 10.1111/j.1558-5646.2010.00961.x.

Alexander, H.J., and F. Breden. 2004. Sexual selection and extreme morphological divergence in the Cumaná guppy: A possible case of incipient speciation. *Journal of Evolutionary Biology* 17: 1238–1254. doi: 10.1111/j.1420-9101.2004.00788.x

Almeida, O.G., A. Miranda, P.C. Hubbard et al. 2005. Urine as a social signal in the Mozambique tilapia (*Oreochromis mossambicus*). *Chemical Senses* 30: i309–i310. doi: 10.1093/chemse/bjh238

Amorim, M.C.P., J.M. Simoes, P.J. Fonseca, and G.F. Turner. 2008. Species differences in courtship acoustic signals among five Lake Malawi cichlid species (*Pseudotropheus* spp.). *Journal of Fish Biology* 72: 1355–1368. doi: 10.1111/j.1095-8649.2008.01802.x

Amundsen, T., and E. Forsgren. 2003. Male preference for colourful females affected by male size in a marine fish. *Behavioral Ecology and Sociobiology* 54: 55–64. doi: 10.1007/s00265-003-0593-4

Andersson M. 1994. *Sexual Selection*. Princeton: Princeton University Press.

Arendt, J.D., D.N. Reznick, and A. Lopez-Sepulcre. 2014. Replicated origin of female-biased adult sex ratio in introduced populations of the Trinidadian guppy (*Poecilia reticulata*). *Evolution* 68: 2343–2356. doi: 10.1111/evo.12445

Barata, E.N., P.C. Hubbard, O.G. Almeida, A. Miranda, and A.V.M. Canário. 2007. Male urine signals social rank in the Mozambique tilapia (*Oreochromis mossambicus*, Peters 1852). *BMC Biology* 5: 54. doi: 10.1186/1741-7007-5-54

Bartáková, V., M. Reichard, K. Janko et al. 2013. Strong population genetic structuring in an annual fish, *Nothobranchius furzeri*, suggests multiple savannah refugia in southern Mozambique. *BMC Evolutionary Biology* 13: 196. doi: 10.1186/1471-2148-13-196

Bass, A.H., and J.R. McKibben. 2003. Neural mechanisms and behaviors for acoustic communication in teleost fish. *Progress in Neurobiology* 69: 1–26. doi: 10.1016/S0301-0082(03)00004-2

Bateman, A.J. 1948. Intra-sexual selection in *Drosophila*. *Heredity* 2: 349–368.

Beaugrand, J.P., D. Payette, and C. Goulet. 1996. Conflict outcome in male green swordtail fish dyads (*Xiphophorus helleri*): Interaction of body size, prior dominance/subordination experience, and prior residency. *Behaviour* 133: 303–319. doi: 10.1163/156853996X00161

Belote, D.F., and W.J.E.M. Costa. 2002. Reproductive behavior patterns in the neotropical annual fish genus *Simpsonichthys* Carvalho, 1959 (Cyprinodontiformes, Rivulidae): Description and phylogenetic implications. *Boletim do Museu Nacional, Nova Série, Rio de Janeiro* 489: 1–10.

Belote, D.F., and W.J.E.M. Costa. 2003. Reproductive behavior of the Brazilian annual fish *Cynolebias albipunctatus* Costa and Brasil, 1991 (Teleostei, Cyprinodontiformes, Rivulidae): A new report of sound production in fishes. *Arquivos do Museu Nacional, Rio de Janeiro* 61: 241–244.

Belote, D.F., and W.J.E.M. Costa. 2004. Reproductive behavior patterns in three species of the South American annual fish genus *Austrolebias* Costa, 1998 (Cyprinodontiformes, Rivulidae). *Boletim do Museu Nacional, Nova Série, Rio de Janeiro* 514: 1–7.

Benson, K., and A.L. Basolo. 2006. Male–male competition and the sword in male swordtails, *Xiphophorus helleri*. *Animal Behaviour* 71: 129–137. doi: 10.1016/j.anbehav.2005.05.004

Berglund, A., A. Bisazza, and A. Pilastro. 1996. Armaments and ornaments: An evolutionary explanation of traits of dual utility. *Biological Journal of the Linnean Society* 58: 385–399. doi: 10.1111/j.1095-8312.1996.tb01442.x

Bergstrom, C.T., and L.A. Real. 2000. Towards a theory of mutual mate choice: Lessons from two-sided matching. *Evolutionary Ecology Research* 2: 493–508.

Blažek, R., M. Polačik, and M. Reichard. 2013. Rapid growth, early maturation and short generation time in African annual fishes. *EvoDevo* 4: 24. doi: 10.1186/2041-9139-4-24

Bleckmann, H. 1993. Role of the lateral line in fish behavior. In: *The Behavior of Teleost Fish*, ed. T.J. Pitcher, 201–246. London: Chapman & Hall.

Boake, C.R.B. 2002. Sexual signaling and speciation, a microevolutionary perspective. *Genetica* 116: 205–214. doi: 10.1023/A: 1021224207727

Boake, C.R.B. 2005. Sexual selection and speciation in Hawaiian *Drosophila*. *Behavior Genetics* 35: 297–303. doi: 10.1007/s10519-005-3221-4

Borg, Å.A., E. Forsgren, and T. Amundsen. 2006. Seasonal change in female choice for male size in the two-spotted goby. *Animal Behaviour* 72: 763–771. doi: 10.1016/j.anbehav.2005.11.025

Boul, K.E., W.C. Funk, C.R. Darst, D.C. Cannatella, and M.J. Ryan. 2007. Sexual selection drives speciation in an Amazonian frog. *Proceedings of the Royal Society of London, Series B* 264: 399–406. doi: 10.1098/rspb.2006.3736

Bridge, A.P., R.W. Elwood, and J.T. Dick. 2000. Imperfect assessment and limited information preclude optimal strategies in male–male flights in the orb-weaving spider *Metellina mengei*. *Proceedings of the Royal Society of London, Series B* 267: 273–279. doi:10.1098/rspb.2000.0997

Burley, N.T., and V.S. Foster. 2006. Variation in female choice of mates: Condition influences selectivity. *Animal Behaviour* 72: 713–719. doi: 10.1016/j.anbehav.2006.01.017

Butlin, R., and M. Ritchie. 1994. Behaviour and speciation. In: *Behaviour and Evolution*, eds. P.J. Slater, and T.R. Halliday, 43–79. Cambridge: Cambridge University Press.

Candolin, U. 1999. Male–male competition facilitates female choice in sticklebacks. *Proceedings of the Royal Society of London, Series B* 266: 785–789. doi: 10.1098/rspb.1999.0706

Candolin, U. 2003. The use of multiple cues in mate choice. *Biological Reviews* 78: 575–595. doi: 10.1017/S1464793103006158

Candolin, U. 2004. Opposing selection on a sexually dimorphic trait through female choice and male competition in a water boatman. *Evolution* 58: 1861–1864. doi: 10.1111/j.0014-3820.2004.tb00470.x

Charnov, E.L. 1993. *Life History Invariants*. Oxford: Oxford University Press.

Costa, W.J.E.M. 2001. The neotropical annual fish genus *Cynolebias* (Cyprinodontiformes: Rivulidae): Phylogenetic relationships, taxonomic revision and biogeography. *Ichthyological Exploration of Freshwaters* 12: 333–383.

Costa, W.E.J.M. 2006. The South American annual killifish genus *Austrolebias* (Teleostei: Cyprinodontiformes: Rivulidae): Phylogenetic relationships, descriptive morphology and taxonomic revision. *Zootaxa* 1213: 1–162.

Costa, W.J.E.M., T.P.A. Ramos, L.C. Alexandre, and R.T.C. Ramos 2010. *Cynolebias parnaibensis*, a new seasonal killifish from the Caatinga, Parnaíba River basin, northeastern Brazil, with notes on sound producing courtship behavior (Cyprinodontiformes: Rivulidae). *Neotropical Ichthyology* 8: 283–288. doi: 10.1590/S1679-62252010000200006

Coté, I.M., and W. Hunte. 1989. Male and female choice in the redlip blenny: Why bigger is better. *Animal Behaviour* 38: 78–88. doi: 10.1016/S0003-3472(89)80067-3

Cox, C.R., and B.J. LeBoeuf. 1977. Female incitation of male competition: A mechanism in sexual selection. *The American Naturalist* 111: 317–335.

Coyne, J.A., and H.A. Orr. 2004. *Speciation*. Sunderland: Sinauer Associates, Inc.

Crapon de Caprona, M.D., and M.J. Ryan. 1990. Conspecific mate recognition in swordtails, *Xiphophorus nigrensis* and *X. pygmaeus*: Olfactory and visual cues. *Animal Behaviour* 39: 290–296. doi: 10.1016/S0003-3472(05)80873-5

Culumber, Z.W., and G.G. Rosenthal. 2013. Mating preferences do not maintain the tailspot polymorphism in the platyfish, *Xiphophorus variatus*. *Behavioral Ecology* 24: 1286–1291. doi: 10.1093/beheco/art063

Danley, P.D., M. Husemann, and J. Chetta. 2012. Acoustic diversity in Lake Malawi's rock-dwelling cichlids. *Environmental Biology of Fishes* 93: 23–30. doi:10.1007/s10641-011-9886-z

Darwin, C. 1871. *The Descent of Man, and Selection in Relation to Sex*. London: Murray.

Desjardins, J.K., and R.D. Fernald. 2008. How do social dominance and social information influence reproduction and the brain? *Integrative and Comparative Biology* 48: 596–603. doi: 10.1093/icb/icn089

Dorn, A., Z. Musilová, M. Platzer, K. Reichwald, and A. Cellerino. 2014. The strange case of East African annual fish: Aridification correlates with diversification for a savannah aquatic group? *BMC Evolutionary Biology* 14: 210. doi: 10.1186/s12862-014-0210-3

Dorn, A., E. Ng'oma, E. Janko et al. 2011. Phylogeny, genetic variability and colour polymorphism of an emerging animal model: The short-lived annual *Nothobranchius* fishes from southern Mozambique. *Molecular Phylogenetics and Evolution* 61: 739–749. doi: 10.1016/j.ympev.2011.06.010

Doutrelant, C., and P.K. McGregor. 2000. Eavesdropping and mate choice in female fighting fish. *Behaviour* 137: 1655–1669. doi: 10.1163/156853900502763

Edward, D.A., and T. Chapman. 2011. The evolution and significance of male mate choice. *Trends in Ecology & Evolution* 26: 647–654. doi: 10.1016/j.tree.2011.07.012

Endler, J.A., and A.L. Basolo. 1998. Sensory ecology, receiver biases and sexual selection. *Trends in Ecology & Evolution* 13: 415–420. doi: 10.1016/S0169-5347(98)01471-2

Enquist, M., and O. Leimar. 1983. Evolution of fighting behavior: Decision rules and assessment of relative strength. *Journal of Theoretical Biology* 102: 387–410. doi: 10.1016/0022-5193(83)90376-4

Enquist, M., O. Leimar, T. Ljungberg, Y. Mallner, and N. Segerdahl. 1990. A test of the sequential assessment game: Fighting in the cichlid fish *Nannacara anomala*. *Animal Behaviour* 40: 1–14. doi: 10.1016/S0003-3472(05)80660-8

Fabra, A. 2011. Comportamiento agonístico en machos de *Austrolebias reicherti*. Degree dissertation, Universidad de la República, Uruguay.

Fekete, D.M. 2003. Rocks that roll zebrafish. *Science* 302: 241–242. doi: 10.1126/science.1091171

Fernald, R.D. 2009. Social regulation of reproduction: What changes and why? *Hormones, Brain and Behavior* 1: 683–691.

Fisher, H.S., S. Mascuch, and G.G. Rosenthal. 2009. Multivariate male traits misalign with multivariate female preferences in the swordtail fish, *Xiphophorus birchmanni*. *Animal Behaviour* 78: 265–269. doi: 10.1016/j.anbehav.2009.02.029

Fisher, H.S., and G.G. Rosenthal. 2006. Hungry females show stronger mating preferences. *Behavioral Ecology* 17: 979–981. doi: 10.1093/beheco/arl038

Fisher, H.S., B.B.M. Wong, and G.G. Rosenthal. 2006. Alteration of the chemical environment disrupts communication in a freshwater fish. *Proceedings of the Royal Society of London, Series B* 273: 1187–1193. doi: 10.1098/rspb.2005.3406

Fisher, R.A. 1930. *The Genetical Theory of Natural Selection.* Oxford: Clarendon Press.

Forsgren, E., T. Amundsen, Å.A. Borg, and J. Bjelvenmark. 2004. Unusually dynamic sex roles in a fish. *Nature* 429: 551–554. doi: 10.1038/nature02562

García, D., M. Loureiro, and B. Tassino. 2008. Reproductive behavior in the fish *Austrolebias reicherti* Loureiro and García 2004 (Cyprinodontiformes: Rivulidae). *Neotropical Ichthyology* 6: 243–248. doi: 10.1590/S1679-62252008000200012

García, G. 2006. Multiple simultaneous speciation in killifishes of the *Cynolebias adloffi* species complex (Cyprinodontiformes, Rivulidae) from phylogeography and chromosome data. *Journal of Zoological Systematics and Evolutionary Research* 44: 75–87. doi: 10.1111/j.1439-0469.2005.00346.x

García, G., M. Loureiro, N. Berois et al. 2009. Pattern of differentiation in the annual killifish genus *Austrolebias* (Cyprinodontiformes: Rivulidae) from a biosphere reserve site in South America: A multidisciplinary approach. *Biological Journal of the Linnean Society* 98: 620–635. doi: 10.1111/j.1095-8312.2009.01303.x

Gonçalves, C., U.P. Souza, and M.V. Volcan. 2011. The opportunistic feeding and reproduction strategies of the annual fish *Cynopoecilus melanotaenia* (Cyprinodontiformes: Rivulidae) inhabiting ephemeral habitats on southern Brazil. *Neotropical Ichthyology* 9: 191–200. doi: 10.1590/S1679-62252011000100019

Gray, S.M., and J.S. McKinnin. 2007. Linking color polymorphism maintenance and speciation. *Trends in Ecology & Evolution* 22: 71–79. doi: 10.1016/j.tree.2006.10.005

Haas, R. 1976a. Sexual selection in *Nothobranchius guentheri* (Pisces-Cyprinodontidae). *Evolution* 30: 614–622.

Haas, R. 1976b. Behavioral biology of the annual killifish, *Nothobranchius guentheri*. *Copeia* 1976: 80–91.

Hack, M.A. 1997. Assessment strategies in the contests of male crickets, *Acheta domesticus* (L). *Animal Behaviour* 53: 733–747. doi: 10.1006/anbe.1996.0310

Herdman, E.J.E., C.D. Kelly, and J.J. Godin. 2004. Male mate choice in the guppy (*Poecilia reticulata*): Do males prefer larger females as mates? *Ethology* 110: 97–111. doi: 10.1111/j.1439-0310.2003.00960.x

Higashi, M., G. Takimoto, and N. Yamamura. 1999. Sympatric speciation by sexual selection. *Nature* 402: 523–526. doi: 10.1038/990087

Hsu Y., R.L. Earley, and L.L. Wolf. 2006. Modulation of aggressive behavior by fighting experience: Mechanisms and contest outcomes. *Biological review of the Cambridge Philosophical Society* 81: 33–74. doi: 10.1017/S146479310500686X

Hunt, J., R. Brooks, and M.D. Jennions. 2005. Female mate choice as a condition-dependent life-history trait. *The American Naturalist* 166: 79-92. doi: 10.1086/430672

Hurd, P.L. 1997. Cooperative signaling between opponents in fish fights. *Animal Behaviour* 54: 1309–1315. doi: 10.1006/anbe.1997.0531

Jennions, M.D., and M. Petrie. 1997. Variation in mate choice and mating preferences: A review of causes and consequences. *Biological Review* 72: 283–327. doi: 10.1111/j.1469-185X.1997.tb00015.x

Kirkpatrick, M. 1987. The evolutionary forces acting on female mating preferences in polygynous animals. In: *Sexual Selection: Testing the Alternatives*, eds. J.W. Bradbury, and M.B. Andersson, 67–82. Chichester: Wiley.

Kodric-Brown, A. 1998. Sexual dichromatism and temporary colour changes in the reproduction of fishes. *American Zoologist* 38: 70–81. doi: 10.1093/icb/38.1.70

Kodric-Brown, A., and U. Strecker. 2001. Responses of *Cyprinodon maya* and *C. labiosus* females to visual and olfactory cues of conspecific and heterospecific males. *Biology Journal of the Linnean Society* 74: 541–548. doi: 10.1111/j.1095-8312.2001.tb01411.x

Kokko, H., R. Brooks, M.D. Jennions, and J. Morley. 2003. The evolution of mate choice and mating biases. *Proceedings of the Royal Society of London, Series B* 270: 653–664. doi: 10.1098/rspb.2002.2235

Kokko, H., and R.A. Johnstone. 2002. Why is mutual mate choice not the norm? Operational sex ratios, sex roles, and the evolution of sexually dimorphic and monomorphic signalling. *Philosophical Transactions of the Royal Society of London, Series B* 357: 319–330. doi: 10.1098/rstb.2001.0926

Kokko, H., and J. Mappes. 2005. Sexual selection when fertilization is not guaranteed. *Evolution* 59: 1876–1885. doi: 10.1111/j.0014-3820.2005.tb01058.x

Kraaijeveld, K., F.J. Kraaijeveld-Smit, and M.E. Maan. 2011. Sexual selection and speciation: the comparative evidence revisited. *Biological Reviews* 86: 367–377. doi: 10.1111/j.1469-185X.2010.00150.x

Lande, R. 1981. Models of speciation by sexual selection on polygenic traits. *Proceedings of the National Academy of Sciences USA* 78: 3721–3725.

Lehtonen, T.K., and K. Lindström. 2008. Repeatability of mating preferences in the sand goby. *Animal Behaviour* 75: 55–61. doi: 10.1016/j.anbehav.2007.04.011

Levesque H.M., D. Scaffid, C.N. Polkinghorne, and P.W. Sorensen. 2011. A multi-component species identifying pheromone in the goldfish *Journal of Chemical Ecology* 37: 219–227. doi: 10.1007/s10886-011-9907-6

Libran, F. 2011. Comportamiento reproductivo en los peces anuales *Austrolebias affinis* Amato 1986. Degree dissertation, Universidad de la República, Uruguay.

Lobel, P.S. 1998. Possible species specific courtship sounds by two sympatric cichlid fishes in Lake Malawi, Africa. *Environmental Biology of Fishes* 52: 443–452. doi: 10.1023/A: 1007467818465

Loureiro, M. 2004. Sistemática y biogeografía de los peces anuales de la subtribu Cynolebiatina (Cyprinodontiformes: Rivulidae: Cynolebiatinae). PhD dissertation, Universidad de la República, Uruguay.

Loureiro, M., and G. García. 2008. *Austrolebias reicherti* Loureiro & García, a valid species of annual fish (Cyprinodontiformes: Rivulidae) from southwestern Laguna Merín basin. *Zootaxa* 1940: 1–5.

Macías Garcia, C., and E. Ramirez. 2005. Evidence that sensory traps can evolve into honest signals. *Nature* 434: 501–505. doi: 10.1038/nature03363

MacLaren, R.D., and D. Daniska. 2008. Female preferences for dorsal and body size in *Xiphophorus helleri*: further investigation of the LPA bias in Poesiliid fishes. *Behaviour* 145: 897–913. doi: 10.1163/156853908784089289

Maan, M.E., and O. Seehausen. 2011. Ecology, sexual selection and speciation. *Ecology Letters* 14: 591–602. doi: 10.1111/j.1461-0248.2011.01606.x

Maruska, K.P., and R.D. Fernald. 2012. Contextual chemosensory urine signaling in an African cichlid fish. *The Journal Experimental Biology* 215: 68–74. doi: 10.1242/jeb.062794

Maruska, K.P., U.S. Ung, and R.D. Fernald. 2012. The African cichlid *Astatotilapia burtoni* uses acoustic communication for reproduction: sound production, hearing, and behavioral significance. *PLoS ONE* 7: e37612. doi: 10.1371/journal.pone.0037612

Maynard Smith, J., and D. Harper. 1988. The evolution of aggression: Can selection generate variability? *Philosophical Transactions of the Royal Society of London, Series B* 319: 557–570.

Maynard Smith, J., and G.A. Parker. 1976. The logic of asymmetric contests. *Animal Behaviour* 24: 159–175.

Mayr, E. 1963. *Animal Species and Evolution*. Cambridge: Harvard University Press.

McLennan, D.A., and M.J. Ryan. 1997. Responses to conspecific and heterospecific olfactory cues in the swordtail *Xiphophorus cortezi*. *Animal Behaviour* 54: 1077–1088. doi: 10.1006/anbe.1997.0504

McLennan, D.A., and M.J. Ryan. 1999. Interspecific recognition and discrimination based upon olfactory cues in northern swordtails. *Evolution* 53: 880–888. doi: 10.2307/2640728

Milinski M., S. Griffiths, K.M. Wegner et al. 2005. Mate choice decisions of stickleback females predictably modified by MHC peptide ligands. *Proceedings of the National Academy of Sciences USA* 102: 4414–4418. doi: 10.1073/pnas.0408264102

Montgomerie, R., and R. Thornhill. 1989. Fertility advertisement in birds: a means of inciting male-male competition. *Ethology* 81: 209–220. doi: 10.1111/j.1439-0310.1989.tb00767.x

Moore, A.J., and P.J. Moore. 1999. Balancing sexual selection through opposing mate choice and male competition. *Proceedings of the Royal Society of London, Series B* 266: 711–716. doi: 10.1098/rspb.1999.0694

Moore, P.J., and A.J. Moore. 2001. Reproductive aging and mating: The ticking of the biological clock in female cockroaches. *Proceedings of the National Academy of Sciences USA* 98: 9171–9176. doi: 10.1073/pnas.161154598

Moretz, J.A. 2003. Aggression and RHP in the northern swordtail fish, *Xiphophorus cortezi*: the relationship between size and contest dynamics in male–male competition. *Ethology* 109: 995–1008. doi: 10.1046/j.0179-1613.2003.00938.x

Morris, M.R., M. Mussel, and M.J. Ryan. 1995. Vertical bars on male *Xiphophorus multilineatus*: a signal that deters rival male and attracts females. *Behavioral Ecology* 6: 274–279. doi: 10.1093/beheco/6.3.274

Morris, M.R., P.F. Nicoletto, and E. Hesselman. 2003. A polymorphism in female preference for a polymorphic male trait in the swordtail fish *Xiphophorus cortezi*. *Animal Behaviour* 65: 45–52. doi: 10.1006/anbe.2002.2042

Moskalik, B., and G.W. Uetz. 2011. Female hunger state affects mate choice of a sexually selected trait in a wolf spider. *Animal Behaviour* 81: 715–722. doi: 10.1016/j.anbehav.2010.12.016

Nelissen, M.H.J. 1991. Communication. In: *Cichlid Fishes: Behaviour, Ecology and Evolution*, ed. M.H.A. Keenleyside, 225–240. London: Chapman Hall.

Ng'oma, E., M. Groth, R. Ripa, M. Platzer, and A. Cellerino. 2014. Transcriptome profiling of natural dichromatism in the annual fishes *Nothobranchius furzeri* and *Nothobranchius kadleci*. *BMC Genomics* 15: 754. doi: 10.1186/1471-2164-15-754

Owens, I.P.F., and D.B.A. Thompson. 1994. Sex differences, sex ratios and sex roles. *Proceedings of the Royal Society of London, Series B* 258: 93–99. doi: 10.1098/rspb.1994.0148

Panhuis, T.M., R. Butlin, M. Zuk, and T. Tregenza. 2001. Sexual selection and speciation. *Trends in Ecology & Evolution* 16: 364–371. doi: 10.1016/S0169-5347(01)02160-7

Parker, G.A. 1974. Assessment strategy and evolution of fighting behavior. *Journal of Theoretical Biology* 47: 223–243. doi: 10.1016/0022-5193(74)90111-8

Passos, C. 2013. *Austrolebias*: un modelo para explorar la selección sexual. PhD dissertation, Universidad de la República, Uruguay.

Passos, C., F. Reyes, B. Tassino, G.G. Rosenthal, and A. González. 2013b. Female annual killifish *Austrolebias reicherti* (Cyprinodontiformes, Rivulidae) attend to male chemical cues. *Ethology* 119: 891–897. doi: 10.1111/eth.12129

Passos, C., B. Tassino, M. Loureiro, and G.G. Rosenthal. 2013a. Intra- and intersexual selection on male body size in the annual killifish *Austrolebias charrua*. *Behavioral Process* 96: 20–26. doi: 10.1016/j.beproc.2013.01.008

Passos, C., B. Tassino, F. Reyes, and G.G. Rosenthal. 2014. Seasonal variation in female mate choice and operational sex ratio in wild populations of an annual fish, *Austrolebias reicherti*. *PLoS ONE* 9: e101648. doi: 10.1371/journal.pone.0101649

Pelabon, C., A.A. Borg, J. Bjelvenmark, E. Forsgren, I. Barber, and T. Amundsen. 2003. Do male two-spotted gobies prefer large fecund females? *Behavioral Ecology* 14: 787–792. doi: 10.1093/beheco/arg074

Plenderleith, M., C. van Oosterhout, R.L. Robinson, and G.F. Turner. 2005. Female preference for conspecific males based on olfactory cues in a Lake Malawi cichlid fish. *Biology Letters* 1: 411–414. doi: 10.1098/rsbl.2005.0355

Polačik, M., and M. Reichard. 2009. Indirect fitness benefits are not related to male dominance in a killifish. *Behavioral Ecology and Sociobiology* 63: 1427–1435. doi: 10.1007/s00265-009-0798-2

Polačik, M., and M. Reichard. 2011. Asymmetric reproductive isolation between two sympatric annual killifish with extremely short lifespans. *PLoS One* 6: e22684. doi: 10.1371/journal.pone.0022684

Pomiankowski, A., and Y. Iwasa. 1998. Runaway ornament diversity caused by Fisherian sexual selection. *Proceedings of the National Academy of Sciences USA* 96: 5106–5111.

Prenter, J., P.W. Taylor, and R.W. Elwood. 2008. Large body size for winning and large swords for winning quickly in swordtail males, *Xiphophorus helleri*. *Animal Behaviour* 75: 1981–1987. doi: 10.1016/j.anbehav.2007.12.008

Pryke, S.R., and S.C. Griffith. 2006. Red dominates black: agonistic signalling among head morphs in the colour polymorphic Gouldian finch. *Proceedings of the Royal Society B: Biological Sciences* 273: 949–957. doi: 10.1098/rspb.2005.3362

Ptacek, M.B., and J. Travis. 1997. Mate choice in the sailfin molly, *Poecilia latipinna*. *Evolution* 51: 1217–1231.

Qvarnström, A. 2001. Context-dependent genetic benefits from mate choice. *Trends in Ecology & Evolution* 16: 5–7. doi: 10.1016/S0169-5347(00)02030-9

Qvarnström, A., and E. Forsgren. 1998. Should females prefer dominant males? *Trends in Ecoloiy & Evolution* 13. 498–501. doi: 10.1016/S0169-5347(98)01513-4

Qvarnström, A., T. Pärt, and B. Sheldon. 2000. Adaptive plasticity in mate preference linked to differences in reproductive effort. *Nature* 405: 344–347. doi: 10.1038/35012605

Rafferty, N.E., and J.W. Boughman. 2006. Olfactory mate recognition in a sympatric species pair of threespined sticklebacks. *Behavioral Ecology* 17: 965–970. doi: 10.1093/beheco/arl030

Reding, L.P., J.P. Swaddle, and H.A. Murphy. 2013. Sexual selection hinders adaptation in experimental populations of yeast. *Biology Letters* 9: 20121202. doi: 10.1098/rsbl.2012.1202

Reichanbacher, B., and M. Reichard. 2014. Otoliths of five extant species of the annual killifish *Nothobranchius* from the East African savannah. *PLoS One*, accepted. doi: 10.1371/journal.pone.0112459

Reichard, M., S.C. Le Comber, and C. Smith. 2007. Sneaking from a female perspective. *Animal Behaviour* 74: 679–688. doi:10.1016/j.anbehav.2007.03.005

Reichard, M., and M. Poláčik. 2010. Reproductive isolating barriers between colour-differentiated populations of an African annual killifish, *Nothobranchius korthausae* (Cyprinodontiformes). *Biological Journal of the Linnean Society* 100: 62–72. doi: 10.1111/j.1095-8312.2010.01406.x

Reichard, M., M. Poláčik, R. Blažek, and M. Vrtílek. 2014. Female bias in the adult sex ratio of African annual fishes: Interspecific differences, seasonal trends and environmental predictors. *Evolutionary Ecology* 28: 1105–1120. doi: 10.1007/s10682-014-9732-9

Ritchie, M.G. 2007. Sexual selection and speciation. *Annual Review Ecology, Evolution, and Systematics* 38: 79–102. doi: 10.1016/S0169-5347(01)02160-7

Rosenthal, G.G. 2000. Design considerations and techniques for constructing video stimuli. *Acta ethologica* 3: 49–54. doi: 1 0.1007/s102110000024

Rosenthal, G.G. 2012. Individual mating decisions and hybridization. *Journal of Evolutionary Biology* 26: 252–255. doi: 10.1111/jeb.12004

Rosenthal, G.G., X.F. de la Rosa Reyna, S. Kazianis et al. 2003. Dissolution of sexual signal complexes in a hybrid zone between the swordtails *Xiphophorus birchmanni* and *Xiphophorus malinche* (Poeciliidae). *Copeia* 2: 299–307. doi: 10.1643/0045-8511(2003)003

Rosenthal, G.G., and C.S. Evans. 1998. Female preference for swords in *Xiphophorus helleri* reflects a bias for large apparent size. *Proceedings of the National Academy of Sciences USA* 95: 4431–4436.

Rosenthal, G.G., J.N. Fitzsimmons, K.U. Woods, G. Gerlach, and H.S. Fisher. 2011. Tactical release of a sexually-selected pheromone in a swordtail fish. *PLoS ONE* 6: e16994. doi: 10.1371/journal.pone.0016994

Rosenthal, G.G., and P.S. Lobel. 2006. Communication. In: *Behaviour and Physiology of Fish*, Vol. 24, eds. K. Sloman, S. Balshine, and R. Wilson, 39–78, New York: Academic Press.

Ryan, M.J. 1998. Sexual selection, receiver biases, and the evolution of sex differences. *Science* 281: 1999–2003. doi: 10.1126/science.281.5385.1999

Ryan, M.J., D.K. Hewes, and W.E.J. Wagner. 1990. Sexual selection on alleles that determine body size in the swordtail *Xiphophorus nigrensis*. *Behavioral Ecology and Sociobiology* 26: 231–237. doi: 10.1007/BF00178316

Sargent, R.C., V.N. Rush, B.D. Wisenden, and H.Y. Yan. 1998. Courtship and mate choice in fish: Integrating behavioral and sensory ecology. *American Zoologist* 38: 82–96. doi: 10.1093/icb/38.1.82

Satou, M., H.A Takeuchi, J. Nishii et al. 1994a. Behavioral and electrophysiological evidences that the lateral line is involved in the inter-sexual vibrational communication of the himé salmon (landlocked red salmon, *Oncorhynchus nerka*). *Journal of Comparative Physiology A* 174: 539–549. doi: 10.1007/BF00217373

Satou, M., H.A. Takeuchi, K. Takei, T. Hasegawa, T. Matsushima, and N. Okumoto. 1994b. Characterization of vibrational and visual signals which elicit spawning behavior in the male himé salmon (landlocked red salmon). *Journal of Comparative Physiology A* 174: 527–537. doi: 10.1007/BF00217372

Schalk, C.M., C.G. Montaña, and M.E. Libson. 2014. Reproductive strategies of two Neotropical killifish, *Austrolebias vandenbergi* and *Neofundulus ornatipinnis* (Cyprinodontiformes: Rivulidae) in the Bolivian Gran Chaco. *Revista de Biología Tropical* 62: 109–117. http://dx.doi.org/10.15517/rbt.v62i1.6567.

Schluter, D., and T. Price. 1993. Honesty, perception and population divergence in sexually selected traits. *Proceedings of the Royal Society of London, Series B* 253: 117–122. doi: 10.1098/rspb.1993.0089

Seehausen, O., Y. Terai, I.S. Magalhaes et al. 2008. Speciation through sensory drive in cichlid fish. *Nature* 455: 620–626. doi: 10.1038/nature07285

Seehausen, O., and J.J.M. van Alphen. 1998. The effect of male coloration on female mate choice in closely related Lake Victoria cichlids (*Haplochromis nyererei* complex). *Behavioral Ecology and Sociobiology* 42: 1–8. doi: 10.1007/s002650050405

Seehausen, O., J.J.M. van Alphen, and F. Witte. 1997. Cichlid fish diversity threatened by eutrophication that curbs sexual selection. *Science* 277: 1808–1811. doi: 10.1126/science.277.5333.1808

Sinervo, B., and C.M. Lively. 1996. The rock-paper-scissors game and the evolution of alternative male strategies. *Nature* 380: 240–243. doi: 10.1038/380240a0

Sorensen, P.W., and T.R. Hoye. 2010. Pheromones in Vertebrates. In: *Comprehensive Natural Products II Chemistry and Biology*, eds. L. Mander, and H.W. Lui. Oxford: Elsevier.

Taylor, P.W., and R.W. Elwood. 2003. The mismeasure of animal contests. *Animal Behaviour* 65: 1195–1202. doi: 10.1006/anbe.2003.2169

Turelli, M., N.H. Barton, and J.A. Coyne. 2001. Theory and speciation. *Trends in Ecology & Evolution* 16: 330–342. doi: 10.1016/S0169-5347(01)02177-2

Valenzano, D.R., J. Kirschner, R.A. Kamber et al. 2009. Mapping loci associated with tail color and sex determination in the short-lived fish *Nothobranchius furzeri*. *Genetics* 183: 1385–1395. doi: 10.1534/genetics.109.108670

Vaz-Ferreira, R., and B. Sierra. 1972. Caracteres etológicos genéricos y específicos en los peces del género *Cynolebias* Steindachner. *Boletín de la Sociedad Zoológica del Uruguay* 2: 22–35.

Vaz-Ferreira, R., B. Sierra, and S. Scaglia. 1964. Eco-etología de la reproducción en los peces del género *Cynolebias* Steindachner, 1876. *Apartados de los Archivos de la Sociedad de Biología de Montevideo* 26: 44–49.

Verzijden, M.N., C. ten Cate, M.R. Srevedio, G.M. Kozak, J.W. Boughman, and E.I. Svensson. 2012. The impact of learning on sexual selection and speciation. *Trends in Ecology & Evolution* 27: 511–519. doi: 10.1016/j.tree.2012.05.007

Verzijden, M.N, J. van Heusden, N. Bouton, F. Witte, C. ten Cate, and H. Slabbekoorn. 2010. Sounds of male Lake Victoria cichlids vary within and between species and affect female mate preferences. *Behavioral Ecology* 21: 548–555. doi: 10.1093/beheco/arq018

Vrtílek, M., and M. Reichard 2015. Highly plastic resource allocation to growth and reproduction in females of an African annual fish. *Ecology of Freshwater Fish* doi:10.1007/s10682-014-9732-9.

West-Eberhard, M.J. 1983. Sexual selection, social competition and speciation. The *Quarterly Review of Biology* 58: 155–183.

Wildekamp, R.H. 2004. *A World of Killies: Atlas of the Oviparous Cyprinodontiform Fishes of the World.* Vol. 4. Elyria: American Killifish Association.

Wong, B.B.M. 2004. Superior fighters make mediocre fathers in the Pacific blue-eye fish. *Animal Behaviour* 67: 583–590. doi: 10.1016/j.anbehav.2003.08.015

Wong, B.B.M., and U. Candolin. 2005. How is female mate choice affected by male competition? *Biological Review of the Cambridge Philosophical Society* 80: 559–571. doi: 10.1017/S1464793105006809

Wong, B.B.M., H.S. Fisher, and G.G. Rosenthal. 2005. Species recognition by male swordtails via chemical cues. *Behavioral Ecology* 16: 818–822. doi: 10.1093/beheco/ari058

Zahavi, A. 1975. Mate selection—A selection for a handicap. *Journal of Theoretical Biology* 53: 205–214. doi: 10.1016/0022-5193(75)90111-3

CHAPTER 13

Comparative Anatomy and Proliferative Zones of Adult *Austrolebias* Brain

Gabriela Casanova Larrosa, Juan Carlos Rosillo, Silvia Olivera-Bravo, and Anabel S. Fernández

CONTENTS

The brain has diversified and advanced in evolution more than any other organ; the variety of nervous systems and behaviors among animal species is thus available for our exploitation. Comparative neuroscience is likely to reach insights so novel as to constitute revolutions in understanding the structure, functions, ontogeny, and evolution of nervous systems. This promise requires pursuit on a wide front, in respect to disciplines and in respect to the species, stages, and states compared. It also requires deliberate concentration on the differences among animals, in addition to the prevailing concern for the basic and common. Neglect of these challenges would be costly. Without due consideration of the neural and behavioral correlates of differences between higher taxa and between closely related families, species, sexes, and stages, we cannot expect to understand our nervous systems or ourselves.

Bullock 1984

13.1 INTRODUCTION

Teleosts are presently, by far, the most numerous group of bony fishes, encompassing about 28,000 species and thus representing the most abundant group of vertebrates (Nelson, 2006). As the earliest teleosts lived long before the earliest land vertebrates, they have had a longer time to differentiate and adapt during phylogeny and have indeed made use of this opportunity. Teleosts show a great variability in form and function, and are adapted to a variety of niches in both salt and freshwater. Consequently, in the long temporal window since their origin, the teleost central nervous system experimented greatly with variations depending on both phylogenetic and functional diversity (Meek and Nieuwenhuys, 1998).

Among vertebrates, annual fishes represent one of the most remarkable extremophiles. They share several features with other fish models; however, they exhibit unique traits related to their peculiar life cycle. They have unique adaptive strategies in reproduction and maintenance of embryos in extreme environmental situations. These attributes make them a useful model to study diverse topics in developmental biology using a comparative and evolutionary approach (Berois et al., 2012).

Herein we provide a description and comparison of the overall gross neuroanatomy among three species of annual fishes: *Austrolebias affinis*, *A. charrua*, and *A. reicherti*. Further, we focus our studies on the similarities and differences of the proliferative cell zones and their effects on neurogenic capacity in the brain of these three species. Analyses of the brain of *Austrolebias* provide new perspectives to the understanding of the variability of brain sizes and shapes exhibited by this vertebrate group. Moreover, comparative studies of areas of cell proliferation within the major cerebral divisions provide new tools for understanding the differential relative growth of the cerebral divisions and its relation to different sensorial modalities. Our research contributes to and is enriched by a multidisciplinary approach to understanding the evolutionary biology of the genus *Austrolebias* as undertaken by different research teams in Uruguay.

13.2 PHENOTYPE DIFFERENCES AMONG THE THREE STUDIED SPECIES

The phylogenetic history of the three species may be related to significant diversity in body characteristics. Molecular and morphological studies support that *A. charrua* and *A. reicherti* are sister taxa within the *A. adloffi* species group (Loureiro and García, 2008). Diversification in the *A. adloffi* species group has been triggered by events of explosive speciation that likely occurred during the Late Pleistocene (1.25 mya to 450,000 years ago; García et al., 2009, 2014). *Austrolebias charrua* and *A. reicherti* are similar in size but differ in (1) the origin and the number of rays of dorsal fins; (2) coloration patterns, for example, spots on the fins, and number, thickness, and separation of body vertical bands; and (3) the supraorbital bands, which is well-developed in *A. reicherti* but moderately developed in *A. charrua* (Loureiro and García, 2008).

Austrolebias affinis belongs to the *A. alexandri* species group (Costa, 2006; García et al., 2014); consequently, it is distantly related to the other two species included in this study. This is a smaller species, and its body coloration consists of white spots spread over the entire body with no vertical bands. Moreover, *A. reicherti* and *A. charrua* can occur in sympatry, whereas *A. affinis* is allopatric with respect to both the other species, occurring geographically distant with no possibility of geographic merging (Figure 13.1).

In spite of the three species belong to the same genus, genetic studies proposed that this group participates in current speciation events with the consequent genome expansion (García et al., 2014). The explosive speciation that the clade presents is reflected in the varied phenotype and also in the brain anatomy (Figure 13.2).

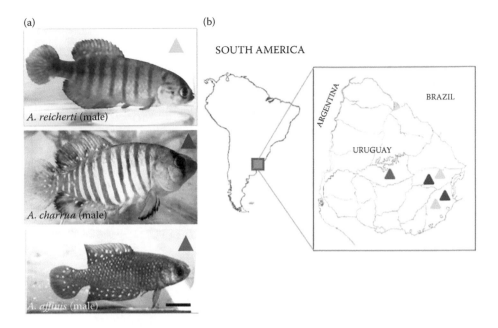

Figure 13.1   (a) Phenotype of adult males from each *Austrolebias* species studied. Note the differences in size and body shape, as well as the presence and pattern of bars or dots. (b) Geographical distribution and sites of collection of the three studied species in Uruguay. *Austrolebias affinis* (magenta) found in the department of Durazno, *A. reicherti* (green triangle) and *A. charrua* (red) cohabit the departments of Rocha and Treinta y Tres.

13.3 VERTEBRATE BRAIN AND THE ENVIRONMENT

The brains of living vertebrates are a reflection of the very diverse niches occupied by the different species that comprise each major taxon. The variation in complexity and relative brain size that exists across all living vertebrate groups and individual species is a direct function of the available niches and the adaptations of various species that successfully occupy them (Butler, 2009).

Bony fishes represent the largest radiation of vertebrates; they exhibit an enormous taxonomic and morphological diversity and occupy all available aquatic habitats. This diversity offers a great opportunity to relate ecological adaptations to morphology, including brain anatomy and sensory systems. Moreover, the anatomical variation among closely related species may represent trends favoring alternative evolutionary pressures toward functional specializations (sensory) of related structures as adaptations to environmental variation (Kotrschal et al., 1998).

Fish brains exhibit the arrangement of subdivisions typical of most vertebrates: forebrain, midbrain, and hindbrain. However, the anatomy of fish brains has greater variation when compared to other vertebrates (Northcutt and Davis, 1983; Meek and Nieuwenhuys, 1998). This variation can be partially explained by their phylogenetic history along with functional specializations that result from their interaction with the environment (Kotrschal et al., 1998). The brain of fishes follows the overall embryological and developmental patterns of other vertebrates with few unique differences. The formation of the nervous system begins at the early stages of the embryo development. A thickening of the ectoderm forming the neural plate will be transformed during neurulation into the neural tube. Subsequently, the most anterior part of neural tube will give rise to the brain vesicles, whereas the posterior portion develops into the spinal cord. In the later development, it is possible to observe the formation of the three main vesicles: forebrain or telencephalon, midbrain or mesencephalon, and hindbrain or rhombencephalon. These primary swellings will subdivide,

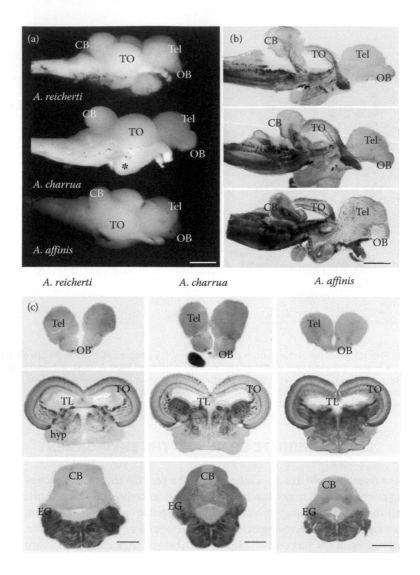

Figure 13.2   (a) Lateral views of the gross anatomy of *Austrolebias* brain showing the commonalities and differences in sizes and shapes. The differences in the cerebellum and olfactory bulbs (OB) are more pronounced when comparing *A. affinis* with the other two closely related species. The asterisk points to the hypothalamic lobes bulging from the ventral portion of *A. reicherti* brain. Prominent optic nerves are seen in all the species studied. (b) Corresponding vibratome parasagittal sections show the differences in brain with greater detail. (c) Unstained vibratome transverse sections showing the internal brain anatomy at different levels: telencephalon (upper), mesencephalon (middle), and rhombencephalon (bottom). Note the particular anatomical features of each species, such as the smaller size of the torus longitudinalis (TL) and the broader appearance of tectum opticum (TO) in *A. affinis*. *A. reicherti,* and *A. charrua* both present prominent OB and thinner TO. Different cerebellum (CB) profiles are also shown. Pictures show the myelinated fibers appearing dark, which allows the recognition of structures and regions even without any staining. Other abbreviations: Tel, telencephalon; OB, olfactory bulb; EG, eminentia granularis. Scale bar: 1 mm. (Reprinted from *Neuroscience*, 189, Fernández, A.S. et al., Proliferation zones in the brain of adult fish *Austrolebias* (Cyprinodontiform: Rivulidae): A comparative study, Figures 1 and 2, 12–24, Copyright (2011), with permission from Elsevier.)

and the forebrain gives rise to the paired telencephalic vesicles and the diencephalon, whereas the rhombencephalon will subdivide into the metencephalon and the myelencephalon. The relative development of these subdivisions is closely related to the phylogeny and adaptability of each species and has been suggested as the underlying cause of the wide variety in fish brains (Sanes et al., 2006). The development of the cerebral hemispheres of ray-finned fishes is radically different from that of all other vertebrates. The hemispheres increase in size and originally may have tended to an everted condition in which the medial hemispheres are separated by an extended sheet of choroid plexus (Aboitiz and Montiel, 2007). This process results in the eversion of the dorsal telencephalon, and the degree of eversion varies considerably among ray-finned fishes; consequently, the pallium (or dorsal telencephalon) shows significant structural diversity.

13.4 COMPARATIVE BRAIN ANATOMY OF THE STUDIED SPECIES

Among teleosts, *Austrolebias* is an annual genus belonging to the order Cyprinodontiformes (Atherinomorpha). Brain anatomy of Cyprinodontiformes has been reported for a few genera: *Fundulus heteroclitus* (Peter et al., 1975), *Xiphophorus helleri* (Anken and Rahmann, 1994), and *Nothobranchius furzeri* (D'Angelo, 2013); and also for *Oryzias latipes* (Ishikawa et al., 1999), a member of the order Beloniformes (Atherinomorpha), the closest relatives to Cyprinodontiformes (Saeed et al., 1994; Allen, 1998; Setiamarga et al., 2008). We reported the first study on the brain anatomy of *Austrolebias* (Fernández et al., 2011), and based on our results, we present here (1) an overview of its gross neuroanatomy, (2) a comparative analysis of the three species, and (3) a discussion of shared features and differences in a developmental/evolutionary context.

Overall, the brain anatomy of *Austrolebias* shares common features with various other groups of teleosts. These features do not show particular sensory or motor specialization, beyond those related to vision, which resulted in a relatively large tectum opticum (TO; Meek and Nieuwenhuys, 1998).

The adult brain of *Austrolebias* is characterized by a well-balanced development of the three major brain divisions (i.e., prosencephalon, mesenenphalon, and rhomboencephalon) and extends 4 mm into the rostrocaudal axis and 2 mm in the dorsoventral axis Figure 13.2. The overall anatomy of *A. charrua* and *A. reicherti* brains is very similar. They are characterized by the following traits: (1) prominent, rounded, and sessile olfactory bulbs (OB), placed anterior and ventral to the hemispheres of the telencephalon; (2) an everted and well-developed telencephalon; (3) a diencephalon with paired and large inferior hypothalamus lobes bulging out on the ventral surface of the brain; (4) two prominent tectal halves located dorsally to the midbrain tegmentum and diencephalon; (5) a large rhombencephalon with a single and prominent cerebellum that differs in size among species (Figure 13.2); and (6) a large medulla oblongata. The cranial nerves seen in the images are representative of the general patterns found in other described teleosts. In contrast, *A. affinis* shows a compact brain with more ventral and less defined OB, large telencephalic areas relative to the TO, and a smaller cerebellum that just passes the level of the TO. Regarding the relative propor tions between the size of telencephalic regions and TO among the species, *A. affinis* shows a large telencephalon, whereas in *A. reicherti* and *A. charrua* both regions are of the approximately equal size (Figure 13.2).

Regarding internal anatomy, transverse histological sections, at the same anatomical level, showed structural variations among the species (Figure 13.2). At the level of the telencephalon, *Austrolebias charrua* and *A. reicherti* have a notable increase in the size in dorsal-ventral axis, a more extended ventricular region, and broader interhemispheric distance relative to *A. affinis*. Furthermore, both species have a rounded and prominent OB, whereas *A. affinis* has triangular, smaller OB (Figure 13.2). In the mesencephalon there is the torus longitudinalis (TL), a small and unique medial structure found in ray-finned fishes that relays motor inputs to the TO. The TL is different among the three studied species. It is the largest and bilobed (appears with the shape of an inverted heart) in *A. reicherti*, whereas in the other two species it is a single medial lobe, prominent in *A. charrua*, and small in *A. affinis*.

The TO consists of relatively expanded lobes on each side of the brain with the TL located between both halves and the third or tectal ventricle extending laterally into each tectal lobe. The TO consists of various layers of cell and fibers, with dendrites of many neurons oriented in a radial way and afferent fibers coursing through it and in parallel to the surface. The tectal wall is thickest in *A. affinis* (Figure 13.2). Transverse sections of the hindbrain revealed clear differences in the size, shape, and proportion of different cerebellar formations. *Austrolebias reicherti* has a wider corpus cerebelli, whereas *A. charrua* has the most prominent eminentia granularis and *A. affinis* has the smallest cerebellum but a larger ventricular area.

Whole brain 3D reconstructions, made by stacking serial transverse sections, showed similar overall brain volumes in the three studied species. However, a previous study has shown species-specific differences in relative volumes among brain regions. *Austrolebias reicherti* has the most prominent TO and cerebellum, *A. charrua* has the largest OB, and *A. affinis* the smallest TL (Fernández et al., 2011). The differences found in brain anatomy among the studied species could be correlated to habitat differences, since visual conditions for different teleostean species may vary dramatically, particularly depending on water turbidity and depth (Gruber et al., 1990).

13.5 CELL PROLIFERATIVE ZONES AND NEUROGENESIS: BrdU RESULTS AND 3D BRAIN RECONSTRUCTIONS

Cell proliferation zones in the brain of adult teleosts have long been recognized (Kirsche, 1967). However, little is known about location and neuroanatomical description of fish brain cell proliferation and neurogenic zones and, among the teleosts, information is limited to *Apteronotus leptorhynchus*, *Gasterosteus aculeatus*, *Danio rerio*, and *Gymnotus omarorum* (Zupanc and Horschke, 1995; Ekström et al., 2001; Zupanc, 2006; Olivera-Pasilio et al., 2014). Early investigations of adult neurogenic niches in fishes, using classical histological approaches (Kirsche, 1967) and [³H]thymidine auto-radiography (Rahmann, 1968; Kranz and Richter, 1970), identified major areas of mitotic activity in the brain of teleosts. Analysis of the types of newly generated cells, particularly study of cell proliferation in the TO and the continued formation of photoreceptors in the retina (Meyer, 1978; Johns, 1982), evidenced that newly born cells in the teleostean adult brain are capable of neuronal differentiation (Raymond and Easter, 1983). This finding strongly evidenced the existence of adult neurogenesis in teleost fishes. Subsequent progress and studies on adult neurogenesis in teleost fishes include (1) mapping and quantitative analysis of proliferative activity in the brain of several teleost species (Zupanc and Horschke, 1995; Zikopoulos et al., 2000; Ekström et al., 2001; Zupanc et al., 2005; Grandel et al., 2006; Fernández et al., 2011); (2) demonstrated long-term persistence, neuronal differentiation, and integration of newly born neurons into already existing neural networks (Zupanc and Zupanc, 1992; Zupanc et al., 1996, 2005); (3) isolation and characterization of the stem cells that generate new neurons and glial cells (Hinsch and Zupanc, 2007); (4) potential of teleost fish to regenerate nervous tissue after injury by formation of new neurons (Zupanc and Zupanc, 2006); and (5) the identification of several proteins associated with neuronal regeneration (Zupanc and Zupanc, 2006). However, there are only few studies on adult neurogenesis in annual fishes: on the African *Nothobranchius* (Tozzini et al., 2012) and in *Austrolebias* among Neotropical annual fishes (Fernández et al., 2011).

13.6 BrdU METHOD TO IDENTIFY PROLIFERATING CELLS

We employed the most widely used marker to study cellular proliferation and identification of potentially neurogenic areas, 5-bromo-2′-deoxyuridine (BrdU). BrdU is a synthetic nucleotide analogue of thymidine. After injection, cells in the S phase of the cellular cycle incorporate BrdU to the DNA strand during replication. Subsequent detection, via immunohistochemistry, allows the identification of proliferating cells. Moreover, its combination with other antibodies allows determination of the lineage of newly generated cells (Rakic, 1974; Nowakowski and Hayes, 2000). Checking for location on the tissues as well as the type and number of labeled cells, it is possible to establish a chronology of cell proliferation and to evaluate its contribution to tissue composition.

13.7 BrdU APPLIED TO THE STUDY OF *AUSTROLEBIAS* BRAIN

BrdU was injected intraperitoneal to adult *Austrolebias* males that were processed 24 h later (Fernández et al., 2011). The brain of the studied species showed many proliferative zones with a concentration of BrdU-positive cells close to the ventricular walls. The pattern of cell proliferation in each region was similar in the three species. In the forebrain, the highest concentration of BrdU-positive cells was found ventrally on the rostral areas of the OB and on the transition region between the OB and the telencephalic lobes. In the latter structures, most of the BrdU-positive cells formed a narrow band contiguous to and following the overall shape of the ventricle. In the midbrain, the zones with the largest concentration of BrdU-positive nuclei were detected on the periphery of the TL and on its boundary with the medial poles of the TO. Whereas in the TO, BrdU-labeled cells were concentrated on the most superficial layer of the periventricular gray zone, equally distributed throughout the rostrocaudal axis. In the hindbrain, as in other teleosts, most of the BrdU-labeled cells were found on the granular layer of the corpus cerebelli, with few positive cells on the granular cell layer. In the caudal pole, BrdU cells were mostly concentrated on the region adjacent to the wall of the ventricle's granular layer. A peculiar distribution of BrdU nuclei that were arranged radially in the valvula cerebelli was found in *A. reicherti* (Fernández et al., 2011). This arrangement of BrdU nuclei is similar to, but more prominent than, the chains of "radial spokes" previously reported in the molecular layer up to the dorsal tip of the corpus cerebelli in zebrafish (Grandel et al., 2006).

Mapping and 3D reconstruction of proliferative zones of the *Austrolebias* showed a common pattern of distribution of BrdU-labeled cells characterized by the contiguity of the proliferation zones along the rostrocaudal axis throughout the brain. From the OB up to the medulla oblongata the concentration of BrdU-positive nuclei was located very close to the ventricular regions. Interestingly, cell proliferation in the cerebellum of *A. charrua* was concentrated on the medial portion, whereas in the other two species proliferation areas were distributed widely, forming lateral bands toward the periphery (Figure 13.3). This is in agreement with previous studies that reported cell proliferation zones located in the proximity of ventricles in a wide range of representative species of fishes, amphibians, reptiles, birds, and mammals (Kirsche, 1967; García-Verdugo et al., 2002; González-Granero et al., 2011). Furthermore, our 3D reconstruction of proliferation zones demonstrated that the contiguity of proliferative cells along the entire rostrocaudal brain axis in the three studied species is similar to that suggested in the mapping of some transverse sections of zebrafish brain (Hinsch and Zupanc, 2007).

13.8 NEUROGENESIS IN *AUSTROLEBIAS* BRAIN

Identification of proliferative cell lineages can be obtained by combining BrdU recognition with specific cell type markers. Double immunostaining for BrdU with the glial markers S-100 and glial acidic fibrillar protein (GFAP) showed that in the brain of *Austrolebias*, most of the BrdU-positive cells also coexpressed S-100. The analysis of colocalization of BrdU with the general neuronal marker HuC/D (Marusich and Weston, 1992) allowed identifying neurogenesis in *Austrolebias* as early as 24 h after BrdU injection. Newborn neurons were found in the OB, dorsal Tel, TO, TL, and CB in the three *Austrolebias* studied species. In Tel, TO, and TL, the newly born neurons were

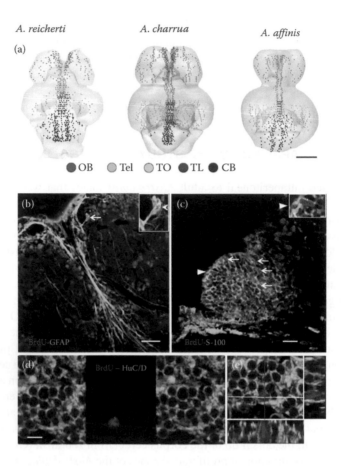

A. reicherti   A. charrua   A. affinis

● OB   ◑ Tel   ○ TO   ● TL   ● CB

Figure 13.3 (a) Three-dimensional reconstruction and topographical representation of *Austrolebias* brain cell proliferative zones. Brains were fully and serially cut and then reconstructed with the BioVis 3D software. Dorsal views of 3D brain reconstructions in which BrdU-labeled nuclei have a contiguous distribution along the medial line throughout the rostrocaudal axis and its adjacence to the ventricle wall. The colored dots represent the places where 5–8 BrdU nuclei were found. The pattern of colors in the circles refers to the regions. OB, olfactory bulb; Tel, telencephalic ventricles; TO, tectum opticum; TL, torus longitudinalis; CB, cerebellum. Scale bar: 1 mm. (b–c) Gliogenesis in *Austrolebias* brain. (b) Immunocytochemistry for the glial marker GFAP (green) and BrdU in a transverse section showing the medial line of the ventral zone of the tectal ventricle. A double-labeled cell is indicated with a white arrow in the panoramic view and in the higher-magnification inset. (c) Olfactory bulb transverse hemisection showing extensive double-labeling against the early glial marker S-100 (green) and BrdU (red). Colocalizations are indicated with white arrows. Arrowheads indicate one double-labeled cell at panoramic and higher magnification (inset), respectively. (d–e) Neurogenesis in *Austrolebias* brain. (d) Double immunostaining against the early neuronal marker HuC/D (green) and BrdU (red) in the granular zone of the cerebellum. An apparent colocalization was found in one newly born granular cell. (e) Stack from multiple confocal sections to demonstrate colocalization in the cell shown in (d). The stack of 6 μm contains 14 orthogonal planes, which analysis of the intersection point demonstrates to be the colocalization of both markers in the cell. Scales bars: (b) 40 μm, (c) 20 μm, (d, e) 15 μm. (Reprinted from *Neuroscience*, 189, Fernández, A.S. et al., Proliferation zones in the brain of adult fish *Austrolebias* (Cyprinodontiform: Rivulidae): A comparative study, Figures 8 and 9, 12–24, Copyright (2011), with permission from Elsevier.)

located close to the ventricular areas. Our results are significantly earlier than those reported for zebrafish, in which the earliest double-labeled cells (BrdU/HuC/D) were found in the telencephalon 3 days after BrdU administration and later (Adolf et al., 2006). Interestingly, zebrafish lifespan is about 3.5 years (Hinsch and Zupanc, 2007), whereas *Austrolebias* are annual fishes with a lifespan of about 9 months. Thus, the faster appearance of double-labeled proliferative cells with neuronal markers in *Austrolebias* could reflect an overall earlier or accelerated neurogenesis program in fishes with a shorter lifespan. In zebrafish, the proportion of double BrdU/HuC/D-labeled cells increased over 2 weeks (Adolf et al., 2006). Accordingly, our preliminary results show that at 1 week after BrdU injection, the population of BrdU/HuC/D-positive cells increased notably. Unlike mammals, fish body and brain grow throughout the lifespan. This continuous brain growth includes constitutive neurogenesis and a high rate of neuron turnover largely throughout the lifespan (Zupanc, 2011). The brain of the electric fish *Gymnotus omarorum* showed many proliferative zones adjacent to the ventricular walls (Olivera-Pasilio et al., 2014). Thus, allometries may account for lifelong changes in brain morphology.

13.9 QUANTITATIVE ANALYSIS OF CELL PROLIFERATION

Mitotic activity was found in all major divisions of the brain of adult *Austrolebias* with *A. reicherti* having the largest number of BrdU-positive cells. The cerebellum exhibited the highest value of cell proliferation in the three species. Quantitative analyses for other regions showed differences in OB, TO, and TL. *Austrolebias reicherti* and *A. charrua* showed similar values of cell proliferation in the OB and TO, but the TL of *A. charrua* has BrdU-positive cells that are about 50% of those found in *A. reicherti*. The brain of *A. affinis* showed the lowest proliferation value on the OBs and the highest in the TO. The TO of *Austrolebias* exhibited high mitotic activity, particularly on a specific area at the caudal pole, similar to previous reports for zebrafish (Zupanc et al., 2005). Proliferation in this region seems to have functional relevance for these macroptic fishes, given that visual input plays an important role during courtship (García et al., 2008). Among *Austrolebias* species, the TL was the region of the brain with the greatest anatomical differences, including size, shape, and proliferation rate. These differences may have functional importance, since the TL is functionally related to the TO, the eye movements, and visuotopic mapping (Butler and Hodos, 2005). As expected, the cerebellum, as the region that develops later (Hinsch and Zupanc, 2007), exhibited the largest number of mitotic cells in *Austrolebias*. This phenomenon has been reported for other teleosts and is linked to the late development of the cerebellum during postnatal life and parallel to the continuous fish brain growth throughout life (Zupanc and Horschke, 1995; Ekström et al., 2001; Zupanc, 2006). In summary, quantitative proliferation values found in *Austrolebias* are in agreement with data reported for zebrafish that exhibits abundant proliferation in the telencephalon, in the mantle zone of mesencephalon (TO and TL; Zupanc et al., 2005; Zupanc, 2006), and in other regions of the brain (Hinsch and Zupanc, 2007). The distribution of BrdU-labeled nuclei also resembles that reported for stickleback (Ekström et al., 2001). However, our results in *Austrolebias* differ from those reporting cell proliferation restricted to the ventricular zones in zebrafish (Grandel et al., 2006).

We analyzed cell proliferation in relation to the volume of each brain region, given that size of different brain regions varied. The volume of each region was calculated using 3D reconstruction by BioVis 3D software. Volume-normalized proliferation indexes were obtained by dividing the number of BrdU-positive nuclei by the volume of each brain region. These indexes were used as an initial approach to understanding the significance that cell proliferation may have in each brain zone and species. The TL shows the highest volume-normalized proliferation index of all brain regions analyzed: the TL of *A. affinis* has the highest index, almost three times that of *A. charrua*. Olfactory bulb indexes were the second largest, with *A. reicherti* having the highest value and *A. affinis* the lowest. The TO had the lowest normalized indexes among brain regions. The three species showed similar cerebellar proliferation indexes. Overall, except for the TL, other brain regions in *A. charrua* showed values intermediate to those of *A. affinis* and *A. reicherti*. It seems that the volume-normalized proliferation indexes found in OB and TO in the three species has inverse correlation. Whereas *A. reicherti* has the highest index value in the OB and the lowest in the TO, *A. affinis* has exactly the opposite (Figure 13.4). Differences in these values among species may be

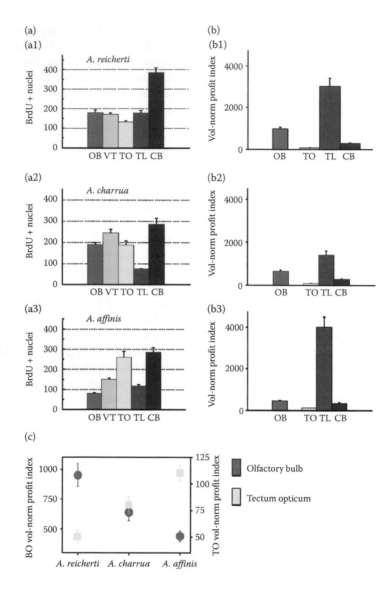

Figure 13.4 Quantitative analysis of total and volume-normalized cell proliferation in brain regions. (a) Whole BrdU-positive nuclei in olfactory bulb (OB), telencephalic ventricle zone (VT), tectum opticum (TO), torus longitudinalis (TL), and cerebellum (CB) in the three species studied. *Austrolebias reicherti* (a1) showed similar values in all regions except for the CB. *Austrolebias charrua* (a2) and *A. affinis* (a3) showed more heterogeneous patterns. TL and OB were the regions with less cell proliferation. (b) Volume-normalized proliferation indexes shown by *A. reicherti* (b1), *A. charrua* (b2), and *A. affinis* (b3) brain. The index was calculated as the ratio between the number of proliferating cells and the volume of each region determined with BioVis 3D. Values were calculated only for the regions that could be delimited in 3D brain reconstructions. The TL had the highest index value in the three species. OB was the second-most proliferative region but had different values among species. The CB showed the minor proliferation index in all species. (c) Inverse correlation of volume-normalized proliferation indexes in OB (pink) and TO (cyan). Whereas OB of *A. reicherti* had the highest index and TO the lowest, the inverse of that shown for *A. affinis*. *Austrolebias charrua* exhibited an intermediate behavior. In all cases, descriptive statistical analysis was made with Sigma Stat 2.0. Data were plotted using Origin 8.0. (Reprinted from *Neuroscience*, 189, Fernández, A.S. et al., Proliferation zones in the brain of adult fish *Austrolebias* (Cyprinodontiform: Rivulidae): A comparative study, Figure 7, 12–24, Copyright (2011), with permission from Elsevier.)

functionally important. *Austrolebias affinis* showed high volume-normalized proliferation indexes in vision-related areas (TO and TL), but lowest values in OBs. Conversely, *A. reicherti* exhibited inverse patterns; this may point to differences in the hierarchy of sense modalities between these species. Furthermore, it may suggest that proliferation capacity may be correlated to phylogenetic relationship and/or shared habitat. Volume-normalized proliferation indexes of regions involved in different sensorial modalities suggest an evolutionary trend favoring olfaction. In this context, *A. reicherti* has highest proliferation indexes in OB and lowest in the TO, whereas *A. affinis* showed the inverse pattern. Indeed, vision is important in *Austrolebias*, because males are elaborately colored, and both coloration and size play an important role in sexual courtship (García et al., 2008; Passos et al., 2014). In addition, recent behavioral studies suggest that olfaction directs *A. reicherti* females to find and select males (Passos, 2013), suggesting that olfaction seems at least equally as important as vision in this species.

Recently, we have described the centrifugal olfacto-retinal system of *Austrolebias*, a constant component of the CNS that appears to exist in all vertebrates and is part of the terminal nerve (TN) complex (Rosillo et al., 2013). TN allows the integration of different sensory modalities, and its anatomical variability may have functional and evolutionary significance. We suggest that in *Austrolebias*, the olfacto-retinal branch of the TN is an important anatomical link that allows the functional interaction between olfactory and visual systems. The olfacto-retinal TN branch of *Austrolebias* is well developed and complex, consisting of different neuronal groups. Moreover, the presence of constitutive gonadotropin-releasing hormone (GnRH)-positive neurons indicates a role in sexual and reproductive behaviors (Kawai et al., 2009). Furthermore, *Austrolebias* presents a rostral group of neurons that is the main component of the link between retinal and olfactory pathways that has not been previously reported in other teleosts and may suggest a higher hierarchy of TN *Austrolebias*. The circuital organization of the pathway changes depending on the group of vertebrate analyzed (Kawai et al., 2009). Small differences may be related to the hierarchy of each sensory modality, even among closely related species such as the three species of *Austrolebias* studied. Consequently, some of the observed differences in the olfacto-retinal pathway in *Austrolebias* may be explained by ecological differences in habitats. The ponds inhabited by *Austrolebias* differ in vegetation, with varying light conditions, particularly depending on the turbidity and depth of the water (Zucker and Dowling, 1987), which may influence the sensory modalities used. Thus, the proliferative and neurogenic activity of each brain region involved in different sensory modalities can reflect the functional activity of this area.

13.10 CHARACTERISTICS OF THE NEUROGENIC NICHES

During embryogenesis, the folding of the neural ectoderm gives rise to the CNS primordium, and its cavities will constitute the future ventricular brain regions. In adults the "ventricular zone" is formed by the layer of cells that lines the brain ventricles. Cells present in these zones maintain their proliferative and neurogenic capacity similar to what happens during embryo neurogenesis (Gross, 2000; García-Verdugo et al., 2002). Thus, these zones are called "matrix zones" or "neurogenic niches" and constitute the embryonic reservoirs from which differentiation can result in neuronal or glial cells, according to the system's needs. The fine structure of these proliferative sites has been extensively studied in mammals (Alvarez-Buylla et al., 2001) and is less known in reptiles and birds (García-Verdugo, 2002; González-Granero et al., 2011). Neurogenic niches are formed by a variety of types of cells. In mammals, these niches contain four different cell types: multiciliated ependymal cells, (type E), astrocytes (type B), transit-amplifying cells (type C), and migrating neuroblasts (type A) (Doetsch et al., 1999). A mix of radial glial cells, ependymal multiciliated cells, and proliferative neuroblasts has also been described in the brains of adult canary birds (Alvarez-Buylla and

Temple, 1998) and reptiles (García-Verdugo et al., 2002). The ventricular zone of the telencephalon in zebrafish has three types of cells: GFAP+ radial glial cells, PCNA+ proliferative cells, and PSA-NCAM+ neuroblasts, revealing similarities to and differences from the ventricular neurogenic niches of mammal brains (Sawada and Sawamoto, 2013).

In mammals the cells proposed as the "neural stem cell" (NSC) are the type B astrocytes, whereas Pax6-positive radial glial cells (Götz and Barde, 2005) are proposed as NSCs in reptiles and birds. Both cell types preserve some features of the monociliated cells that constitute the neural tube. However, a current controversial issue regarding neurogenic niches is the determination of the structural characteristics that identify the NSCs. Neural stem cells are defined as cells that possess the ability to self-renew and generate the three major cell types in the CNS (neurons, astrocytes, and oligodendrocytes in mammals; Gage, 2000; Temple, 2001; Okano, 2002). NSCs originate from the pseudostratified monociliated epithelium of the neural tube and preserve the primary cilium projecting into the ventricle lumen as a structural landmark (Mori et al., 2005; Chapouton et al., 2007; Kazanis et al., 2008; Rodríguez, 2012) that is essential for retaining pluripotent capacities (Mizradeh et al., 2011; Wang et al., 2009).

Another remarkable difference among adult vertebrates is related to the number of the NSC reservoirs. Whereas in adult mammalian brain neurogenic niches are restricted to the subventricular and subgranular sites, many of those niches are reported in different sites of brain ventricular walls in nonmammalian vertebrates (García-Verdugo et al., 2002; Sawada and Sawamoto, 2013). Similar to their avian relatives, reptiles contain only one big primary neurogenic compartment that extends through the whole periventricular zone of the lateral ventricular walls. Moreover, proliferation in the adult reptilian brain follows a pattern similar to embryonic brain development (Lindsey and Tropepe, 2006).

Studies in teleost fishes described cells with proliferative and neurogenic capacities in the neighborhood of the ventricles in many regions of the CNS, but the ultrastructure of those sites has been studied only in a few species (Zupanc, 2006; Rosillo et al., 2010; Sawada and Sawamoto, 2013). A comparative study showed that these mitotically active regions have both characteristics common to all groups and some characteristics unique to each group studied (Doetsch and Scharff, 2001). In *Austrolebias* there are numerous proliferative zones lining the ventricles of the telencephalon, the TO, and the TL (Fernández et al., 2011). Such regions exhibit different cell populations similar to those observed in other neurogenic niches (Rosillo, 2010).

13.11 SUBCELLULAR MORPHOLOGICAL ANALYSIS OF THE TELENCEPHALIC VENTRICULAR EPITHELIUM OF *AUSTROLEBIAS*

A transmission electron microscopy (TEM) analysis of the ventricular zone of the telencephalon (TVZ) identified in at least three well-differentiated cell types: (1) Type I cells lining the ventricles that present with heterochromatic nuclei (with variable profiles) and an electron-dense cytoplasm due to the abundant free ribosomes and rough endoplasmic reticulum. Most of these cells have a single cilium oriented toward the ventricular lumen. (2) Type II cells bordering the ventricular lumen intermingled among type I cells or just a line behind it. Their cytoplasm is lighter than type I cells and has a remarkable development of the cytoskeleton with large areas occupied by bundles of filaments that suggest a glial nature. The nuclear envelope has indentations, and the chromatin is homogeneous with a moderate degree of condensation. Some type II cells show a single cilium. (3) Type III cells are arranged one or two cell lines away from the ventricular lumen. These cells have an abundant and electron-lucid cytoplasm with a large, spherical, and euchromatic nucleus with one or two prominent nucleoli, strongly suggesting that type III cells probably are newly formed neurons. The immunocytochemistry images obtained with the antineuronal HuC/D antibody confirmed the neuronal identity of these cells (Rosillo et al., 2010).

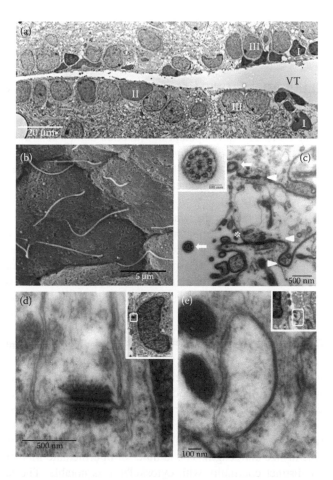

Figure 13.5 TEM and SEM ultrastructural features of *A. charrua* ventricular zone of the telencephalon (TVZ). (a) Dorsoventral panoramic view obtained from thin-section overlapped TEM pictures, evidencing three well-differentiated cell types: type I cells characterized by the most electron-dense nucleus of variable profiles and dark cytoplasm enriched in ribosomes; type II cells that are close to the ventricle and show nuclei with intermediate electron density and size. These cells in general are bigger than I type. Type III cells are young neurons bearing electron-lucid and round nuclei with one or two nucleoli. (From Rosillo, J.C. et al. 2010. *Microscopica Acta* (Figure 2d) 19: 152–159. With permission.) (b) SEM topographic view of TVZ evidencing that all of the cells that line the wall bear a single cilium. All cilia present are similar in length and appearance and have a thinner end. (c) TEM micrograph showing different cell membrane specializations, including microvilli (*), different intercellular junctions (arrowhead), and cilium (arrow). The inset shows the typical microtubule structure of a single cilium consisting of the arrangement of 9 + 2 pairs of microtubules. (d) TEM micrograph showing the presence of a prominent desmosome interacting with long intermediate filaments of cell cytoskeleton. The white box in the inset shows the location of the desmosome. (e) TEM of gap junction connecting neighboring cell membranes. The box in the inset shows the panoramic location of the intercellular communication zone.

TEM analysis of the ciliar axoneme of type I and II cells (Figure 13.5) showed that it is composed of nine doublets of outer microtubules surrounding a central pair in the characteristic "9 + 2" pattern. The minus axonemal microtubule ends are bound to the basal body located in the cytoplasm (Figure 13.5). This basal body has a structure similar to a centriole (9 triplets + 0) and acts as an organizing center for microtubules in axoneme formation (Dawe et al., 2006). The cilium ranges in length from 8 to 10 μm. We did not find multiciliated cells in the ventricular wall of the telencephalon of *Austrolebias* (Rosillo et al., 2010). This finding was contrary to the abundance of multiciliated cells reported for the ventricular walls of mammals and the few ones described in reptiles and zebrafish (Sawada and Sawamoto, 2013).

13.12 INTERCELLULAR JUNCTIONAL COMPLEXES AT NEUROGENIC NICHES

Most of the cell junctions have been described in different regions of vertebrate CNS, mainly in mammals (Peters et al., 1991). The development of "junction complexes" among cells of the brain proliferative zones is particularly interesting due to their potential roles in the dynamics of cell division in neurogenic niches. Moreover, apically localized cell junctions and the apicobasal polarity complexes have been implicated in the asymmetric cell division of neural proliferative cells (Farkas and Huttner, 2008; Hatakeyama, 2014).

In *Austrolebias* brain, we found that the ventricular cells form an intricate network of processes that overlap and interdigitate each other, showing a significant development of tight, gap, and adherens junctions, as wells as desmosomes, that bind and connect cells to each other (Rosillo et al., 2010; Casanova et al., 2014). Tight junctions are a network of transmembrane proteins always located between adherens junctions and the ventricular lumen (Grupp et al., 2010). We observed that the membranes of neighboring cells in the ventricles of *Austrolebias* have tight junctions forming points of cell-to-cell adhesion that seal the more apical intercellular space of the ventricular wall. Similarly, junctional complexes in the glial membranes were found on the ventricular surfaces of the TO and telencephalon of zebrafish brain (Grupp et al., 2010). In this species, radial glial cells contacting the ventricle bear tight junctions and form a kind of barrier between the ventricular cerebrospinal fluid and the neuronal compartment. Close to the tight junctions, it is often possible to observe the presence of adherens junctions. In *Austrolebias*, adherens junctions were found among almost all the cells that line the ventricular lumen of the telencephalon. TEM analysis showed (Figure 13.5) that each adherens junction is a 30-nm-thick plaque of dense material placed in the cytoplasmic face of both plasma membranes (Casanova et al., 2014). In *Austrolebias*, ventricular desmosomes are well developed and variable in size and length. The number of intermediate filaments anchored to each desmosome and their distinct continuity with cytoskeleton is notable. The lateral membranes of adjacent cells often interdigitate and sometimes attach through desmosome-like junctions (Casanova et al., 2014). In addition, to have complexes that join together, the cells of the ventricular walls connect with each other via gap junctions. Gap junctions consist of transmembrane proteins (i.e., connexins, Cx in the text) that form complexes called connexons that play a role coordinating cells and the exchange of molecules and ions, enhancing the chemical and electrical coupling among cells. In neurogenic precursors, functional coupling via gap junctions has been reported as a basic and phylogenetically conserved mechanism (Nakase and Naus, 2004). This coupling appears to involve both neural precursors and radial glia (Bittman et al., 1997). In mammals, the cells of the ventricle walls express Cx43 and Cx26 (Bittman and Lo Turco, 1999), as also occurs in the central channel of the spinal cord of turtles, another neurogenic niche (Russo et al., 2006). In adult *Austrolebias*, glial cells are often connected to each other through gap junctions that express Cx43. Furthermore, these gap junctions are generally located close to desmosomes and to the ventricular surface, extending along large surfaces. Gap junctions are also present among cell interdigitations (Casanova et al., 2014).

High proliferation and the presence of undifferentiated cells bearing a single cilium that projects toward the ventricular lumen, together with cell heterogeneity found in the TVZ of *Austrolebias*, are consistent with the characteristics reported for other neurogenic niches (García-Verdugo et al., 2002; Sawada and Sawamoto, 2013). Moreover, the appearance and distribution of the three cell types described here are consistent with those reported on the wall of the third ventricle of the brain in the teleost *Eigenmannia* sp. (Zupanc, 1992). In addition, the presence of electron-dense nuclei, with marked indentations (equivalent to the *Austrolebias* type I cells), in *Eigenmannia* sp. also suggest high mitotic activity (Zupanc and Zupanc, 1992). Type II cells show ultrastructural characteristics that are comparable with those S100+ cells described by Zupanc (1992) as "intermediate cells." Their characteristics suggest that they could be proposed to be NSC s of the *Austrolebias* TVZ. Finally, the characteristics of type III cells coincide with those previously identified as "young and migrating neurons" (Zupanc, 2001).

13.13 CONCLUSIONS

The analysis of the comparative brain anatomy of the genus *Austrolebias* allows us to conclude that these annual fishes share common features with other teleosts that show no particular specializations beyond those related to vision. However, the *Austrolebias* species studied showed anatomical and cell proliferation differences across the main brain divisions. In addition to having comparable brain volumes, *A. reicherti* and *A. charrua* showed the greatest similarity in terms of size, shape, and arrangement of the larger cerebral divisions. These similarities are consistent with their phylogenetic relationship and/or because they inhabit similar environments. However, both species showed significant size and shape differences in the TL. *Austrolebias affinis*, a more distantly related species, has significant differences in the TL but also in the OB, TO, and cerebellum. These differences could be attributed to variations in the environments. Visual conditions for different teleost species may vary dramatically depending on water turbidity and depth and the hierarchy of sensory modalities used. Using BrdU, we identified, quantified, and mapped cell proliferation throughout the entire brain of the three analyzed species. Significant cell proliferation was found in all brain subdivisions, showing remarkable continuity along the rostrocaudal axis and proximity to ventricle walls. Patterns of proliferation areas were similar to those reported for other teleosts. However, the brain regions related to different sensory modalities exhibited different proliferation indexes depending on phylogeny or environment. The largest variations in cell proliferation indexes were found in the OB and TO. Whereas *A. reicherti* presents the highest values in the OB and the lowest in the TO, *A. affinis* exhibits the inverse pattern. Indeed, vision is important in *Austrolebias* for social communication, sexual courtship, and mating, but other studies suggest that olfaction seems at least equally important in *A. reicherti*. In summary, the different proliferation indexes may suggest a high plasticity that probably reflects plastic adaptations to diverse environmental pressures.

Coimmunolabeling of BrdU with cell lineage markers showed that adult *Austrolebias* have gliogenesis and early neurogenesis throughout all brain regions. We suggest that the rapid appearance of BrdU+ cells bearing neuronal markers in *Austrolebias* could reflect a relatively rapid neurogenesis program related to their short lifespan. Adult neurogenesis could enable the dynamic remodeling of mature neuronal circuits by adding new neurons, contributing to structural and functional plasticity in the adult brain. In this context, adult neurogenesis seems a fundamental mechanism for experience-dependent plasticity, which is the most important feature of the adult brain. Recent reports have indicated that environmental stimuli, such as exercise, exploration, enriched environments, learning, and memory, improve the rate of the generation of new neurons (Kempermann, 2011). Adult neurogenesis could also have an important role in the injured brain in regenerating lost neurons from adult NSCs.

The complex biological events that occur in neurogenesis are only possible due to the particular features of neurogenic zones. Studies performed in the proliferative zones of *Austrolebias* showed that neurogenic niches have characteristics in common with other vertebrates, including the variety of cell types. In *Austrolebias*, we recognized glial and neuronal cells in distinct differentiation stages and also undifferentiated cells that could not be classified in any lineage. The presence of a single cilium in most cells lining the ventricular lumen seems a unique feature for *Austrolebias*. In addition, the absence of multiciliated cells contacting the ventricular lumen, reported in neurogenic niches of mammals and reptiles, seems also unique in these fishes. Another particular feature found in *Austrolebias* is the existence of many types of intercellular junctions, including desmosomes and gap junctions. This degree of organization suggests a highly dynamic zone where the intercellular junction may play a role in cell movements during mitotic activity. On the other hand, anchoring junctions could be important in the maintenance of the everted epithelia of *Austrolebias* forebrain.

REFERENCES

Aboitiz, F. and J. Montiel. 2007. *Origin and Evolution of the Vertebrate Telencephalon, with Special Reference to the Mammalian Neocortex.* Berlin: Springer, Science & Business Media.

Adolf, B., P. Chapouton, C.S. Lam et al. 2006. Conserved and acquired features of adult neurogenesis in the zebrafish telencephalon. *Developmental Biology* 295: 278–293. doi: 10.1016/j.ydbio.2006.03.023

Alberts, B., A. Johnson, J. Lewis, M. Raff, K. Roberts, and P. Walter. 2008. Cell junctions, cell adhesion, and the extracellular matrix. In: *The Cell.* 5th edition. 1131–1204. New York: Garland Science.

Allen, G.R. 1998. *Encyclopedia of Fishes.* J.R. Paxton and W.N. Eschmeyer (eds.). 153–156. San Diego, CA: Academic Press. ISBN 0-12-54766.

Alvarez-Buylla, A. 1992. Neurogenesis and plasticity in the CNS of adult birds. *Experimental Neurology* 115: 110–114. doi: 10.1016/0014-4886(92)90232-F

Alvarez-Buylla, A. and J.M. García-Verdugo. 2002. Neurogenesis in adult subventricular zone. *The Journal of Neuroscience* 22: 629–634. doi: 0270-6474/02/220629-06$15.00/0

Alvarez-Buylla, A., J.M. García-Verdugo, and A.D. Tramontin. 2001. A unified hypothesis on the lineage of neural stem cells. *Nature Reviews Neuroscience* 2: 287–293. doi: 10.1038/35067582.

Alvarez-Buylla, A., C.Y. Ling, and N. Nottebohm. 1992. High vocal center growth and its relation to neurogenesis, neuronal replacement and song acquisition in juvenile canaries. *Journal of Neurobiology* 23: 396–406. doi: 10.1002/neu.480230406

Alvarez-Buylla, A. and S. Temple. 1998. Stem cells in the developing and adult nervous system. *Journal of Neurobiology* 36: 105–110. doi: 10.1002/(SICI)1097-4695(199808)36: 2<105: : AID-NEU1>3.0.CO;2-5

Alvarez-Buylla, A., M. Theelen, and F. Nottebohm. 1990. Proliferation "Hot Spots" in adult avian ventricular zone reveal radial cell division. *Neuron* 5: 101–109. doi: 10.1016/0896-6273(90)90038-H

Anken, R.H. and Rahmann, H. 1994. *Brain Atlas of the Adult Swordtail Fish Xiphophorus helleri and of Certain Developmental Stages.* Gustav Fisher Verlag, Stuttgart, Jena, New York.

Banerjee, S., A.D. Souza, and M.A. Bhat. 2006. Organization and function of septate junctions: An evolutionary perspective. *Cell Biochemistry and Biophysics* 46: 65–77. doi: 10.1385/CBB: 46: 1: 65

Berois, N., M.J. Arezo, N.G. Papa, and G.A. Clivio. 2012. Annual fish: Developmental adaptations for an extreme environment. WIREs *Developmental Biology* 1: 595–602. doi: 10.1002/wdev.39

Bittman, K.S. and J.J. Lo Turco. 1999. Differential regulation of connexin 26 and 43 in murine neocortical precursors. *Cerebral Cortex* 9: 188–195. doi: 10.1093/cercor/9.2.188

Bittman, K.S., D.F. Owens, A.R. Kriegstein, and J.J. Lo Turco. 1997. Cell coupling and uncoupling in the ventricular zone of developing neocortex. *Journal of Neuroscience* 17: 7037–7044. NII Article ID (NAID): 10016130012 0270-6474/97/177037-08$05.00/0.

Bullock, T.H. 1984. Comparative neuroscience holds promise for quiet revolutions. *Science* 225: 473–478.

Butler, A.B. 2009. *Evolution of Vertebrate Brains.* Fairfax, VA, USA: George Mason University, Elsevier Ltd.

Butler, A.B. and W. Hodos. 2005. Overview of the midbrain. In: *Comparative Vertebrate Neuroanatomy: Evolution and Adaptation.* N.J. Hoboken (ed.). 57–66. Hoboken, NJ: Wiley.

Casanova, G., J.C. Rosillo, S. Olivera, and A. Fernández. 2014. Ultrastructural features of progenitors cells in the *Austrolebias* brain. *Paper Presented at the 18th International Microscopy Congress.* Prague, Check Republic.

Chapouton, P., R. Jagasia, and L. Bally-Cuif. 2007. Adult neurogenesis in non-mammalian vertebrates. *BioEssays* 29: 745–757. doi: 10.1002/bies.20615

Costa, W.E.J.M. 2006. The South American annual killifish genus *Austrolebias* (Teleostei: Cyprinodontiformes: Rivulidae): Phylogenetic relationships, descriptive morphology and taxonomic revision. *Zootaxa* 1213: 1–162.

D'Angelo, L. 2013. Brain atlas of an emerging teleostean model: *Nothobranchius furzeri. The Anatomical Record* 296: 681–691. doi: 10.1002/ar.22668

Dawe, H.R., H. Farr, and K. Gull. 2006. Centriole/basal body morphogenesis and migration during ciliogenesis in animal cells. *Journal of Cell Science* 120: 7–15. doi: 10.1242/jcs.03305

Doetsch, F. 2003. The glial identity of neural stem cells. *Nature Neuroscience* 6: 1127–1134. doi: 10.1038/nn1144

Doetsch, F., I. Caille, D.A. Lim, J.M. García-Verdugo, and A. Álvarez-Buylla. 1999. Subventricular zone astrocytes are neural stem cells in the adult mammalian brain. *Cell* 97: 703–716. doi: 10.1016/s0092-8674(00)80783-7

Doetsch, F. and C. Scharff. 2001. Challenges for brain repair insights from adult neurogenesis in birds and mammals. *Brain, Behavior and Evolution* 58: 306–322. doi: 10.1159/000057572

Ekström, P., C.M. Johnsson, and L.M. Ohlin. 2001. Ventricular proliferation zones in the brain of an adult teleost fish and their relation to neuromeres and migration (secondary matrix) zones. *Journal of Comparative Neurology* 436: 92–110. doi: 10.1002/cne.1056

Farkas, L.M. and W.B. Huttner. 2008. The cell biology of neural stem and progenitor cells and its significance for their proliferation versus differentiation during mammalian brain development. *Current Opinion in Cell Biology* 20: 707–715. doi: 10.1016/j.ceb.2008.09.008

Fernández, A.S., J.C. Rosillo, G. Casanova, and S. Olivera-Bravo. 2011. Proliferation zones in the brain of adult fish *Austrolebias* (Cyprinodontiform: Rivulidae): A comparative study. *Neuroscience* 189: 12–24. doi: 10.1016/j.neuroscience.2011.05.063

Gage, F.H. 2000. Mammalian neural stem cells. *Science* 287: 1433–1438. doi: 10.1126/science.287.5457.1433

García, G., F.Y. Álvarez-Valín, and N. Gómez. 2002, Mitochondrial genes: Signals and noise in the phylogenetic reconstruction of the annual killifish genus *Cynolebias* (Cyprinodontiformes, Rivulidae). *Biological Journal of the Linnean Society* 76: 49–59. doi: 10.1111/j.1095-8312.2002.tb01713.x

García, G., S. Claramunt, A.I. Lalanne. 2004. Genetic differentiation among annual fishes of the genus *Cynolebias* (Cyprinodontiformes, Rivulidae) in a biosphere reserve site from Uruguay. *Environmental Biology of Fishes* 70: 247–256. doi: 10.1023/b: ebfi.0000033339.55585.f6

García, G., V. Gutiérrez, N. Ríos et al. 2014. Burst speciation processes and genomic expansion in the neotropical annual killifish genus *Austrolebias* (Cyprinodontiformes, Rivulidae). *Genética* 142: 87–98. doi: 10.1007/s10709-014-9756-7

García, G., M. Loureiro, N. Berois et al. 2009. Pattern of differentiation in the annual killifish genus *Austrolebias* (Cyprinodontiformes: Rivulidae) from a biosphere reserve site in South America: A multidisciplinary approach. *Biological Journal of the Linnean Society London* 98: 620–635. doi: 10.1111/j.1095 8312.2009.01303.x

García, D., M. Loureiro, and B. Tassino. 2008. Reproductive behavior in the annual fish *Austrolebias reicherti* Loureiro & García 2004 (Cyprinodontiformes, Rivulidae). *Neotropical Ichthyology* 6: 243–248. doi: 10.1590/S1679-62252008000200012

García-Verdugo, J.M., S. Ferron, N. Flames, L. Collado, E. Desfilis, and E. Font. 2002. The proliferative ventricular zone in adult vertebrates: A comparative study using reptiles, birds, and mammals. *Brain Research Bulletin* 57: 765–775. doi: 10.1016/s0361-9230(01)00769-9

García-Verdugo, J.M., S. Llahi, I. Ferrer, and C. Lopez-García. 1989. Postnatal neurogenesis in the olfactory bulbs of a lizard. A tritiated thymidine autoradiographic study. *Neuroscience Letters* 98: 247–252. doi: 10.1016/0304-3940(89)90408-4

González-Granero, S., M. Lezameta, and J.M. García-Verdugo. 2011. Adult neurogenesis in reptiles. In: *Neurogenesis in Adult Brain I Neurobiology*. T. Seki, K. Sawamoto, J.M. Parent, and A. Alvarez-Buylla (eds.). Tokyo, Dordrecht, Heidelberg, London, New York: Springer. doi: 10.1007/978-4-431-53933-9_6

Götz, M. and Y.A. Barde. 2005. Radial glial cells: Defined and major intermediates between embryonic stem cells and CNS neurons. *Neuron* 46: 369–372. doi: 10.1016/j.neuron.2005.04.012

Grandel, H., J. Kaslin, J. Ganz, I. Wenzel, and M. Brand. 2006. Neural stem cells and neurogenesis in the adult zebrafish brain: Origin, proliferation dynamics, migration and cell fate. *Developmental Biology* 295: 263–277. doi: 10.1016/j.ydbio.2006.03.040

Gross, C.G. 2000. Neurogenesis in the adult brain: Death of the dogma. *Nature Reviews Neuroscience* 5: 308–315. doi: 10.1038/35036235

Gruber, S.H., E.R. Loew, and W.N. Mc Farland. 1990. Rod and cone pigments of the Atlantic guitarfish, *Rhinobatos lentiginosus* Garman. *Journal of Experimental Zoology* 5 (Suppl.): 85–87. doi: 10.1002/jez.1402560512

Grupp, L., H. Wolburg, and A.F. Mack. 2010. Astroglial structures in the zebrafish brain. *Journal of Comparative Neurology* 518: 4277–4287. doi: 10.1002/cne.22481

Hatakeyama, J., Y. Wakamatsu, A. Nagafuchi, R. Kageyama, R. Shigemoto, and K. Shimamura. 2014. Cadherin-based adhesions in the apical endfoot are required for active Notch signaling to control neurogenesis in vertebrates. *Development* 14: 1671–1682. doi: 10.1242/dev.102988

Hinsch, K. and G.K.H. Zupanc. 2007. Generation and long-term survival in the adult zebrafish brain: A quantitative analysis. *Neuroscience* 146: 679–696. doi: 10.1016/j.neuroscience.2007.01.071

Ishikawa, Y., M. Yoshimoto, and H. Ito. 1999. A brain atlas of a wild-type inbred strain of the medaka, *Oryzias latipes*. *The Fish Biology Journal of Medaka* 10: 1–26. doi: 10.1159/000006577

Johns, P.R. 1982. Formation of photoreceptors in larval and adult goldfish. *Journal of Neuroscience* 2: 178–198.

Kawai, T., Y. Oka, and H. Eisthen. 2009. The role of the terminal nerve and GnRH in olfactory system neuromodulation. *Zoological Science* 26: 669–680. doi: 10.2108/zsj.26.669

Kazanis, I., J. Lathia, L. Moss, and C.F. Constant. 2008. *The Neural Stem Cell Microenvironment*. D. Scadden (ed.), StemBook, The Stem Cell Research Community.

Kempermann, G. 2011. Regulation of adult neurogenesis by environment and learning. In *Neurogenesis in the Adult Brain I. Neurobiology*. T. Seki, K. Sawamoto, J.M. Parent, and A. Alvarez-Buylla (eds.). 271–284. Tokyo, Dordrecht, Heidelberg, London, New York: Springer. doi: 10.1007/978-4-431-53933

Kirsche, W. 1967. Über postembryonale Matrixzone im Gehirn vershiedener Vertebraten und deren Beziehung zur Hirnbauplahnlehre. *Zeitschrift für Mikroskopisch-anatomische Forschung* 77: 313–406.

Kotrschal, K., M.J. Van Staaden, and R. Huber. 1998. Fish brains: Evolution and environmental relationships. *Reviews in Fish Biology and Fisheries* 8: 373–408.

Kranz, D. and W. Richter. 1970. Autoradiographische Untersuchungen über die Lokalisation der Matrixzonen des Diencephalons von juvenilen und adulten *Lebistes reticulatus* (Teleostei). *Zeitschrift für Mikroskopisch-anatomische Forschung* 82: 42–66.

Lane, N. 1984. A comparison of the construction of intercellular junctions in the CNS of vertebrate and invertebrates. *Trends in Neurosciences* 7: 95–99. doi: 10.1016/s0166-2236(84)80165-4

Lindsey, B.W. and V. Tropepe. 2006. A comparative framework for understanding the biological principles of adult neurogenesis. *Progress in Neurobiology* 80: 281–307.

Loureiro, M. and G. García. 2008. *Austrolebias reicherti* Loureiro & García, a valid species of annual fish (Cyprinodontiformes: Rivulidae) from Uruguay. *Zootaxa* 1940: 1–15. ISSN 1175-5334 (online edition).

Marusich, M.F. and J.A. Weston. 1992. Identification of early neurogenic cells in the neural crest lineage. *Developmental Biology* 149: 295–306. doi: 10.1016/0012-1606(92)90285-o

Meek, J. and R. Nieuwenhuys. 1998. Holosteans and teleosts. In: *The Central Nervous System of Vertebrates*. R. Nieuwenhuys, H.J. Ten Donkelaar and C. Nicholson (eds.). 759–937. Berlin: Springer-Verlag. doi: 10.1007/978-3-642-18262-4_15

Meyer, R.L. 1978. Evidence from thymidine labeling for continuing growth of retina and tectum in juvenile goldfish. *Experimental Neurology* 59: 99–111. doi: 10.1016/0014-4886(78)90204-2

Mirzadeh, Z., Y.G. Han, J.M. García-Verdugo, and A. Alvarez-Buylla. 2011. Chapter 12: Epithelial organization of adult neurogenic germinal niches. In: *Neurogenesis in the Adult Brain I: Neurobiology*. T. Seki et al. (eds.). USA: Springer. doi: 10.1007/978-4-431-53933-9_12

Mori, T., A. Buffo, and M. Gotz. 2005. The novel roles of glial cells revisited: The contribution of radial glia and astrocytes to neurogenesis. *Current Topic in Developmental Biology* 69: 67–99. doi: 10.1016/s0070-2153(05) 69004-7

Mori, M., Y. Rikitake, K. Mandai, and Y. Takai. 2014. Roles of nectins and nectin-like molecules in the nervous system. *Advances in Neurobiology* 8: 91–116. doi: 10.1007/978-1-4614-8090-7_5

Morshead, C.M., B.A. Reynolds, C.G. Craig et al. 1994. Neural stem cells in the adult mammalian forebrain: A relatively quiescent subpopulation of subependymal cells. *Neuron* 13: 1071–1082. doi: 10.1016/0896-6273(94)90046-9

Nakase, T. and C.C.G. Naus. 2004. Gap junctions and neurological disorders of the central nervous system. *Biochimica et Biophysics Acta—Biomembranes* 1662: 149–158. The Connexins. Review. doi: 10.1016/j.bbamem.2004.01.009

Nelson, J. 2006. *Fishes of the World*. 4th edition. New Jersey: Wiley and Sons, Inc. doi: 10.1007/s11160-006-9004-z

Nieuwenhuys, R. and J. Meek. 1990. The telencephalon of actinopterygian fishes. In: *Comparative Structure and Evolution of Cerebral Cortex, Part I*. E.G. Jones and A. Peters (eds.). 31–73. New York: Plenum. doi: 10.1007/978-1-4757-9622-3_2

Northcutt, R.G. 1983. Evolution of the optic tectum in ray-finished fishes. In: *Fish Neurobiology: Higher Brain Areas and Function, Vol. 2*. R.E. Davis and R.G. Northcutt (eds.). 1–42. Ann Arbor, MI: The University of Michigan Press.

Nowakowski, R.S. and N.L. Hayes. 2000. New neurons: Extraordinary evidence or extraordinary conclusion. *Science* 5: 771. doi: 10.1126/science.288.5467.771a

Olivera-Pasilio, V., D.A. Peterson, and M.E. Castelló. 2014. Spatial distribution and cellular composition of adult brain proliferative zones in the teleost, *Gymnotus omarorum*. *Frontiers in Neuroanatomy* 8: 88. doi: 10.3389/fnana.2014.00088

Okano H. 2002. Stem cell biology of the central nervous system. *Journal of Neuroscienc Resesearch* 69(6): 698–707. Review.

Passos, C. 2013. *Austrolebias*: un modelo para explorar la selección sexual. PhD thesis. Faculty of Sciences, UdelaR. Uruguay.

Passos, C., B. Tassino, F. Reyes, and G.G. Rosenthal. 2014. Seasonal variation in female mate choice and operational sex ratio in wild populations of an annual fish. *Austrolebias reicherti*. *Plos One* 9: 1–5.

Peter, R.E., M.J. Macey, and V.E. Gill. 1975. A stereotaxic atlas and technique for forebrain nuclei of the killifish, *Fundulus heteroclitus*. *Journal of Comparative Neurology* 159: 103–127. doi: 10.1002/cne.901590107

Peters, A., S.L. Palay, and H. deF. Webster. 1991. *Fine Structure of the Nervous System: Neurons and Their Supporting Cells*. 3rd edition. USA: Oxford University Press.

Rahmann, H. 1968. Autoradiographische Untersuchungen zum DNS-Stoffwechsel (Mitose-Häufigkeit) im ZNS von *Brachydanio rerio* HAM. BUCH. (Cyprinidae, Pisces). *Journal für Hirnforschung* 10: 279–284.

Rakic, P. 1974. Neurons in rhesus monkey visual cortex: Systematic relation between time of origin and eventual disposition. *Science* 183: 425–427. doi: 10.1126/science.183.4123.425

Raymond, P.A., S.S. Easter, J.A. Burnham, and M.K., Powers. 1983. Postembryonic growth of the optic tectum in goldfish. II. Modulation of cell proliferation by retinal fiber input. *Journal of Neuroscience* 3: 1092–1099. doi: 0270-6474/83/0305-1092$02.00/O

Rodríguez, E.M., M.M. Guerra, K. Vío et al. 2012. A cell junction pathology of neural stem cells leads to abnormal neurogenesis and hydrocephalus. *Biological Research* 45: 231–241. doi: 10.4067/S0716-97602012000300005

Rosillo, J.C. 2010. Zonas proliferativas y neurogénicas en el cerebro de peces adultos del género *Austrolebias*. MSc thesis, Faculty of Sciences, UdelaR. Uruguay.

Rosillo, J.C., G. Casanova, S. Olivera, and A. Fernández. 2010. Cell heterogenity of the telencephalic ventricular zone: A neurogenic brain region of *Austrolebias charrua*. *Microscopica Acta* 19: 152–159.

Rosillo, J.C., S. Olivera-Bravo, G. Casanova, J.M.D. García-Verdugo, and A.S. Fernández. 2013. Olfacto-retinalis pathway in *Austrolebias charrua* fishes: A neuronal tracer study. *Neuroscience* 253: 304–315. doi: 10.1016/j.neuroscience.2013.08.047

Russo, R.E., C. Reali, M. Radmilovich, A. Fernández, and O. Trujillo-Cenóz. 2008. Connexin 43 delimits functional domains of neurogenic precursors in the spinal cord. *The Journal of Neuroscience* 28: 3298–3309. doi: 10.1523/jneurosci.5736-07.2008

Saeed, B., W. Ivanstoff, L.E.L.M. Crowley. 1994. Systematic relationships of Atheriniform families within division I of the series Atherinomorpha (*Acanthopterygii*) with relevant historical perspectives. *Journal of Ichthyology* 34: 27–72.

Sanes, H.D., A.R.A. Thomas, and H. William. 2006. *Development of the Nervous System*. San Diego: Academic Press, Elsevier.

Sawada, M. and K. Sawamoto. 2013. Mechanisms of neurogenesis in the normal and injured adult brain. *The Keio Journal of Medicine* 62: 13–28. doi: 10.2302/kjm.2012-0005-re

Setiamarga, D.H., M. Miya, Y. Yamanoue, K. Mabuchi, T.P. Satoh, J.G. Inoue, and M. Nishida. 2008. Interrelationships of Atherinomorpha (medakas, flyingfishes, killifishes, silversides, and their relatives): The first evidence based on whole mitogenome sequences. *Molecular Phylogenetics and Evolution* 49(2): 598–605. doi: 10.1016/j.ympev.2008.08.008

Temple, S. 2001. The development of neural stem cells. *Nature* 414: 112–117. doi: 10.1038/35102174

Togashi, H., T. Sakisaka, and Y. Takai. 2009. Special focus: Synapse-Glia interactions, cell adhesion molecules in the central nervous system. *Cell Adhesion and Migration* 3: 29–35. doi: 10.4161/cam.3.1.6773

Tozzini, E.T., M. Baumgart, G. Battistoni, and A. Cellerino. 2012. Adult neurogenesis in the short-lived teleost *Nothobranchius furzeri*: Localization of neurogenic niches, molecular characterization and effects of aging. *Aging Cell* 11: 241–251. doi: 10.1111/j.1474-9726.2011.00781.x

Wang, X., J.W. Tsai, J.H. Imai, W.N. Lian, R.B. Vallee, and S.H. Shi. 2009. Asymmetric centrosome inheritance maintains neural progenitors in the neocortex. *Nature* 461: 947–955. doi: 10.1038/nature08435

Zikopoulos, B., M. Kentouri, and C.R. Dermon. 2000. Proliferation zones in the adult brain of a sequential hermaphrodite teleost species (*Sparus aurata*). *Brain Behavior and Evolution* 56: 310–322.

Zucker, C.L. and J.E. Dowling. 1987. Centrifugal fibers synapse on dopaminergic interplexiform cells in the teleost retina. *Nature* 330: 166–168. doi: 10.1038/330166a0

Zupanc, G.K.H. 2001. A comparative approach towards the understanding of adult neurogenesis. *Brain Behavior and Evolution* 58: 246–249. doi: 10.1159/000057568

Zupanc, G.K.H. 2006. Neurogenesis and neuronal regeneration in the adult fish brain. *Journal of Comparative Physiology A* 192: 649–670. doi: 10.1007/s00359-006-0104-y

Zupanc, G.K.H. 2011. Adult neurogenesis in teleost fish. In: *Neurogenesis in the Adult Brain I. Neurobiology.* T. Seki et al. (eds.). 137–167. Tokyo, Dordrecht, Heidelberg, London, New York: Springer. doi: 10.1007/978-4-431-53933-9_1

Zupanc, G.K.H., K. Hinsch, and F.H. Gage. 2005. Proliferation, migration, neuronal differentiation and long-term survival of new cells in the adult zebrafish brain. *Journal of Comparative Neurology* 488: 290–319. doi: 10.1002/cne.20571

Zupanc, G.K.H. and I. Horschke. 1995. Proliferation zones in the brain of adult gymnotiform fish: A quantitative mapping study. *Journal of Comparative Neurology* 353: 213–233. doi: 10.1002/cne.903530205

Zupanc, G.K.H. and M.M. Zupanc. 1992. Birth and migration of neurons in the central posterior/prepacemaker nucleus during adulthood in weakly electric knife fish (*Eigenmannia* sp.). *Proceedings of the National Academy of Sciences of the United States of America* 89: 9539–9543. doi: 10.1073/pnas.89.20.9539

Zupanc, G.K.H. and M.M. Zupanc. 2006. New neurons for the injured brain: Mechanisms of neuronal regeneration in adult teleosteost fish. *Journal of Regenerative Medicine* 1: 207–216. doi: 10.2217/17460751.1.2.207

Zupanc, M.M., I. Horschke, R. Ott, and G.B. Rascher. 1996. Postembryonic development of the cerebellum in gymnotiform fish. *Journal of Comparative Neurology* 370: 443–464. doi: 10.1002/(sici)1096-9861(19960708)370: 4<443: : aid-cne3>3.3.co;2-8

CHAPTER 14

Genome of the Annual Fishes *Austrolebias* Dynamized by Transposable Elements

Verónica Gutiérrez, Néstor Ríos, and Graciela García

CONTENTS

14.1 INTRODUCTION

Repetitive DNA sequences form a large portion of the genomes of eukaryotes. The major classes of tandem-repeat noncoding sequences are *satellite DNA sequences*, very highly repetitive sequences typically organized as large clusters of up to 100 megabases (Mb), in the heterochromatic regions of chromosomes; *minisatellite sequences*, moderately repetitive sequence arrays as long as ~15 base pairs (bp), found in euchromatic regions of the genomes of vertebrates, fungi, and plants; *microsatellite sequences*, arrays of short nucleotide repeats (2–5 bp), found in vertebrate, insect, and plant genomes (copy numbers are variable within a population, typically with mean array sizes of about 100 bp but with multiple array sizes classes distributed around the mean); and *transposable elements* (TEs), another class of moderately repetitive, mobile, and dispersed sequences (Charlesworth et al., 1994).

14.2 CLASSIFICATION OF EUKARYOTIC TEs

First discovered in maize by Barbara McClintock (McClintock, 1950), TEs are sequences of DNA that move within their host genomes from one genomic position to another. They are major constituents of repetitive DNA in many prokaryotes and eukaryotes, could be the source of spontaneous mutations, and can promote important chromosomal rearrangements such as deletions, inversions, and translocations (Capy et al., 1998).

The TEs are divided into two different classes according to their mechanism of transposition (Figure 14.1). Class I elements, named *retrotransposons*, are transposed via an intermediate

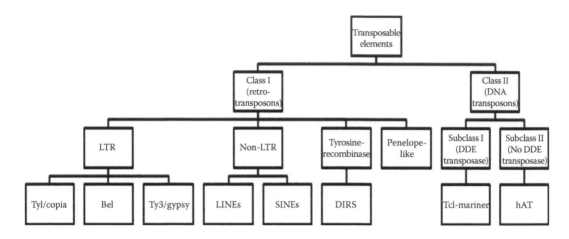

Figure 14.1 Classification of eukaryotic TE. (Adapted from Capy, P. et al. 1998. *Dynamics and Evolution of Transposable Elements*. Molecular Biology Intelligence Unit. Georgetown, Texas, USA: Landes Bioscienece; Eickbush, T.H., and V.K. Jamburuthugoda. 2008. *Virus Research* 134: 221–234.)

RNA that is reverse transcribed and integrated into a new genomic site. Each complete replication cycle produces one new copy, so they are often the major contributors to the repetitive fraction in large genomes (Wicker et al., 2007). Class II, called *DNA transposons*, uses only DNA as direct transposition intermediate. They move by a conservative cut-and-paste mechanism, where the excision of the donor element is followed by its reinsertion elsewhere in the genome (Capy et al., 1998).

Within class I, retrotransposons are classified as long terminal repeats (LTRs) and non-LTRs, according to the presence or absence of LTRs encompassing a central region that contains one or two open reading frames (ORFs) corresponding to the *gag-pol* regions of retroviruses. The *gag* gene specifies a polyprotein that is processed into three mature proteins (matrix, capsid, and nucleocapsid) corresponding to the components of the viral capsides. The *pol* gene encodes all the proteins required for transposition (protease, Pr; reverse transcriptase, RT; Rnase, H; and integrase, Int). The LTRs contain all the *cis*-regulatory elements required for the initiation of transcription and polyadenylation of the intermediate RNA, as well as regulatory boxes that control the expression of retroelements. Based on the phylogeny of their RT domains, the LTR retrotransposons are divided into major lineages historically referred to as the Ty1/copia group, the Bel group, and the Ty3/gypsy group. Ty1 and Ty3 are well-characterized elements from *Saccharomyces cerevisiae*, while Copia, Bel, and Gypsy are elements from *Drosophila melanogaster* (Eickbush and Jamburuthugoda, 2008).

The Ty1/copia and Ty3/gypsy groups of elements have extremely broad distributions in animals, plants, and fungi, whereas the Bel class has only been reported in animals (Eickbush and Jamburuthugoda, 2008). The abundance of these elements is usually low in fungi, high in plants, and highly variable in animals (Eickbush and Jamburuthugoda, 2008).

The non-LTR retrotransposons are highly abundant in eukaryotes. These elements have neither inverted nor tandem terminal repeats, instead ending most frequently with a poly (A) tail at their 3′ ends, while their 5′ ends often contain variable deletions. They encode ORFs that were usually disrupted by mutations. The highly abundant insertions identified in mammals are called long interspersed nucleotide elements (LINEs) to differentiate them from short interspersed nucleotide elements (SINEs). The insertions of the LINEs appeared similar to the reverse transcription mechanism used by processed pseudogenes and SINEs (Eickbush and Jamburuthugoda, 2008).

Most autonomous DNA transposons encode a transposase that possess a protein domain containing an acidic amino acid triad (DDE or DDD) that catalyzes the cut-and-paste transposition reaction and is responsible for removing the TE from its original position and inserting it into a new site in the genome (Curcio and Derbyshire, 2003). Initially, these TEs are classified according to the presence of DDE motif in the sequence of the transposase and the nature of the target insertion sites (Capy et al., 1998). In vertebrates, numerous superfamilies of DNA transposons, for example, Mariner-Tc1, hAT, and P, have been identified (Figure 14.1) (Feschotte and Pritham, 2007).

14.3 TRANSPOSABLE ELEMENTS IN TELEOST FISH

Almost all TEs described in eukaryotes have been found in the genome of teleost fishes (Aparicio et al., 2002). More than 30 families of phylogenetically ancient retrotransposons (generated before separation of tetrapods and fishes) were identified in the genomes of *Takifugu rubripes*, *Tetraodon nigroviridis*, and *Danio rerio*. The overall repeat content in the genome of the African annual fish *Nothobranchius furzeri* amounts to approximately 45% (the highest among the analyzed fish species) of which approximately 6.3% are retrotransposons and 1.5% DNA transposons (Reichwald et al., 2009). Numerous retrotransposons present in vertebrates, but absent from mammalian genomes, were identified in the genome of different teleost fish species. For example, the LTR retrotransposons Ty3/gypsy-like, Ty3/copia, tyrosine recombinase-encoding elements, BEL-like, Penelope-like, and the non-LTR retrotransposons with restriction enzyme-like endonuclease are found in fishes but absent from or have no functional equivalent in mammals (Volff, 2005). Taken together, as many as 16–23 clades of retrotransposons have been detected in different fish species, while only six clades are present in mouse and human genomes. A similar situation was detected for some major families of DNA TE. Although evidence for frequent and recent activity has been provided for numerous families of fish TEs, their copy number is generally lower in fishes than in mammals, suggesting that mobile sequences apparently undergo a higher turnover in teleost fish genomes (Volff, 2005).

14.4 ISOLATION AND CHARACTERIZATION OF TEs IN GIANT *AUSTROLEBIAS* GENOMES

Among Neotropical fishes the South American killifish genus *Austrolebias* (Cyprinodontiformes: Rivulidae) constitutes an excellent model for studying the genomic evolutionary processes underlying speciation events. Recently, the genome size has been determined for 16 *Austrolebias* species, revealing an unexpectedly high average DNA content of about 5.95 ± 0.45 picograms per diploid cell (mean *C*-value of about 2.98 pg; García et al., 2014). However, the genome sizes reported for other rivulid killifish, including *Cynopoecilus melanotaenia* (2.72 ± 0.06 pg/diploid cell) the sister group to *Austrolebias*, are smaller and fall within the average range of other diploid actinopterygian fishes (García et al., 2014). Cytological analyses revealed that *Austrolebias* species are diploids (García et al., 1993, 1995, 2001). So there is no reason to believe they arise from polyploid ancestors and that the giant genome sizes detected could be associated with the great genome instability previously reported in *Austrolebias* (García, 2006). However, it could explain the high morphological diversity described among *Austrolebias* species (Loureiro and de Sá, 1998). Previous phylogenetic analyses, based on mitochondrial genes, proposed that *Austrolebias* underwent sudden cladogenetic events since the Quaternary and that several such events occurred within different species groups, associated with high levels of chromosomal divergence (García et al., 2002; García, 2006).

A comparative analysis of the repetitive DNA content in the *A. charrua* and *C. melanotaenia* genomes was performed using 454-Roche technology to elucidate the origin of the *Austrolebias* giant genome sizes. Approximately 1% of the whole genome of *A. charrua* contains 45% of repetitive elements, while they constitute 25% of the 1.8% of the *C. melanotaenia* genome. Hence, the

Table 14.1 Percentage of Reads Homologous to Some Transposable Elements Detected in Rivulidae Genomes

Transposable Element		*A. charrua*	*C. melanotaenia*
Retrotransposons	Babar	4.04	0.79
	Rex	5.15	3.19
	Jockey	4.63	4.06
DNA transposon	Tc1/mariner	0.74	0.29

genomic proportion of repetitive DNA in the *A. charrua* genome is approximately twice that of *C. melanotaenia*. These data could be suggesting that repetitive elements may have played a role in the differentiation and speciation in Rivulidae (García et al., 2015).

Almost all classes of repetitive DNA are present in the genomes of Rivulidae. In both species analyzed, retroelements make up most of the repetitive DNA (Table 14.1). Retrotransposons Ty3/gypsy and Babar, as well as the DNA transposon Tc1/mariner and the tRNA fraction of the genome of *A. charrua*, largely exceed those found in the genome of *C. melanotaenia*. Conversely, the non-LTR retrotransposons belonging to the Rex, Jockey-like, Penelope, and SINEs classes/groups, as well as ribosomal DNA components, are predominant in the genome of *C. melanotaenia*. Other tandem-repeat elements, such as microsatellites and minisatellites, are more abundant in the *A. charrua* genome than in the *C. melanotaenia* genome, whereas satellite DNA was scarce in *C. melanotaenia* and not detected in the partial genome of *A. charrua* analyzed (García et al., 2015).

14.5 EVOLUTION OF TEs IN THE GENOME OF ANNUAL FISHES

To understand the evolution of TEs, researchers frequently compare TEs' phylogenies with those of their host species (Capy et al., 1998). If the topologies of the resulting trees are concordant, the hypothesis of vertical transmission is the most plausible. Horizontal transfer assumes asexual transmission of genetic material between distantly related species that coexist in time and space. Under this scenario, a "vector" acts as a vehicle to transport the genetic information between genomes. Furthermore, vertical transmission assumes TE transmission through a common ancestor and requires (1) existence of ancestral polymorphisms, (2) different rates of evolution of TEs among host species, and/or (3) random sampling of TEs during speciation of host species (Capy et al., 1998).

Since some TEs belonging to class I and II were detected in Rivulidae, phylogenetic analyses of these TEs found in a wide range of species, from insects to vertebrates, could shed light about their evolutionary history. Clustering of some TEs in *A. charrua* and *C. melanotaenia* were obtained through the Neighbor-joining (NJ) method using the p-distance method to compute similarity distances (Figure 14.2). The relationships among partial *Babar* sequences revealed two monophyletic lineages. One of them, with a bootstrap node support of 100, involved partial *Babar* RT sequences from *Austrolebias* obtained from GenBank. The other one, with a bootstrap node support of 100, encompassed the *Babar* sequence from *Tetraodon nigroviridis* and from *A. charrua* and *C. melanotaenia*, which are similar to *Babar* sequences obtained from GenBank (Figure 14.2a). This analysis suggests the presence of two different lineages of the non-LTR retrotransposon *Babar* in *Austrolebias*, each one comprising different copies of this retrotransposon. On the other hand, the analysis of the Rivulidae sequences homologous to the non-LTR retrotransposon Rex (Figure 14.2b) showed the presence of three monophyletic lineages. The first one, with a bootstrap node support of 88, involved Rex-3 sequences from many Teleost fishes from GenBank, two sequences from *A. charrua*, and one from *C. melanotaenia*. The second group, with a bootstrap node support of 100, includes a sequence of *A. charrua* and a sequence of *C. melanotaenia*. The last group, with

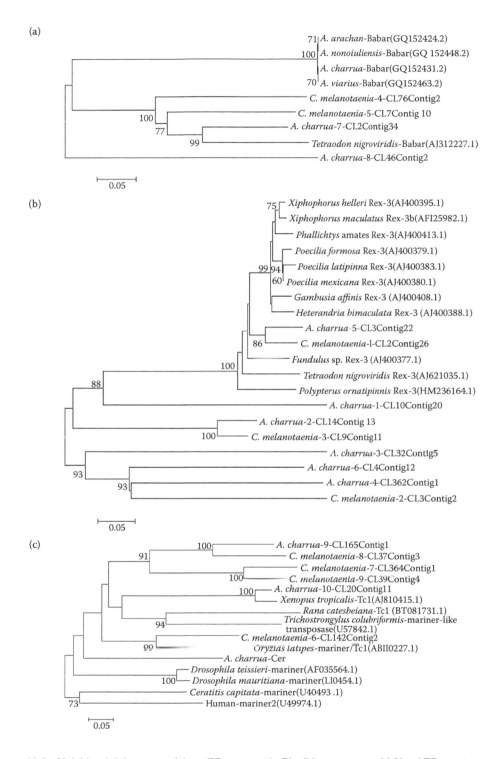

Figure 14.2 Neighbor-joining trees of three TEs present in Rivulidae genomes. (a) Non-LTR retrotransposon *Babar*, (b) non-LTR retrotransposon *Rex-3*, and (c) DNA transposon *Tc1/mariner*. The evolutionary distances were computed using the p-distance method and are in the units of the number of base differences per site. The number at each node represents bootstrap values above 60% recovered in 1000 replicates. GenBank accession numbers are between brackets.

a bootstrap node support of 100, consists only of homologous sequences to Rex-3 from *A. charrua* and *C. melanotaenia* (Figure 14.2b). Hence, the resulting relationships for Rex sequences revealed the presence of at least three different lineages of this retrotransposon in Rivulidae. Finally, the analysis of the transposon Tc1-like/mariner (Figure 14.2c) shows three different lineages: one consisting of sequences of *A. charrua* and *C. melanotaenia* homologous to this TE; a second lineage, with bootstrap node support of 100, consisting of an *A. charrua* sequence and the Tc-1 sequence from *Xenopus tropicalis* from GenBank; and the third grouping consisting of a *C. melanotaenia* sequence and the mariner/Tc1 GenBank sequence from *Oryzias latipes*. The analyses performed for TEs present in genomes of Rivulidae revealed the existence of more than one lineage of them in both species analyzed, suggesting that these TEs could be under divergent evolution.

Overall, our preliminary comparative analyses using repetitive sequences using NGS showed a high lineage divergence within different TEs classes and families. This suggests a possible vertical transference mechanism for these TEs copies among the genomes of Rivulidae.

14.6 TEs DYNAMIZING GENOMES IN THE SPECIATION

Transposable elements are generally active in germ cells and their movement during germline divisions has various effects. Most obviously, they produce mutations caused by the insertion of a TE into a functional gene or its regulatory region. Transposable element activity is also associated with an elevated rate of chromosomal rearrangements. Fixation of different rearrangements, like translocations and inversions, in different populations could result in reproductive isolation (Hurst and Schilthuizen, 1998; Volff, 2005). Their activity has a greater impact in cases of hybridizations, where the presence of gonadal dysgenesis suggests TEs as a potential source of reproductive isolation. In most TE-mediated speciation scenarios, allopatric populations are invaded by a range of different TEs that result in sterility of all hybrids (Hurst and Schilthuizen, 1998). Hence, TEs might be able to contribute to pre and postmating reproductive isolation and therefore might be involved in speciation (Hurst and Schilthuizen, 1998; Hurst and Werren, 2001; Volff, 2005).

Currently, little is known regarding the role of TEs during speciation in fishes or about their possible activation in fish hybrids. In *Salmo salar* and other species, analyses of DNA sequences for the presence of TE families revealed that the sudden timing of transposon replication coincides with the diversification of *Salmo*, *Oncorhynchus*, and *Salvelinus* and subsequent speciation in *Salmoninae*. This analysis agrees with previous suggestions that laterally injected massive transposon movement played a role in chromosomal reorganization at different times during speciation (de Boer et al., 2007). However, phylogenetic analysis of several retrotransposons from various fish species has revealed the presence of multiple waves of retrotransposition that might have been associated with speciation events (Volff et al., 2001). Clearly, the multiple active lineages of TEs present in the genomes of fishes might be predisposed to rapid speciation (Volff, 2005). Recently, comparative gene topologies of the non-LTR retrotransposon *Babar*, partial regions of the conserved intron-2 and exon-3 of the *recombination activating gene-1* gene (*RAG-1*), and mitochondrial *cytochrome b* gene (*cyt-b*) yielded interesting evolutionary features about TEs and speciation events in the genus *Austrolebias* (Gutiérrez and García, unpublished data). The non-LTR retrotransposon *Babar* and *cyt-b* gene topologies are similar and differ from that based on the *RAG-1* gene. Although the non-LTR retrotransposon *Babar* as well as the others gene based topologies do not resolve the basal polytomy in *Austrolebias* (V. Gutiérrez and G. García, unpublished data), could be suggesting a sudden radiation in *Austrolebias*. Molecular evolutionary patterns revealed purifying selection (Ka/Ks < 1), since the three genes evolved under similar selective constraints, suggesting that *Babar* RT copies could be active during the evolution of *Austrolebias* (V. Gutiérrez and G. García, unpublished data).

Therefore, TE lineages found in *Austrolebias* suggest that they could be an interesting model system to explore interrelationships among sequence proliferation, chromosomal divergence, ecological plasticity in unstable environments, population divergence, and speciation.

REFERENCES

Aparicio, S., J. Chapman, E. Stupka et al. 2002. Whole-genome shotgun assembly and analysis of the genome of *Fugu rubripes*. *Science* 297: 1301–1310. doi: 10.1126/science.1072104

Capy, P., C. Bazin, D. Hiquet et al. 1998. *Dynamics and Evolution of Transposable Elements*. Molecular Biology Intelligence Unit. Georgetown, Texas, USA: Landes Bioscienece.

Charlesworth, B., P. Sniegowski, and W. Stephan. 1994. The evolutionary dynamics of repetitive DNA in eukaryotes. *Nature* 371: 215–220. doi: 10.1038/371215a0

Curcio, M.J., and K.M. Derbyshire. 2003. The outs and ins of transposition: From mu to kangaroo. *Nature Reviews Molecular Cell Biology* 4: 865–877. doi: 10.1038/nrm1241

De Boer, J.G., R. Yazawa, W.S. Davidson et al. 2007. Bursts and horizontal evolution of DNA transposons in the speciation of pseudotetraploid salmonids. *BMC Genomics* 8: 422. doi: 10.1186/1471-2164-8-422

Eickbush, T.H., and V.K. Jamburuthugoda. 2008. The diversity of retrotransposons and the properties of their reverse transcriptases. *Virus Research* 134: 221–234. doi: 10.1016/j.virusres.2007.12.010

Feschotte, C., and E.J. Pritham. 2007. DNA transposons and the evolution of eukaryotic genomes. *Annual Review of Genetics* 41: 331. doi: 10.1146/annurev.genet.40.110405.090448

García, G. 2006. Multiple simultaneous speciation in killifishes of the *Cynolebias adloffi* species complex (Cyprinodontiformes, Rivulidae) from phylogeography and chromosome data. *Journal of Zoological Systematics and Evolutionary Research* 44: 75–87. doi: 10.1111/j.1439-0469.2005.00346.x

García, G., F. Alvarez-Valin, and N. Gomez. 2002. Mitochondrial genes: Signals and noise in the phylogenetic reconstruction of the annual killifish genus *Cynolebias* (Cyprinodontiformes, Rivulidae). *Biological Journal of the Linnean Society* 76: 49–59. doi: 10.1111/j.1095-8312.2002.tb01713.x

García, G., V. Gutiérrez, N. Ríos et al. 2014. Burst speciation processes and genomic expansion in the neotropical annual killifish genus *Austrolebias* (Cyprinodontiformes, Rivulidae). *Genetica* 142: 87–98. doi: 10.1007/s10709-014-9756-7

García, G., A.I. Lalanne, G. Aguirre et al. 2001. Chromosome evolution in the annual killifish genus *Cynolebias* and mitochondrial phylogenetic analysis. *Chromosome Research* 9: 437–448. doi: 10.1023/A:1011664009509

García, G., N. Ríos, and V. Gutiérrez. 2015. Next-generation sequencing detects repetitive elements expansion in giant genomes of annual killifish genus *Austrolebias* (Cyprinodontiformes, Rivulidae). *Genetica* 143: 353–360. doi: 10.1007/s10709-015-9834-5

Garcia, G., E. Scvortzoff, and A. Hernández. 1995. Karyotypic heterogeneity in South American annual killifishes of the genus *Cynolebias* (Pisces, Cyprinodontiformes Rivulidae). *Cytologia* 60: 103–110. doi: http://dx.doi.org/10.1508/cytologia.60.103

García, G., E. Scvortzoff, M. Máspoli et al. 1993. Analysis of karyotypic evolution in natural population of *Cynolebias* (Pisces: Cyprinodontiformes, Rivulidae) using banding techniques. *Cytologia* 58: 85–94. doi: http://dx.doi.org/10.1508/cytologia.58.85

Hurst, G.D., and M. Schilthuizen. 1998. Selfish genetic elements and speciation. *Heredity* 80: 2–8. doi: 10.1046/j.1365-2540.1998.00337.x

Hurst, G.D., and J.H. Werren. 2001. The role of selfish genetic elements in eukaryotic evolution. *Nature Reviews Genetics* 2: 597–606. doi: 10.1038/35084545

Loureiro, M., and R.O. de Sá. 1998. Osteological analysis of the killifish genus *Cynolebias* (Cyprinodontiformes: Rivulidae). *Journal of Morphology* 238: 245–262. doi: 10.1002/(SICI)1097-4687(199811)238:2<245::AID-JMOR5>3.0.CO;2-F

McClintock, B. 1950. The origin and behavior of mutable loci in maize. *Proceedings of the National Academy of Sciences* 36: 344–355. doi: 10.1073/pnas.36.6.344.

Reichwald, K., C. Lauber, I. Nanda et al. 2009. High tandem repeat content in the genome of the short-lived annual fish *Nothobranchius furzeri*: A new vertebrate model for aging research. *Genome Biol* 10: R16. doi: 10.1186/gb-2009-10-2-r16

Volff, J. 2005. Genome evolution and biodiversity in teleost fish. *Heredity* 94: 280–294. doi: 10.1038/sj.hdy.6800635

Volff, J.-N., C. Körting, A. Meyer et al. 2001. Evolution and discontinuous distribution of Rex3 retrotransposons in fish. *Molecular Biology and Evolution* 18: 427–431.

Wicker, T., F. Sabot, A. Hua-Van et al. 2007. A unified classification system for eukaryotic transposable elements. *Nature Reviews Genetics* 8: 973–982. doi: 10.1038/nrg2165

CHAPTER 15

Comparative Phylogeographic Patterns in *Austrolebias* from Different South American Basins

Graciela García, Verónica Gutiérrez, Néstor Ríos, and Rafael O. de Sá

CONTENTS

15.1 INTRODUCTION

The Neotropical region is among the world richest in biodiversity; however, it still holds poorly known and understudied biota. Furthermore, comparative phylogeographic syntheses of the Southern Hemisphere fauna are needed to understand the current and historical patterns of populations (Beheregaray, 2008). Particularly, eastern South America's coastal areas and river basins were impacted by marine transgression during the Pleistocene and post-Pleistocene resulting in habitat modifications and fragmentations (Sprechman, 1978).

South American killifish (Cyprinodontiformes: Rivulidae) are unique among Neotropical fishes; they represent a speciose group that includes taxa with annual life cycles. The distribution of annual species is linked to the geological history of different basins in the region, making them an excellent model for comparative phylogeographic studies. However, they have been the focus of only a handful of phylogeographic studies (García, 2006; García et al., 2009, 2012). South American rivulids have been shown to be morphologically and karyologically variable (Vaz-Ferreira and Melgarejo, 1984; Costa, 1990; Loureiro and de Sá, 1998; García, 2006). A recent study suggested that the distribution patterns of the cynolebiasines are congruent with the tectonic and geomorphological history of South America and resulted from a series of vicariant events (Costa, 2010). For example, the split between basal lineages (endemic to eastern Brazilian coastal plains) and the rest of cynolebiasines was correlated to an early vicariant event occurring just after the separation of South America from Africa (at about 90 mya). Overall, that biogeographical analysis exemplifies the potential of annual killifish as a model group for comparative phylogeographic analysis to understand current and historical patterns of speciation.

Historically, taxonomic literature on Neotropical annual fishes commonly used the terms "species groups" and "species complexes" somewhat interchangeably (Huber, 1998; Costa, 2006). Overall, the increased use of molecular systematics over the last 25 years, and particularly the development of DNA-based species delimitation methods, had assisted in uncovering previously unknown genetic diversity in several groups as well as allowing recognition of some morphological "cryptic" species, increasing our understanding of biodiversity (Camargo et al., 2006; Gehring et al., 2012). In this context, herein we refer to "species groups" as a clade of formally described species (e.g., the *Austrolebias bellottii* species group consists of *A. bellottii*, *A. vandenbergi*, *A. melanoorus*, and *A. univentripinnis*). Whereas, when we refer to a "species complex," we mean those species whose identity across a distributional range as a single taxonomic unit may be questionable based on known intraspecific variation, for example, morphological, morphometric, and genetic (e.g., the *A. bellottii* species complex consists of all the populations currently included under this nomina).

Among Neotropical annual killifish, the genus *Austrolebias* consists of 38 currently recognized species distributed in association with two river basins: the Paraná-La Plata basin (PLPB; includes the Paraná, Paraguay, and Uruguay rivers) and the Patos-Merín drainage (PMD) system (Costa, 2006). Some species occur in only one system (e.g., *A. bellottii* in PLPB, *A. adloffi* in PMD), but a few occur in both (e.g., *A. affinis*). Five species groups are recognized within *Austrolebias* (García et al., 2002; Costa, 2006). The present study focuses on a comparative phylogeographic analysis of the *A. bellottii* species complex and the *A. adloffi* species group.

The *A. bellottii* species group currently contains four species: *A. bellottii* (Steindachner, 1881), *A. melanoorus* (Amato, 1986), *A. univentripinnis* (Costa and Cheffe, 2005), and *A. vandenbergi* (Huber, 1995). *A. bellottii* (Steindachner, 1881) occurs across the PLPB, *A. melanoorus* inhabits the upper Río Negro basin (main tributary of lower Uruguay River), *A. univentripinnis* is known from a single location in the PMD system (Costa, 2006), and *A. vandenbergi* is distributed in the Paraguay and high Paraná river basins. This species group consists of medium-sized and deep-bodied species that are distributed throughout the range of the genus. The morphological variation found in *A. bellotti* species complex resulted in misidentification of specimens and taxonomic problems (Vaz-Ferreira and Sierra, 1973; Costa, 2006). The species occurs in the PLPB associated with the middle and lower Paraná and Uruguay (Figure 15.1a) rivers, which are characterized by alluvial deposits and extensive floodplains (Kawakami de Resende, 2003) and drain into the La Plata estuary. These rivers may have been connected at different sites and geological times (Boneto, 1994), and extensive marine transgressions occurred in this area during the late Miocene (11.8–10.0 mya), generating the Paranean Sea (Lundberg et al., 1998). Recent phylogeographic analysis based on mitochondrial cytochrome *b* (cyt-*b*) gene sequences supports historical allopatric fragmentation resulting in independent vicariant events along both river drainages (García et al., 2012). These vicariant events may have played an important role in the diversification and split of the ancestral *A. bellottii* population.

The *Austrolebias adloffi* species group currently contains eight species: *A. adloffi* (Ahl, 1922), *A. charrua* (Costa and Cheffe, 2001), *A. minuano* (Costa and Cheffe, 2001), *A. nigrofasciatus* (Costa and Cheffe, 2001), *A. reicherti* (Loureiro and García, 2004; Loureiro and García, 2008), *A. arachan* (Loureiro, Azpelicueta and García, 2004), *A. nachtigalli* (Costa and Cheffe, 2006), and *A. viarius* (Vaz-Ferreira, Sierra-de-Soriano and Scaglia-de-Paulete 1964). *Austrolebias arachan* and *A. viarius* were first included in the *A. bellottii* species group by Costa (2006) based

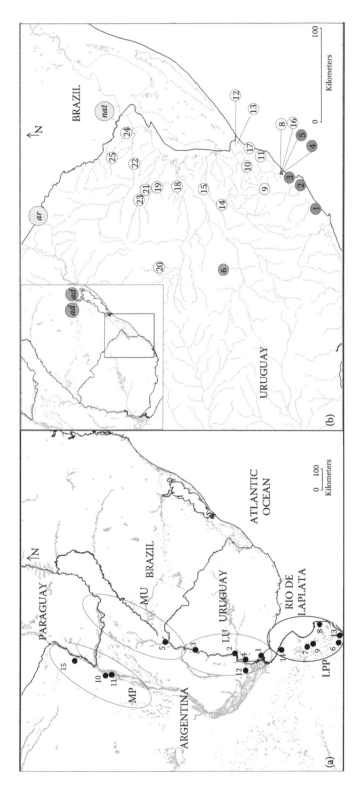

Figure 15.1 Distribution of sampling sites across basins. (a) Number of 15 collecting sites of *A. bellotti* species complex from PLPB grouped into four major regions as follows: LPP, lower Paraná-La Plata basin; LU, lower Uruguay river basin; MU, middle Uruguay river basin; MP, middle Paraná river basin. (b) Collecting sites for the *A. adloffi* species group at PMD. ad: *A. adloffi*; ar: *A. arachan*; nat: *A. nachtigalli*.

on a morphological phylogeny; however, a phylogeographic analysis placed both taxa within the *A. adloffi* species group (García, 2006). The species group also consists of medium-sized and deep-bodied species, but their distribution is confined to the PMD system. These lowland areas include "Bañados del Este" (eastern Uruguay, Figure 15.1b) that are both Biosphere Reserve and Ramsar Sites (Probides, 1999). This is a vast area of Atlantic coastal wetlands with seasonal flooding. The area includes lagoons and several rivers that originated during Quaternary marine transgressions (Montaña and Bossi, 1995; Behling and Negrelle, 2001; Iriondo, 2004).

Previous studies have shown the *A. adloffi* species group as a well-supported clade (Costa, 2006; García, 2006). Historical allopatric fragmentation, with subsequent range expansions and areas of secondary contact, needs to be tested as a possible mechanism of differentiation within the *A. adloffi* species group.

Herein we provide a comparative phylogeographic analysis, based on molecular data, of the *A. adloffi* species group and the *A. bellottii* species complex with the goal of understanding their patterns of diversification. Furthermore, we will explore (1) historical events (particularly during the Quaternary period) that affected the PLPB and PMD systems and impacted the current distribution of these species and (2) the role of the geological history in the diversification of an endemic ichthyofauna. Specifically, we test whether vicariance (instead of dispersal) was the process driving the differentiation of species groups and populations between both hydrological systems. Among molecular markers, the mitochondrial cyt-*b* gene has been commonly used in phylogeographic analyses due to its high performance in resolving phylogenetic relationships among highly related populations and taxa (Avise, 2000; Galtier et al., 2009). Furthermore, previous studies have shown the utility of cyt-*b* in assessing relationships in the *Austrolebias* genus (García et al., 2002, 2009, 2012; García, 2006).

15.2 MATERIAL AND METHODS

15.2.1 Sampling and Molecular Methods

The study included a total of 139 individuals; of these, 58 individuals correspond to *A. bellottii* collected along the PLPB and 81 specimens belong to six species of the *A. adloffi* species group collected from the PMD and southwestern Atlantic coastal lagoons system (Figure 15.1; for locality data see Appendices 15.1 and 15.2). Phylogeographic analysis of the *A. bellottii* species complex included *A. vanderbergi* and *Plesiolebias glaucopterus* as outgroups, whereas analyses of the *A. adloffi* species group included *A. bellottii* and *Plesiolebias glaucopterus* as outgroup taxa. Tissues and voucher specimens are deposited in the Sección Genética Evolutiva (GP), Universidad de la República, Montevideo, Uruguay (Appendices 15.1 and 15.2). All sampling protocols for this scientific study were approved by CNEA (Comisión Nacional de Experimentación Animal) from Uruguay. Genomic DNA was isolated from liver tissue of freshly killed animals fixed in ethanol (95%) with sodium chloride protein precipitation, followed by ethanol precipitation (modified from Medrano et al., 1990). A fragment of 887 base pairs (bp) from the cyt-*b* gene was amplified using the CB3-H or CB6-H and Gludg-L primers (Palumbi et al., 1991) following the PCR profile: 94°C for 1 min, 45°C for 1 min, 72°C for 1 min; 30 cycles. PCR products were cleaned using CONCERT Kit rapid PCR purification System (Life Technologies) and sequenced in both strand directions, using the same amplification primers, in a Perkin-Elmer ABI Prism 377 Automated Sequencer (MACROGEN, Seoul, Korea). Sequence alignments were performed using the CLUSTAL X v. 1.8 software (Thompson et al., 1997). GenBank accession numbers are provided in Appendices 15.1 and 15.2.

15.2.2 Methodology of Analyses

Nucleotide composition and substitution patterns were calculated using MEGA v. 6.0 (Tamura et al., 2013) and DnaSP v. 4.50 (Rozas et al., 2003). The corrected estimates of pairwise sequence divergence were obtained using the two-parameter algorithm (K2P) of Kimura (1980) implemented in MEGA. Within a population, DNA polymorphism was measured calculating the proportion of segregating sites (S), the haplotype diversity (h; Nei 1987:179), and the nucleotide diversity (π; Nei 1987:257) using DnaSP v. 4.50 (Rozas et al., 2003) software package. Tajima's (1989) test implemented in the DnaSP v. 4.50 (Rozas et al., 2003) program was used to test for mutation/drift equilibrium and departure from neutrality. Phylogeographic relationships among mitochondrial cyt-*b* haplotypes from both sampled basins were assessed using two different methodologies. A nonmodel-based method (MP, maximum-parsimony) was implemented in PAUP* 4.0b10 (Swofford, 2002) following an equally weighted MP analysis using heuristic search (MULPARS option, stepwise addition, tree-bisection-reconnection [TBR] branch swapping, 100 replicates). A strict consensus between rival trees was computed to reconcile equally parsimonious topologies. Support for nodes was assessed by bootstrapping with 500 replicates in MP. Also, two model-based approaches were also used, that is, maximum likelihood (ML) and Bayesian inference (BI), implemented in PAUP* 4.0b10 (Swofford, 2002) and BEAST v. 1.5.4 (Drummond and Rambaut, 2007), respectively. In ML and BI analyses, the best-fit nucleotide substitution model for each data set was determined using Modeltest v. 3.7 (Posada and Crandall, 1998), based on the Akaike information criterion comparing 56 DNA substitution models (Akaike, 1974). For the *A. bellottii* data set, the general time-reversible (GTR) model was selected (Rodríguez et al., 1990). The likelihood scores estimated for this model were used as the prior settings for the ML analysis in the data set (–lnL = –5958.98). For the *A. adloffi* species complex, the best fit was the HKY + Γ model (Hasegawa et al., 1985) with gamma distribution (Γ). The gamma-distribution shape parameter value was 0.30. The likelihood scores estimated for these models were used as the prior settings for the ML analysis in the data set (–lnL = –7261.69). A heuristic ML search (with 100 replications of stepwise addition and TBR branch swapping) was implemented in the PAUP*4.0b10 (Swofford, 2002) software package. The robustness of the nodes was assessed after 100 bootstrap replications as implemented in PhyML 3.0 (http://atgc.lirmm.fr/phyml) following the algorithm developed by Guindon et al. (2010). In this case, the NNI (a fast nearest-neighbour edge-interchange search) swapping algorithm option was implemented. Heuristic search (with 100 replicates of stepwise addition and TBR branch swapping) in ML analyses was implemented in PAUP* 4.0b10 (Swofford, 2002). The robustness of the nodes was determined after 1000 bootstrapping replicates, as implemented in PhyML 3.0 (http://atgc.lirmm.fr/phyml; Guindon et al., 2010). In this case, the NNI swapping algorithm option was implemented. Nonparametric bootstrap values above 75% were considered to be robust support for clades (Hillis and Bull, 1993). All trees were rooted using outgroups as described above.

Divergence times of nodes in the reconstructed phylogenies and the age of the most recent common ancestor (tMRCA) were estimated using BEAST v. 1.5.4 software (Drummond and Rambaut, 2007). Input files were generated with Beauti v. 1.5.4 (Drummond and Rambaut, 2007) assuming uncorrelated lognormal trees and a Yule speciation process as prior information. The nucleotide substitution model and its parameter values were selected using Modeltest v. 3.7 (Posada and Crandall, 1998). An uncorrelated lognormal relaxed molecular clock, which allows rate variation among lineages, was implemented using an estimated rate for mitochondrial genome of 2% per million years (Brown et al., 1979) adjusted for ectotherms (Avise, 1994). We performed two independent runs of 10 million generations and trees, and parameters were sampled every 1000 iterations, with a burn-in of 10%. Results of each run were visualized in the Tracer v. 1.5 program (Rambaut and Drummond, 2009) to ensure that stationarity was achieved and that convergence was reached. Each analysis was repeated many times to optimize the operators of parameters until no suggestion message appeared in the log file. The timing of clade divergence and the tMRCA were estimated in million years ago (mya) with a mean and a 95% highest posterior density (lower and upper). Posterior probabilities and the maximum credibility tree were calculated using the TreeAnnotator v. 1.5.4 software (Drummond and Rambaut, 2007).

Genetic structure of *A. bellottii* populations and the *A. adloffi* complex was assessed using the variance components among hierarchical partitions in the data set as implemented in analysis of molecular variance (AMOVA; Excoffier et al., 1992). The Euclidean metric of Excoffier et al. (1992) was used to construct the matrix of pairwise distances. The genetic variation was partitioned into three components: among groups (Φ_{CT}), among populations within groups (Φ_{SC}), and among individuals within populations (Φ_{ST}), disregarding either their original populations or their groups. A cladogram, including the haplotypes from each separated data set was reconstructed using statistical parsimony implemented in the program TCS 1.06 (Clement et al., 2000). The program collapses sequences into haplotypes and calculates the frequencies of the haplotypes in the sample. These frequencies are used to estimate haplotype outgroup probabilities, which correlate with haplotype age (Castelloe and Templeton, 1994). An absolute distance matrix is then calculated for all pairwise comparisons of haplotypes. The probability of parsimony as defined in Templeton et al. (1992) is calculated for pairwise differences until the probability exceeds 0.95. The number of mutational differences associated with the probability just before this 95% cutoff is then the maximum number of mutational connections between pairs of sequences justified by the "parsimony" criterion. These justified connections are then made, resulting in a 95% set of plausible solutions. For the present data sets we have fixed a 95% connection limit of 13 steps between haplotypes to recover the most robust associations.

The long-term gene flow was estimated using a coalescent-based approach implemented in MIGRATE-N v. 3.4.2 (Beerli and Felsenstein, 2001; Beerli, 2006) to assess migration rates and the potential for asymmetry in gene flow among populations. This analysis uses ML or BI to estimate the parameter M (m/μ), mutation-scaled migration rates, and θ, the mutation-scaled effective population size ($2N_e\mu$ for mitochondrial marker), where m is the immigration rate, μ the mutation rate, and N_e the effective population size. The product θM results in the effective number of immigrants per generation. Unlike F statistics, this method allows for the possibility of asymmetrical gene flow between population pairs and differences in the effective population size. Runs for each data set consisted of five replicates of 1,000,000 genealogies each. A random genealogy and parameter settings inferred by an F_{ST}-based method were used as start conditions. Priors for θ and M for the BI were drawn from UNIFORM (0-0.01) and UNIFORM (0-1000) distributions, respectively. We set the burn-in for 10,000 genealogies (number of discarded trees per chain).

15.3 RESULTS

15.3.1 Mitochondrial Cyt-*b* Variation in *Austrolebias*

Pairwise K2P sequence divergence among populations of the *A. bellottii* species complex and among species of the *A. adloffi* species group are shown in Tables 15.1–15.4. *A. bellottii* has a total sequence divergence of 2.8%, SE = 0.022. Table 15.1 shows the level of divergence in each group of *A. bellottii* populations. Remarkably, the highest divergence values were observed in populations from the lower Uruguay River (LU) drainage (*A. bellottii* group B), whereas the lowest one was

Table 15.1 Estimates of DNA Divergence among Groups (A–D) of *Austrolebias bellottii* Species Complex

	Distance	Standard Deviation
A	0.004	0.005
B	0.018	0.010
C	0.000	0.000
D	0.014	0.015

Source: Adapted from Kimura, M. 1980. *Journal of Molecular Evolution* 16: 111–120.

Note: Corrected K2P distances, including both transversions (Tv) and transitions (Ts).

Table 15.2 Estimates of DNA Pairwise Divergence among Groups (A–D) of *Austrolebias bellottii* Species Complex

	A	B	C	D
A		**0.041**	**0.002**	**0.008**
B	0.046		**0.042**	**0.041**
C	0.002	0.044		**0.008**
D	0.009	0.052	0.007	

Source: Adapted from Kimura, M. 1980. *Journal of Molecular Evolution* 16: 111–120.

Note: Corrected K2P distances including both transversions (Tv) and transitions (Ts). Standard deviation (S.D.) in bold.

Table 15.3 Estimates of Intraspecific DNA Divergence in *A. adloffi* Species Groups

	Distance	Standard Deviation
A. viarius	0.031	**0.008**
A. adloffi	0.000	**0.000**
A.arachan	0.012	**0.009**
A. charrua	0.013	**0.004**
A. reicherti	0.020	**0.006**

Source: Adapted from Kimura, M. 1980. *Journal of Molecular Evolution* 16: 111–120.

Note: Corrected K2P distances including both transversions (Tv) and transitions (Ts). Standard deviation (S.D.) in bold.

Table 15.4 Estimates of DNA Pairwise Divergence between Taxa from the *A. adloffi* Species Group

	A. viarius	*A. adloffi*	*A. arachan*	*A. charrua*	*A. reicherti*
A. viarius		**0.039**	**0.036**	**0.028**	**0.036**
A. adloffi	0.141		**0.042**	**0.030**	**0.034**
A. arachan	0.137	0.159		**0.026**	**0.028**
A. charrua	0.102	0.099	0.087		**0.026**
A. reicherti	0.152	0.109	0.103	0.084	

Source: Adapted from Kimura, M. 1980. *Journal of Molecular Evolution* 16: 111–120.

Note: Corrected K2P distances including both transversions (Tv) and transitions (Ts). Standard deviation (S.D.) in bold.

found in middle Paraná (MP, group C) of Argentina. In this context, the lower and middle Uruguay's River populations (LU + MU) of *A. bellottii* are the most divergent in all pairwise comparisons (Table 15.2). Within the *A. adloffi* species group, the highest intraspecific divergence was observed in *A. viarius* and *A. reicherti*, whereas the lowest one was found in *A. adloffi* (Table 15.3). The highest genetic divergence was detected between *A. adloffi* and *A. arachan* (15.9%, SE = 0.042), and the genetically most similar species pairs are *A. charrua* and *A. nachtigalli* (5.7%, SE = 0.008) and *A. charrua* and *A. reicherti* (8.4%, SE = 0.026) (Table 15.4).

15.3.2 Cyt-*b* Gene Genealogy

The recovered topology in the analysis of the *A. bellottii* species complex identified a large monophyletic clade that includes all *A. bellottii* samples from PLPB (Figure 15.2). Clade I consists of two subclades, one grouping samples from the lower Paraná-La Plata basin (LPP, subclade A)

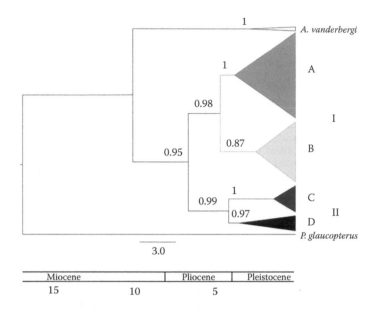

Figure 15.2 Bayesian tree topology generated using the GTR + G model of molecular evolution based on 52 cyt-*b* gene haplotypes (H) of *A. bellottii* species complex from four major regions (see Figure 15.1; Appendix 15.1). Bayesian phylogeographic inference framework implemented in BEAST 1.5.4 and the estimated divergence dates. Numbers above branches are Bayesian posterior probability values for the occurrence of clades. The bottom bar summarizes the time-scale divergence dates in mya. For description of *A. bellotii* clades A–D, see text.

and the other containing a mixture of populations from the LU, MU, and LPP rivers (subclade B). A second clade (clade II) consists of samples of *A. bellottii* from MP (subclade C) and populations from MP + MU + LU rivers (subclade D). *Austrolebias vandenbergi* is the sister species to clades I + II. Node calibration assumed a mutation rate of 2% to capture a plausible time interval (lower and upper estimates). Clade I diverged between 6.5 and 5.5 mya, whereas clade II diverged slightly more recently, between 5.0 and 4.4 mya. The divergence between *A. vandenbergi* and *A. bellottii* was estimated at 7.1–9.0 mya.

The analysis of the *A. adloffi* species group identified a large monophyletic clade, with a high posterior probability, which included the different species from the *A. adloffi* complex included in the analysis (Figure 15.3). *Austrolebias viarius* (clade A) was recovered as the sister species to a large clade consisting of the remaining of the species of the *A. adloffi* species group (clades B, C, D, and E). Within this largest clade, *A. adloffi* (clade b) is the sister taxon to all other species followed by *A. arachan* (clade c) as sister to the remaining species. *Austrolebias reicherti* (clade d) and A. *charrua* (clade e) are recovered as sister taxa, with *A. nachtigalli* imbedded deeply within *A. charrua*.

The time calibration assumes a mutation rate of 2% to capture a plausible time interval (lower and upper estimate) for clades and resulted in a divergence between 9.2–11.8 mya and 9.0–7.2 mya for *A. viarius* and *A. adloffi*, respectively, whereas for *A. arachan* the divergence time corresponds to 7.5–3.7 mya. The most recent clades appeared during the Pleistocene (3.8–1.0 mya) and include *A. charrua*, *A. nachtigalli*, and *A. reicherti*.

15.3.3 Statistical Parsimony Network Based on Cyt-*b* Gene

The 58 samples of *A. bellottii* represent 56 unique haplotypes, and a statistical parsimony analysis revealed two distinct networks (Figure 15.4) and four haplotypes that could not be connected to the network or any other single haplotypes. The larger grouping includes 44 haplotypes clustered in

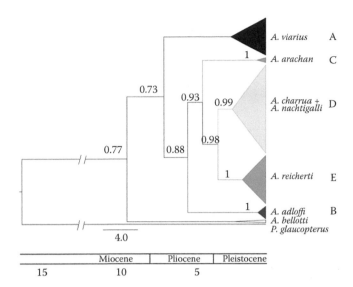

Figure 15.3 Bayesian tree topology generated using the HKY + G model of molecular evolution based on cyt-*b* gene haplotypes (H) from six taxa belonging to *A. adloffi* species group from PMD (see Figure 15.1; Appendix 15.1). Bayesian phylogeographic inference framework implemented in BEAST 1.5.4 and estimated divergence dates. Numbers above branches refer to Bayesian posterior probability of occurrence for clades. The bottom bar summarizes the timescale divergence dates in mya.

three networks. Network A consists of haplotypes exclusively from the LPP area, whereas network B includes haplotypes from the LU + MU + LPP systems, and network C groups samples from the MP area. These three networks connect through sample H28 corresponding to the Gualeguaychu site within the LU area, which represents a geographically equidistant locality from the remaining ones (Figure 15.4). Interestingly, a minor network emerges by connecting ancestral haplotypes from the Gualeguaychu site (H5, H6, and H8 from LU) with a haplotype from the MP (H7) region as well as other detached haplotypes as part of group D. These single and detached ancestral haplotypes were also located in the MU, MP, and LU regions.

The 16 samples of *A. viarius* from the coastal plains of Castillos Lagoon represent 15 unique haplotypes that cluster in three separate networks (Figure 15.5a) and a detached one (H1). Geographically, the network represented by H2–H3 corresponds to samples from slightly higher elevation from the coastal lowland where the other two networks occur. Ancestral haplotypes (H1, H2, and H3) are from relatively distant localities in the upper Cebollati River basin (near the city of Treinta y Tres, 360 m above sea level).

The analysis of the *A. adloffi* species group included a total of 65 samples representing 61 haplotypes and resulted in three distinct networks. Two of these networks consisted exclusively of samples of *A. adloffi* (Figure 15.5b) and *A. arachan* (Figure 15.5c), respectively. The third and largest network includes haplotypes of *A. charrua* (Figure 15.5d), *A. nachtigalli* (Figure 15.5d, H12), and *A. reicherti* (Figure 15.5e). The ancestral haplotype of this large network (H1) corresponds to a sample of *A. charrua* from "La Coronilla" population, a locality close to the Atlantic coastal plains. H1 connects through several step mutations with H18 (also a sample of *A. charrua*) that is geographically central and distantly located from the upper Cebollati River basin. A complex reticulate pattern emerges from alternative links between H1 and H18, and the remaining samples of *A. charrua* (Figure 15.5d). *Austrolebias nachtigalli* (Figure 15.5d, H12) connects with H18 through nine step mutations, whereas 11 step mutations separate H18 (Figure 15.5d, *A. charrua*) from the H1 belonging to *A. reicherti* (Figure 15.5e). Within *A. reicherti*, two haplotypes give raise to two

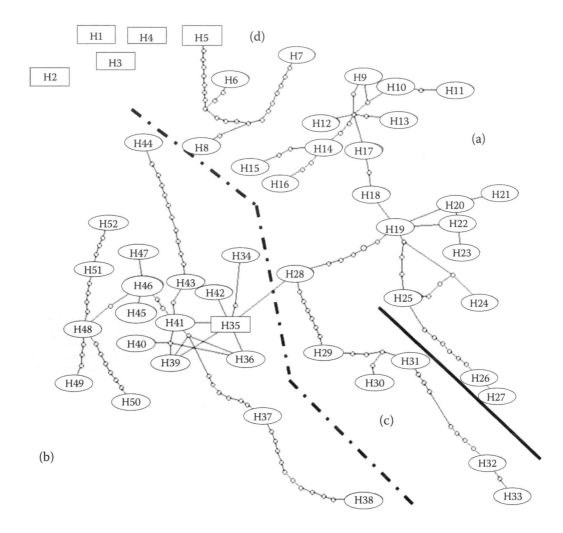

Figure 15.4 Maximum parsimony network for cyt-*b* haplotypes in *A. bellottii* species complex: (a) haplotypes from LPP region; (b) haplotypes from LU and MU regions; (c) haplotypes from MP region;
(d) haplotypes from MP and LU regions. The cladogram was estimated under the 95% statistical
parsimony limits using the program TCS 1.06 (Clement et al., 2000). Ovals include the haplotype number. Solid circles represent hypothetical haplotypes. Dashed-line separates western and eastern Uruguay River drainages; thick line encloses haplotypes from the MP region.
Rectangular polygons represent ancestral haplotypes and the size of the ovals and rectangles
are proportional to the haplotype frequency. Specimens, and the respective haplotype, are listed
in Appendix 15.1.

subnetworks: H2 connects to haplotypes from the lower Cebollati River basin and the H3 links
those from the Tacuarí River basin.

15.3.4 AMOVA, Population Differentiation, and Migration
Estimate Based on Coalescent Approach

Under a two-groups hypothesis, considering populations western and eastern to the Uruguay
River as separate groups, the AMOVA analysis showed that *A. bellottii* has a maximized variance
value (60.93%) among groups. The AMOVA analyses identified five population groups within the
A. adloffi species complex, with most of the genetic variation distributed among groups (84.73%).

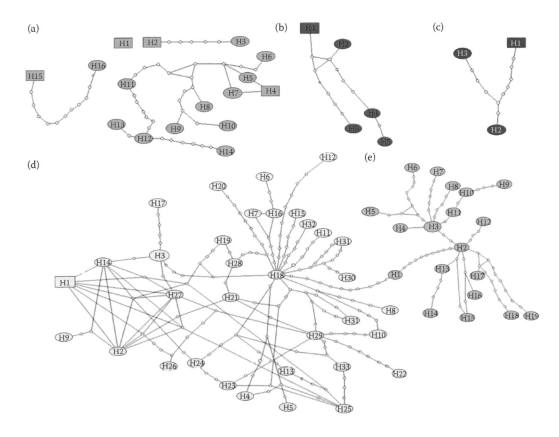

Figure 15.5   Maximum parsimony network for cyt-*b* haplotypes in *A. adloffi* species group: (a) haplotypes of *A. viarius*; (b) haplotypes of *A. adloffi*; (c) haplotypes of *A. arachan*; (d) haplotypes of *A. charrua* and H12 of *A. nachtigalli*; (e) haplotypes of *A. reicherti*. The cladogram was estimated under 95% statistical parsimony limits using the program TCS 1.06 (Clement et al., 2000). Ovals, solid circles, and rectangles as in Figure 15.4. Specimens and the respective haplotype are listed in Appendix 15.1.

The estimates of migration values among populations are shown in Figure 15.6. All the pairwise comparisons exhibited significant asymmetric gene flow rates based on nonoverlapping 95% confidence intervals. *Austrolebias bellottii* from the LU + MU area shows the highest number of immigrant females per generation with19.4 and 39.8 individuals moving toward the LPP and MP areas, respectively (Figure 15.6a). The LPP showed the most restricted gene flow relative to the other areas. The number of females migrating per generation was higher moving from the MU + LU to the MP than from the MU + LU to the LPP.

The five species of the *A. adloffi* species group were included in the pairwise calculation of asymmetric rates of gene flow. Migrate-n Bayesian analysis showed different levels of genetic exchanges among these taxa (Figure 15.6b). The highest number of immigrant females per generation was detected from *A. charrua* to *A. reicherti* (44.3 individuals) and from *A. charrua* to *A. arachan* (61.9 individuals) but less in the reverse direction (15.1 and 5.6 individuals, respectively). In general, all taxa showed restricted female migration toward populations of *A. charrua* (ranging 0.73–5.6) or *A. reicherti* (ranging 0.67–9.0). Remarkably, relatively high female migration was detected from *A. charrua* to *A. viarius* (23.4) and from *A. viarius* to *A. arachan* (39.4 individuals), but lower values were seen in the reverse direction. Interestingly, *A. adloffi*, which has an allopatric distribution relative to *A. arachan*, showed potentially higher genetic exchange with *A. arachan* (20.7 individuals) than with any other species.

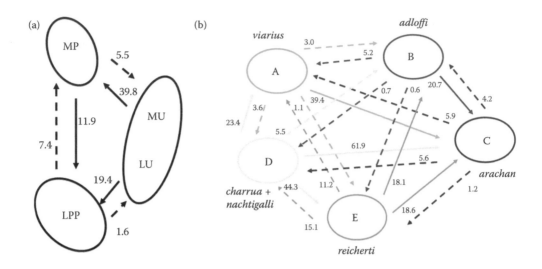

Figure 15.6 Coalescent-based estimates of asymmetric migration from the average of the runs performed in MIGRATE software. Pathway of migrants exchanged between populations based on mtDNA cyt-*b* sequences are shown. (a) migration among *A. bellottii* species complex. Thick arrows indicate number (>10) of female immigrants from LU-MU to MP and LPP; dashed arrows indicate number (<10) of female immigrants from LPP to LU-MU and MP. (b) migration among *A. adloffi* species group: (A) *A. viarius*; (B) *A. adloffi*; (C) *A. arachan*; (D) *A. charrua + nachtigalli*; (E) *A. reicherti*. Thick arrows indicate number (>10) of female immigrants between taxa; dashed arrows indicate number (<10) of female immigrants between taxa. The number next to each arrow is the estimate of the effective number of female immigrants along that pathway per generation.

15.4 DISCUSSION

15.4.1 Population Structure in *A. bellottii* Species Complex

Our analyses detected four different phylogroups of *A. bellottii* from the PLPB. Bayesian analysis suggests that these phylogroups may have emerged from independent allopatric events. An earlier vicariant event would have resulted in the split of western (Uruguay River drainage) and eastern (Paraná–La Plata basin) populations during the late Pliocene-Pleistocene (Figures 15.2 A, B and 15.4). A subsequent (early Pleistocene) split separated the eastern group into northern (MP) and southern (LPP and LU + MU) populations (Figures 15.2 C, D and 15.4). The hypothesized scenarios of differentiation are consistent with a complex Neogene geological history, including different marine ingression events into the PLPB (Iriondo and Kröhling, 2008).

AMOVA and haplotype networks detected a high level of population structure with two major groups occurring in each of the drainages of the Uruguay River basin (i.e., eastern clade B vs. western clades A, C, and D). This suggests that vicariance has played an important role in the diversification of *A. bellottii* populations. Furthermore, the data highlight the role that historical events that affected the hydrological systems played in limiting gene flow and functioning as barriers that favored diversification of populations in this species.

The analysis detected unpredicted high genetic diversity consisting of two lineages among samples from site 12 in Gualeguaychú. One of them was integrated by the haplotype H28, which connects the networks A and C and the network B through the central haplotype H35 belonging to the same site (Figure 15.4). Another detached and ancestral haplotype from this site belonging to network D (H5) links the LU to populations from the MP region. Coalescent estimates of migration rates support these results, since the LU has the highest value of migrating females toward other regions, reinforcing the role of the LU as intermediate populations. A previous study suggested that populations from Gualeguaychú serve as a "genetic corridor" linking both Uruguay River drainages (García et al., 2012). However, present analysis highlighted the role of other isolated populations from Gualeguaychú connecting with the northern MP area and suggesting the occurrence of different vicariant events between northern and southern populations that increased genetic diversity in *A. bellottii*. These results support the idea that an *Austrolebias* ancestor may have occurred in LPP, as previously suggested by Costa (2010).

15.4.2 Population Differentiation in the *A. adloffi* Species Group

Bayesian analysis and statistical parsimony networks detected four major clades corresponding to currently known species of the *A. adloffi* species group (Figures 15.3 and 15.5). *Austrolebias viarius* is the sister group of the remaining taxa of the *A. adloffi* group (Figure 15.3). Although these phylogenetic relationships were previously reported (García, 2006), a morphological study placed *A. viarius* in the *A. bellottii* species group (Huber, 1998). Herein, we found *A. viarius* consisting of three divergent and separate networks, suggesting a high level of intraspecific genetic diversity in the species (Figure 15.5; Table 15.3). In addition, we recovered that all other species of the *A. adloffi* species group (except *A. viarius*) included in the study have evolved relatively recently from an *A. adloffi* ancestor (Figure 15.3). BI, statistical parsimony networks, and the AMOVA results (Figures 15.3 and 15.5) yielded robust support to the species of the *A. adloffi* species group. Considering that *A. adloffi* is allopatric relative to the other species in the group, its differentiation could have occurred through a vicariant event from a large ancestral population that also gave rise to *A. viarius* (García, 2006). Our Bayesian analysis placed this cladogenetic event during the early to mid-Pliocene with subsequent speciation events during the late Pliocene-Pleistocene. These hypothetical cladogenetic events are consistent with the geological history of the Dos Patos-Merín coastal lagoon system, which has undergone geological complex changes since the Pleistocene, as well as post-Pleistocene marine transgression events (Sprechman, 1978; Montaña and Bossi, 1995).

Our statistical parsimony yielded a network, with a 95% degree of confidence, in which *A. charrua* haplotypes (including *A. nachtigalli*) and all *A. reicherti* haplotypes are connected (Figure 15.5). *Austrolebias reicherti* was recognized as a separate species (Loureiro and García, 2004) and subsequently placed in the synonymy of *A. charrua* (Costa, 2006); more recently, based on molecular and morphological data, it was revalidated as a different taxon (Loureiro and García, 2008). The AMOVA results strongly support the genetic distinctiveness between *A. charrua* and *A. reicherti* as separate species. Historic allopatric fragmentation and range expansions, perhaps involving areas of secondary contacts, were proposed as explanation for the differentiation of *A. charrua* and *A. reicherti* (García et al., 2009). The lowlands surrounding the Merín lagoon, close to the southern portion of Tacuarí River (Figure 15.1b), were considered as a potential contact area for these two species (García et al., 2009). Our recovered large network, which groups all *A. charrua* and *A. reicherti* samples, also suggests a lowland ancestral haplotype (Figure 15.1b, collecting site 17), which broadens the proposed area of secondary contact closer to the Atlantic coast than previously suggested (García et al., 2009). Alternatively, the intermediate H1 haplotype of *A. reicherti* found between the range of expansion of both taxa (Figure 15.5d and e) could represent incomplete lineage sorting following the Pleistocene cladogenetic events. However, the genetic divergence (8.4%) between the two species, as well as the extensive haplotype reticulation in the statistical parsimony network and the high structured BEAST topology, suggest secondary contacts between differentiated species.

An asymmetric number of immigrant females per generation was detected for most taxa in the *A. adloffi* species group relative to *A. viarius*. In particular, immigration rates between *A. charrua* versus *A. reicherti* and *A. charrua* versus *A. arachan* (Figure 15.6) agree with previous reports (García et al., 2009). These data may suggest that gene flow is currently a weak cohesive population factor but probably sufficient to maintain genetic exchange among some populations of these taxa.

Parapatric populations with secondary contact between previously allopatric taxa were shown to have variable levels of gene flow depending on the extent of reproductive isolation achieved while in allopatry (Jiggins and Mallet, 2000).

15.4.3 Comparing Phylogeographic Patterns of *Austrolebias* in Different Basins

This study represents the first comparative phylogeographic analysis focused on distinct clades of annual fishes inhabiting these two hydrological systems. The molecular data suggest high to moderate genetic diversity within the *A. bellottii* species complex (A–D), except for MP population (Figures 15.2 and 15.4, clade C) (Table 15.1). The genetic diversity found in populations from the LU site (Figures 15.2 and 15.4, clade B) is comparable with that of *A. reicherti* from the PMD system; furthermore, it is similar to that reported for the *A. adloffi* species group (García, 2006; García et al., 2009).

As expected, genetic pairwise comparisons within the *A. bellottii* species complex (average 4.7%) are lower than those among species in the *A. adloffi* species group (average 10.6%); the latter are similar to those reported among *Rivulus* species (average 15.0%, Ponce de León et al., 2014). In addition, mean divergence estimates within two *R. cylindraceus* lineages range between 0.3% and 0.5% (Ponce de León et al., 2014); we found similar values among clades of *A. bellottii* (0.7%–5.2%). Morphometric analyses of *A. bellottii* reported a high degree of overlap among individuals that can be interpreted as population-level differentiation instead of speciation events (García et al., 2012).

Based on our data we propose that ongoing speciation has not yet reached isolation among divergent lineages within *A. bellottii* in the PLPB as it was attained among different taxa belonging to the *A. adloffi* complex in the PMD system. This could be due in part to different geomorphological events that occurred during the Quaternary in both regions.

Historical allopatric fragmentations, with subsequent range expansions, have been proposed as common processes of *Austrolebias* population differentiation in both basins (García et al., 2009, 2012). However, our data suggest that in the PMD system (1) these events are more recent and (2) the different species of the *A. adloffi* group have differentiated from endemic and isolated populations that were distributed across areas of slightly higher altitude to coastal wetlands environments that originated during Quaternary marine transgressions. Moreover, our study reveals variable levels of gene flow among species in the *A. adloffi* species group (Figure 15.6). A number of examples of speciation occurring without interruption of gene flow have been suggested (Nosil, 2008; Feder and Nosil, 2012; Via 2012). Therefore, speciation with gene flow could also have occurred in these populations, and if so it would represent the first report both in the family Rivulidae and in Cyprinodontiformes.

Finally, in both basins, lowland regions have enhanced opportunities for population mixing and increased genetic exchange. Marine transgressions and regressions within the PLPB during the late Pliocene-early Pleistocene have been suggested as alternative barriers and corridors that prevented complete isolation of the three major phylogroups within *A. bellottii* species complex (García et al., 2012). The hypothetical pattern of populations mixing in the LU from both drainages of the Uruguay River basin could have occurred in the LU drainage since the Neogene (Iriondo and Kröhling, 2008). This study corroborates these cycles of connectivity, perhaps including multiple vicariance and/or dispersal events through populations located in the western lower Uruguay River margin as previously proposed (García et al., 2012). A recent report on nonannual freshwater fish species from the Uruguay River also indicates a high level of connection between the LPP and LU–MU drainages (Zarucki et al., 2010).

Secondary contacts between the differentiated taxa *A. charrua-reicherti* were previously proposed to explain the occurrence of morphological and genetic "intermediate" populations located at the southern portion of Tacuarí River (García et al., 2009). In this sense, April and Turgeon (2006) have reported secondary contact between two subspecies of *Fundulus diaphanous*. These represent distinct evolutionary lineages corresponding to glacial races of Atlantic and Mississippian origins, and the authors proposed that individuals with intermediate phenotypes result from secondary contacts between these races.

Nevertheless, the present work enlarges the secondary contact between both pair of species *A. char-rua-reicherti* near the Atlantic coastal lowland area and lends support to considerable genetic exchange in relation to the sister taxon *A. arachan*. In fact, a considerable number of effective female immigrants between parapatric populations of *A. reicherti* and *A. arachan* were evidenced in the present analysis.

15.5 GENERAL CONCLUSIONS AND CONSERVATION REMARKS

Although past allopatric fragmentations (vicariance events) and range expansion were the common scenarios for *Austrolebias* population differentiation in both basins, present work supports that speciation processes have not reached an endpoint among divergent lineages within the *A. bellottii* taxon through the PLPB, as it was evident among different taxa belonging to the *A. adloffi* species group in PMD system. This could be due, in part, to different geomorphological events that occurred during the Quaternary in both regions. However, in both study areas, concordant patterns of population mixing were detected in lowlands that increased the level of genetic exchange between nearby populations and the biological diversity.

This study supports the occurrence of speciation with flow gene in the *A. adloffi* species group, since limited high levels of female immigrants per generation were detected among all taxa, suggesting an additional possible asymmetry in interspecific crosses.

Taking into account all present findings, we propose that different *A. bellottii* lineages and species of the *A. adloffi* complex would be considered independent evolutionary units in conservation programs of these endemic annual Neotropical fish.

ACKNOWLEDGMENTS

We thank the following colleagues for kindly providing the following fish specimens: L. Malabarba (*A. adloffi*) and W.J.E.M. Costa (*C. melanotaenia*, *C. nachtigalli*) from RS Brazil (1991 and 1997, respectively); V. Etzel and G. Hessfcld (*A. vandenbergi*) from Paraguay (2004); P. Calviño (*A. bellottii*) from Chaco, Salta, and Buenos Aires provinces and from aquarium strains. Graciela García and Verónica Gutiérrez acknowledge the research support of SNI (ANII, Uruguay). R.d.S. was supported by award NSF-DEB 1144692.

Appendix 15.1 Detailed List of Localities, Geographic Areas, GenBank Accession Numbers, Facultad de Ciencias Evolutionary Genetics Laboratory (GP) of *A. bellottii*, *A. vanderbergi*, and *P. glaucopterus* Specimens Used in This Article

Species	River Basin	Locality	Catalog and Haplotype Numbers	GenBank Accession Number
A. bellottii	LU	1 Carmelo Town, Colonia Department, Uruguay	GP363-H45	AF245006.1
A. bellottii	LU	2 Puerto Viejo, San Javier Town, Río Negro Department, Uruguay	GP1452-H40 GP1455-H35 GP1456-H35	HQ149645 HQ149646 HQ149647
A. bellottii	MU	3 Bañado Verocay, Salto Department, Uruguay	GP400-H48	AF245007.1
A. bellottii	MU	3 Salto City, Salto Department, Uruguay	GP1436-H52 GP1442-H48 GP1437-H49 GP1438-H46 GP1440-H44 GP1441-H43	HQ149639 HQ149644 HQ149640 HQ149641 HQ149642 HQ149643

(Continued)

Appendix 15.1 (*Continued*) Detailed List of Localities, Geographic Areas, GenBank Accession Numbers, Facultad de Ciencias Evolutionary Genetics Laboratory (GP) of *A. bellottii*, *A. vanderbergi*, and *P. glaucopterus* Specimens Used in This Article

Species	River Basin	Locality	Catalog and Haplotype Numbers	GenBank Accession Number
A. apaii	LU	4 Villa Soriano Town, Soriano Department, Uruguay	GP1572-H50	HQ149648
A. bellottii	MU	5 Franquia, Bella Unión City, Artigas Department, Uruguay	GP1426-H2 GP1427-H37 GP1430-H30 GP1434-H51	HQ149635 HQ149636 HQ149637 HQ149638
A. bellottii	LPP	6 Viborota City, Buenos Aires Province, Argentina	R298-H22 R299-H9 R300-H18 R301-H18 R302-H26	HQ149667 HQ149668 HQ149669 HQ149670 HQ149671
A. bellottii	LPP	7 Dolores City, Buenos Aires Province, Argentina	R290-H27 R294-H15 R292-H27 R293-H27	HQ149663 HQ149666 HQ149664 HQ149665
A. bellottii	LPP	8 Cabo San Antonio, Buenos Aires Province, Argentina	R308-H14 R309-H17 R310-H13 R311-H20 R312-H23	HQ149672 HQ149673 HQ149674 HQ149675 HQ149676
A. bellottii	LPP	9 R2-K145.2. Buenos Aires Province, Argentina	R324-H10 R325-H21 R326-H11 R327-H12	HQ149677 HQ149678 HQ149679 HQ149680
A. bellottii	MP	10 Puerto Tirol City, Chaco Province, Argentina	R173-H7 R174-H42	HQ149655 HQ149656
A. bellottii	MP	11 R11, 21 K from Río de Oro, Chaco Province, Argentina	R153-H31 R154-H29 R155-H3 R156-H30 R157-H32 R158-H33	HQ149650 HQ149651 HQ149652 FJ 826891 HQ149653 HQ149654
A. bellottii	LU	12 Gualeguaychú City, Entre Ríos Province, Argentina	R175-H28 R176-H41 R178-H41 R179-H34	HQ149657 HQ149658 HQ149659 HQ149660
		12 Gualeguaychú, "Tacural" Locality, Entre Ríos Province, Argentina	R380-H8 R381-H4 R382-H6	
A. bellottii	LPP	13 Mar de Cobo, Buenos Aires Province, Argentina	R120-H19	HQ149649
A. bellottii	LPP	14 La Plata City, Buenos Aires Province, Argentina	R225-H25 R226-H24	HQ149661 HQ149662
A. bellottii	MP	15 Formosa, Argentina	R379-H1	
A. vandenbergi		Typical locality, Paraguay	R116	FJ826901
P. glaucopterus		Next to Cuiabá City, Mato Grosso, Brazil	R006	AF245468.1

Note: Several localities of the catalog map in the same geographic area (circles in Figure 15.1a). Collection codes: R = route, K = kilometer.

Appendix 15.2 Detailed List of Localities, Geographic Areas, GenBank Accession Numbers, Facultad de Ciencias Evolutionary Genetics Laboratory (GP) of *A. adloffi* Species Groups (Circles in Figure 15.1b)

Species	Locality	Catalog and Haplotype Numbers	GenBank Accession Number
A. viarius	1 CH1, R10-K.250 (Rocha Department, Uruguay)	854-H11 855-H5 856-H13 859-H13	AY724408 AY724383
A. viarius	2 CH4, R10-K.266.5 (Rocha Department, Uruguay)	329-H7	
A. viarius	3 CH11, R16-K.2.5 (Rocha Department, Uruguay)	880-H10 881-H9 883-H8 888-H4	AY724383 AY724385 AY724384 AY724382
A. viarius	4 CH23, R10-16 (Rocha Department, Uruguay)	957-H12 937-H7 377-H7	AY724408 AF245456
A. viarius	5 CH49, close to R9-K.272 (Rocha Department, Uruguay)	1231-H6 1237-H16 1239-H15	AY724387 AY724386
A. viarius	6 R8 A° Pirarajá and A° Retamosa. (Rocha Department, Uruguay)	1674-H3 1675-H2 1676-H1 1677-H14	KJ734179 KJ734180 KJ731182 KJ734181
A. arachan	7 PPM, Melo City. (Cerro Largo Department, Uruguay)	657-H1 1678-H2 1679-H3	AF245466 KJ734168 KJ734167
A. charrua	8 CH6, R9-K.272. (Rocha Department, Uruguay)	947-H6	AY724390
A. charrua	9 CH8, R13-16 (Rocha Department, Uruguay)	745-H10 747-H8 746-H9	AY724391 AY724379 AY724380
A. charrua	10 CH28, R14-K.489.5 (Rocha Department, Uruguay)	1015-H26	AY724377
A. charrua	11 CH26, R9-K.302 (Rocha Department, Uruguay)	1031-H16	AY724390
A. charrua	12 CH33, R19-K.6.5-7 (Rocha Department, Uruguay)	362-H15 350-H14	AY724409 AY724390
A. charrua	13 CH32, R9-K.336.5 (Rocha Department, Uruguay)	1008-H22	AY724389
A. charrua	14 CH37, R15-K.151.7 (Rocha Department, Uruguay)	1032-H13	AY724400
A. charrua	15 CH44, R15-K.173 (Rocha Department, Uruguay)	1022-H7	AY724399
A. charrua	16 CH50, close to R9-K.272 (Rocha Department, Uruguay)	1194-H11 122 I-H32 1222-H23 1027-H34 1082-H33	AY724378 AY724381 AY724406 AY724405 AY724410
A. charrua	17 La Coronilla Town (Rocha Department, Uruguay)	2500-H1 2478-H2 2744-H2 1745-H3 1746-H3 1747-H24 1748-H25 1749-H4 1750-H5	KJ734149 KJ734148 KJ734150 KJ734151 KJ734152 KJ734153 KJ734154 KJ734155 KJ734156

(Continued)

Appendix 15.2 (*Continued*) Detailed List of Localities, Geographic Areas, GenBank Accession Numbers, Facultad de Ciencias Evolutionary Genetics Laboratory (GP) of *A. adloffi* Species Groups (Circles in Figure 15.1b)

Species	Locality	Catalog and Haplotype Numbers	GenBank Accession Number
A. charrua	18 CH53, R91 (A° C. del Parao) (Treinta y Tres Department, Uruguay)	1258-H27 1322-H19	
A. charrua	19 CH54, R91 (A° C. del Parao) (Treinta y Tres Department, Uruguay)	1655-H31 1656-H29 1657-H28	KJ734165 KJ734164 KJ734163
A. charrua	20 CH66, Treinta y Tres City (Treinta y Tres Department, Uruguay)	1595 1597-H30	KJ734166
A. reicherti	21 CH55, R91-29 K from La Charqueada (Treinta y Tres Department, Uruguay)	1320-H1 1317-H9 1318-H16 1100-H14 1103-H19 1104-H13	AY724394 AY724392 AY724393 AY724401 AY724403 AY724393
A. reicherti	22 CH42, R18-K.369.5 (Treinta y Tres Department, Uruguay)	2001-H17	KJ734169
A. reicherti	23 CH43, R18 close to Vergara City (Treinta y Tres Department, Uruguay)	1096-H18 1089-H15 1085-H7 1086-H2	AY724398 AY724396 AY724397 AY724395
A. reicherti	24 CH58, Lower Tacuarí basin (Treinta y Tres Department, Uruguay)	1288-H12 1289-H4 1299-H10 1298-H6 1300-H11	KJ734178 KJ734170 KJ734173 KJ734172 KJ734174
A. reicherti	25 CH59, R26 (5 K from Rio Branco City) (Cerro Largo Department, Uruguay)	1301-H3 1290-H3 1302-H5 1303-H8	KJ734175 KJ734171 KJ734176 KJ734177
A. adloffi	BR290, RGS, Brazil Puente Gravataí, RGS, Brazil BR290, RGS, Brazil	308-H1 303-H2 304-H3 300-H4 313-H5	AY724375 AY724372 AY724374 AY724376
A. natchigalli	RG-Pond from BR116-K.15, RGS, Brazil	1186-H12	AY724407

Note: Collection codes: CH = pond, R = route, K = kilometer, A° = stream.

REFERENCES

Akaike, H. 1974. A new look at the statistical model identification. *IEEE Transactions on Automatic Control* 19: 716–723.

April, J., and J. Turgeon. 2006. Phylogeography of the banded killifish (*Fundulus diaphanus*): Glacial races and secondary contact. *Journal of Fish Biology* 69: 212–228. doi: 10.1111/j.1095-8649.2006.01233.x

Avise, J. C. 1994. *Molecular Markers, Natural History and Evolution.* Chapman and Hall, New York.

Avise, J.C. 2000. *Phylogeography: The History and the Formation of Species.* Harvard University Press, Cambridge, MA.

Beerli, P. 2006. Comparison of Bayesian and maximum-likelihood inference of population genetic parameters. *Bioinformatics* 22: 341–345. doi: 10.1093/bioinformatics/bti803

Beerli, P., and J. Felsenstein. 2001. Maximum likelihood estimation of a migration matrix and effective population sizes in *n* subpopulations by using a coalescent approach. *Proceedings of the National Academy of Sciences of the United States of America* 98: 4563–4568. doi: 10.1073/pnas.081068098

Beheregaray, L. 2008. Twenty years of phylogeography: The state of the field and the challenges for the Southern Hemisphere. *Molecular Ecology* 17: 3754–3774. doi: 10.1111/j.1365-294X.2008.03857.x

Behling. H., and R. R. B. Negrelle. 2001. Tropical rain forest and climate dynamics of the Atlantic lowland, southern Brazil, during the late Quaternary. *Quaternary Research* 56: 383–389. doi: 10.1006/qres.2001.2264

Boneto, A. A. 1994. Austral rivers of South America. *Limnology Now: A Paradigm of Planetary Problems.* R. Margalef ed. pp. 425–472. Elsevier Science B. V.

Brown, W. M., M. Jr. George, and A. C. Wilson. 1979. Rapid evolution of animal mitochondrial DNA. *Proceeding of the National Academy of Sciences of the United State of America* 76: 1967–1971.

Camargo, A., R. O. de Sá, and W. R. Heyer. 2006. Phylogenetic analyses of mtDNA sequences reveal three cryptic lineages in the widespread neotropical frog *Leptodactylus fuscus* (Schneider, 1799) (Anura, Leptodactylidae). *Biological Journal of the Linnean Society* 87: 325–341. doi: 10.1111/j.1095-8312.2006.00581.x

Castelloe, J., and A. R. Templeton. 1994. Root probabilities for intraspecific gene trees under neutral coalescent theory. *Molecular Phylogenetics and Evolution* 3: 102–113. doi: 10.1006/mpev.1994.1013

Clement, M., D. Posada, and K. A. Crandall. 2000. TCS: A computer program to estimate gene genealogies. *Molecular Ecology* 9: 1657–1659. doi: 10.1046/j.1365-294x.2000.01020.x

Costa, W. E. J. M. 1990. Analise filogenetica da familia Rivulidae (Cyprinodontiformes, Aplocheiloidei). *Revista Brasileira de Biologia* 50: 65–82.

Costa, W. E. J. M. 2006. The South American annual killifish genus *Austrolebias* (Teleostei: Cyprinodontiformes: Rivulidae): Phylogenetic relationships, descriptive morphology and taxonomic revision. *Zootaxa* 1213: 1–162. http://dx.doi.org/.

Costa, W. E. J. M. 2010. Historical biogeography of cynolebiasine annual killifishes inferred from dispersal vicariance analysis. *Journal of Biogeography* 37: 1995–2004. doi: 10.1111/j.1365-2699.2010.02339.x

Drummond, A. J., and A. Rambaut. 2007. BEAST: Bayesian evolutionary analysis by sampling trees. *BMC Evolutionary Biology* 7: 214. doi: 10.1186/1471-2148-7-214.

Excoffier, L., P. E. Smouse, and J. M. Quattro. 1992. Analysis of molecular variance inferred from metric distances among DNA haplotypes: Application to human mitochondrial DNA restriction data. *Genetics* 131: 479–491.

Feder, J. L., S. P. Egan, and P. Nosil. 2012. The genetics of speciation with gene flow. *Trends in Genetics* 28: 342–350. doi: 10.1016/j.tig.2012.03.009

Galtier, N., B. Nabholz, S. Glémin, and G. Hurst. 2009. Mitochondrial DNA as a marker of molecular diversity: A reappraisal. *Molecular Ecology* 18: 4541–4550.

García, G. 2006. Multiple simultaneous speciation in killifishes of the *Cynolebias adloffi* species complex (Cyprinodontiformes, Rivulidae) from phylogeography and chromosome data. *Journal of Zoological Systematics and Evolutionary Research* 44: 75–87. doi: 10.1111/j.1439-0469.2005.00346.x

García, G., F. Alvarez-Valin, and N. Gómez. 2002. Mitochondrial genes: Signals and noise in phylogenetic reconstruction within killifish genus *Cynolebias* (Cyprinodontiformes, Rivulidae). *Biological Journal of the Linnean Society* 76: 49–59. doi: 10.1111/j.1095-8312.2002.tb01713.x

García, G., V. Gutiérrez, J. Vergara, P. Calviño, A. Duarte, and M. Loureiro. 2012. Patterns of population differentiation in annual killifishes from the Paraná-Uruguay-La Plata basin: The role of vicariance and dispersal. *Journal of Biogeography* 39: 1707–1719. doi: 10.1111/j.1365-2699.2012.02722.x

García, G., M. Loureiro, N. Berois et al. 2009. Pattern of differentiation in the annual killifish genus *Austrolebias* (Cyprinodontiformes; Rivulidae) from a biosphere reserve site in South America: A multidisciplinary approach, *Biological Journal of the Linnean Society* 98: 620–635. doi: 10.1111/j.1095-8312.2009.01303.x

Gehring, P. S., K. A. Tolley, F. S. Eckhardt et al. 2012. Hiding deep in the trees: Discovery of divergent mitochondrial lineages in Malagasy chameleons of the *Calumma nasutum* group. *Ecology and Evolution* 2: 1468–1472.

Guindon, S., J. F. Dufayard, V. Lefort et al. 2010. New algorithms and methods to estimate maximum-likelihood phylogenies: Assessing the performance of PhyML 3.0. *Systematic Biology* 59: 307–21. doi: 10.1093/sysbio/syq010

Hasegawa, M., K. Kishino, and T. Yano. 1985. Dating the human-ape splitting by a molecular clock of mitochondrial DNA. *Journal of Molecular Evolution* 22: 160–174.

Hillis, D. M., and J. J. Bull. 1993. An empirical test of bootstrapping as a method for assessing confidence in phylogenetic analysis. *Systematic Biology* 42: 182–192.

Huber, J. 1998. *Comparison of Old World and New World Tropical Cyprinodonts.* Société Françaised´Ichtyologie pp. 1–109. Mus. Nat. Hist. Nat. Paris (France). I.S.B.N. 2-9507330-4-2.

Iriondo, M. 2004. Large wetlands of South America: A model for Quaternary humid environments. *Quaternary International* 114: 3–9.

Iriondo, M., and D. Kröhling. 2008. *Cambios ambientales en la cuenca del río Uruguay desde dos millones de años hasta el Presente.* M. Iriondo and D. Kröhling eds. Universidad Nacional del Litoral, Santa Fé, Argentina.

Jiggins, C., and J. Mallet. 2000. Bimodal hybrid zones and speciation. *Trends in Ecology and Evolution* 15: 250–255.

Kawakami de Resende, E. 2003. *Migratory Fishes of the Paraguay-Paraná Basin Excluding the Upper Paraná Basin. Migratory Fishes of South America: Biology, Fisheries and Conservation Status.* J. Carolsfeld, B. Harvey, C. Ross and A. Baer eds. Victoria BC: The World Bank—IDRC—World Fisheries Trust.

Kimura, M. 1980. A simple method for estimating evolutionary rate of base substitutions through comparative studies of nucleotide sequences. *Journal of Molecular Evolution* 16: 111–120.

Loureiro, M., and R. O. de Sá. 1998. Osteological analysis of the killifish genus *Cynolebias* (Cyprinodontiformes: Rivulidae). *Journal of Morphology* 238: 245–262. doi: 10.1002/(sici)1097-4687 (199811)238:2<245::aid-jmor5>3.0.co;2-f

Loureiro, M., and G. García. 2004. *Cynolebias reicherti* a new annual fish (Rivulidae:Cynolebiatinae) from southern Laguna Merim basin. *Acta Zoológica Lilloana* 48: 13–25.

Loureiro, M., and G. García. 2008. *Austrolebias reicherti* Loureiro and García, a valid species of annual fish (Cyprinodontiformes: Rivulidae) from Uruguay *Zootaxa* 1940: 1–15.

Lundberg, J. G., L. G. Marshall, J. Guerrero, B. Horton, L. R. Malabarba, and F. Wesselingh. 1998. The stage for neotropical fish diversification: A history of tropical South American rivers. *Phylogeny and Classification of Neotropical Fishes.* L. R. Malabarba, R. E. Reis, R. P. Vari, Z. M. S Lucena and C. A. S Lucena eds. Edipucrs, Porto Alegre.

Medrano, J. F., E. Aasen, and L. Sharrow. 1990. DNA extraction from nucleated red blood cells. *Biotechniques* 8: 43.

Montaña, J. R., and J. Bossi. 1995. Geomorfología de los humedales de la cuenca de la Laguna Merín en el departamento de Rocha. UDELAR. *Revista Facultad de Agronomía.* Montevideo, Uruguay 2: 1–32.

Nei, M. 1987. *Molecular Evolutionary Genetics.* New York, Columbia University Press.

Nosil, P. 2008. Speciation with gene flow may be common. *Molecular Ecology* 17: 2103–2106.

Palumbi, S., A. Martin, S. Romano, W. O. McMillan, L. Stice, and G. Grabowski. 1991. The simple fool's guide to PCR. Department of Zoology and Kewalo Marine Laboratory, University of Hawaii, Honolulu.

Ponce de León, J. L., G. León, R. Rodríguez et al. 2014. Phylogeography of Cuban Rivulus: Evidence for allopatric speciation and secondary dispersal across a marine barrier. *Molecular Phylogenetics and Evolution* 79: 404–414. doi: 10.1111/j.1365-2699.2011.02594.x

Posada, D., and A. Crandall. 1998. Modeltest: Testing the model of DNA substitution. *Bioinformatics* 14: 817–818. doi: 10.1093/bioinformatics/14.9.817.

Probides. 1999. Plan director. *Reserva de Biosfera "Bañados del Este".* p. 159. Rocha, Uruguay.

Rambaut, A., and A. J. Drummond. 2009. Tracerv. 1.5.http://beast.Bio.Ed.Ac.UK/Tracer.

Rodríguez, F., J. L. Oliver, A. Marín, and J. R. Medina. 1990. The general stochastic model of nucleotide substitution. *Journal of Theoretical Biology* 142: 482–501.

Rozas, J., J. C. Sánchez-Delbarrio, X. Messeguer, and R. Rozas. 2003. DnaSP, DNA polymorphism analyses by the coalescent and other methods. *Bioinformatics* 19: 2496–2497. doi: 10.1093/bioinformatics/btg359.

Sprechman, P. 1978. The paleoecology and paleogeography and stratigraphy of the Uruguayan coastal area during the neogene and quaternary. *Zitteliana* 4: 3–72.

Swofford, D. L. 2002. *PAUP* Phylogenetic Analysis Using Parsimony* (*and other Methods). Sunderland, MA.: Sinauer Associates.

Tajima, F. 1989. Statistical method for testing the neutral mutation hypothesis by DNA polymorphism. *Genetics* 123: 585–595.

Tamura, K., G. Stecher, D. Peterson, A. Filipski, and S. Kumar. 2013. MEGA6: Molecular evolutionary genetics analysis version 6.0. *Molecular Biology and Evolution* 30: 2725–2729. doi: 10.1093/molbev/msr121

Templeton, A. R., K. A. Crandall, and C. F. Sing. 1992. A cladistic analysis of phenotypic associations with haplotypes inferred from restriction endonuclease mapping and DNA sequence data. III. Cladogram estimation. *Genetics* 132: 619–633.

Thompson, J. D., T. J. Gibson, F. M. Plewniak, F. Jeanmougin, and D. G. Higgins. 1997. The CLUSTAL_X windows interface: Flexible strategies for multiple sequence alignment aided by quality analysis tools. *Nucleic Acids Research* 25: 4876. doi: 10.1093/nar/25.24.4876

Vaz-Ferreira, R., and A. Melgarejo. 1984. La distribución de las especies del género *Cynolebias* Steindachner 1876, en el Uruguay, con notas sobre *C. alexandri* Castello y López, 1974. *Boletin de la Sociedad Zoológica del Uruguay* 2: 41–46.

Vaz-Ferreira, R., and B. Sierra. 1973. El género *Cynolebias* Steindachner, 1876 (Atheriniformes, Cyprinodontidae): Caracteres, especies y distribución. *Trabajos del V Congreso Latinoamericano de Zoología* 1: 245–260.

Via, S. 2012. Review article: Divergence hitchhicking and the spread of genomic isolation during ecological speciation with gene flow. *Philosophical Transactions of the Royal Society B, Biological Sciences* 367: 451–460. doi: 10.1098/rstb.2011.0260)10.1098/rstb.2011.0260

Zarucki, M., I. González-Bergonzoni, F. Teixeira de Mello et al. 2010. New records of freshwater fish for Uruguay. *Check List* 62: 191–194.

CHAPTER 16

Chromosome Evolution, Divergent Meiotic Pathways, and Speciation in the Family Rivulidae

Graciela García

CONTENTS

16.1 INTRODUCTION

In contrast to chromosomal conservation, chromosomal instability characterizes situations in which multiple closely related taxa (populations, subspecies, and/or species) belonging to a single phylogenetic lineage differ drastically from each other by major chromosomal rearrangements, sometimes resulting in high variability in chromosome number (Talavera et al., 2013).

In general, chromosome numbers vary little within and among teleost groups (Mank and Avise, 2006) and do not differ greatly from a widely proposed ancestral karyotype of 48 acrocentric chromosomes (Mank and Avise, 2006; Kornfield, 1984). This is also the case for some Neotropical rivulid species that include annual species (Elder et al., 1993; García et al., 1993, 1995, 2001). However, among annual killifishes, the chromosome numbers range from 2N = 48 to 2N = 28 with arm numbers (NF) as high as 80 (García et al., 1993, 1995, 2001; Azpelicueta et al., 2009). Robertsonian and non-Robertsonian rearrangements, including additions and/or deletions of heterochromatic regions, have been particularly prominent in this fish group (García et al., 1993, 1995, 2001). Moreover, chromosomal instability may have played a crucial role in the recent diversification of the species within this group (García, 2006; García et al., 2014).

A recent study suggested rapid diversification associated with considerable genomic expansion events in the genus *Austrolebias*, diverging from hypothetical ancestral genomes containing approximately 3.0 pg/diploid nucleus and a basal karyotype constituted by 48 small telocentric chromosomes (García et al., 2014). Moreover, the drastic nuclear DNA–increasing events found in *Austrolebias* (approx. 6.0 pg/diploid nucleus) would be associated with great genome instability and considerable increase in the proportion of transposable elements (TEs). This hypothetical scenario was corroborated through partial genome sequencing of *Austrolebias charrua* and the closely related and syntopic Cynolebiatinae species *Cynopoecilus melanotaenia* (García et al., 2015; Chapter 14). TEs are responsible for rapid genome remodeling through the creation of new regulatory gene networks and chromosome restructuring (Rebollo et al., 2010). In fishes, many bursts of retrotransposition potentially associated with speciation events have been observed for non-LTR retrotransposons (Volff et al., 2001). Ectopic recombination between TEs can also lead to lineage- or species-specific genomic rearrangements. Several hypothetical models of speciation might be consistent with a role of transposition and TE-mediated rearrangements in speciation (Böhne et al., 2008).

In this chapter, chromosome repatterning associated with cladogenetic events in different Cynolebiatinae genera is revisited. At the same time, comparative meiotic processes are presented in order to understand (1) patterns of differentiation among species groups of *Austrolebias* and (2) potential reproductive isolation (RI) barriers among taxa. We hypothesized that a sudden speciation process is associated with bursts of TE transposition that can cause chromosomal rearrangements encompassing meiotic divergent patterns and reinforcing intrinsic postzygotic reproductive barriers within and among clades in *Austrolebias*.

16.2 KARYOTYPE ORTHOSELECTION IN THE FAMILY RIVULIDAE?

The subfamily Cynolebiatinae is an endemic and speciose group of Neotropical fishes with an annual life cycle. Members of this subfamily vary greatly in morphology (Vaz-Ferreira and Melgarejo, 1984; Loureiro and de Sá, 1998; Costa, 1998), and interspecific karyotypic divergence was documented among species from Uruguay, Argentina, and Rio Grande do Sul, Brazil (García et al., 1993, 1995, 2001) (Figure 16.1).

Two major clades within *Austrolebias* could represent different genomic repatterning pathways of the karyotypes of this lineage. Figure 16.2 presents a phylogenetic hypothesis for clade differentiation associated with karyotypic parameters in each taxa.

The analysis revealed that each species group is characterized by a distinct pattern of karyotype evolution (karyotypic orthoselection). Clade I, corresponding to the *Austrolebias alexandri-affinis* species group, is basal and differentiated early in the Miocene (García et al., 2014). This clade is characterized by a 2N = 48, different numbers of NF in each species, and the presence of extralarge acrocentric and subtelocentric chromosomes with additional interstitial C-bands. Clade II is composed of all the remaining species groups included in the analyses and clustered in four subclades (subclades a–d). Herein, we summarize various mechanisms of chromosomal evolution proposed for the different clades (García et al., 1993, 1995, 2001, 2014). The occurrence of pericentric inversions, including heterochromatic addition events, was proposed for clade I and several groups within clade II. Subclade II.a consists of the *A. gymnoventris-luteoflammulatus-quirogai* species group and has a diploid number of 2N = 48 to 2N = 34, respectively, showing NF numbers next to 48. Large karyotypic rearrangements, probably involving centric fusions, were suggested in the evolution of this subclade (García et al., 1995). Subclade II.b consists of large predatory species, whose relationships have shifted over the years, and currently is assigned to the genus *Austrolebias* (Costa, 2006). These species have a noticeable decrease of chromosomal number, 2N = 46 in *A. wolterstorffi* to 2N = 38 in *A. prognathus*. Robertsonian chromosomal rearrangements (i.e., centric fusion events) and drastic changes in constitutive heterochromatin (e.g., telomeric, centromeric, and interstitial C-heterochromatic blocks) may have played an important role in the chromosome evolution in this species group. Members of subclade II.c, consisting of the *A. bellottii-robustus* species group, have diploid numbers of 2N = 48 to 2N = 44 and clear differences in karyotype structure, showing a decrease of acrocentric in favor of subtelocentric chromosomes. All chromosomes exhibited centromeric and telomeric C-bands and major karyotype repatterning probably involving centric fusions and pericentric inversions, including heterochromatic addition

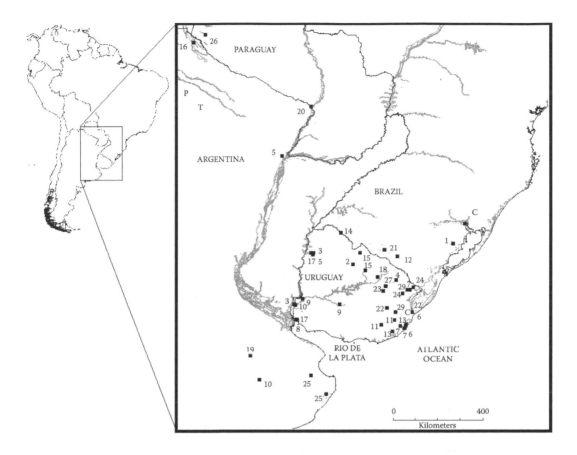

Figure 16.1 Distribution of collection sites for *Austrolebias* species and other Rivulidae genera from natural populations. 1. *A. adloffi*, 2. *A. affinis*, 3. *A. alexandri*, 4. *A. arachan*, 5. *A. bellottii*, 6. *A. charrua*, 7. *A. cheradophilus*, 8. *A. cinereus*, 9. *A. duraznensis*, 10. *A. elongatus*, 11. *A. gymnoventris*, 12. *A. juanlangui*, 13. *A. luteoflammulatus*, 14. *A. luzardoi*, 15. *A. melanoorus*, 16. *A. monstrosus*, 17. *A. nigripinnis*, 18. *A. nioni*, 19. *A. nonoiuliensis*, 20. *A. patriciae*, 21. *A. periodicus*, 22. *A. prognathus*, 23. *A. quirogai*, 24. *A. reicherti*, 25. *A. robustus*, 26. *A. vandenbergi*, 27. *A. vazferreirai*, 28. *A. viarius*, 29. *A. wolterstorffi*, C. *Cynopoecilus melanotaenia*, P. *Papiliolebias hatinne*, T. *Trigonectes balzanii*.

events. It is likely these events have influenced the evolution of this subclade. Finally, subclade II.d, consisting of the *Austrolebias adloffi-viarius* species group, has diploid numbers ranging from 46 to 48 and NF = 48–58, with a predominance of subtelocentric or acrocentric chromosomes and the presence of only one or two biarmed chromosome pairs of medium size. Moreover, high levels of genome instability with different cytotypes among different populations of *A. charrua* from geographically distant ponds have been reported (García, 2006). Thus, chromosome rearrangements like pericentric inversions in isolated populations could have triggered cladogenetic events from a hypothetical highly polymorphic population (García, 2006). Each cladogenetic event in the *A. adloffi* species group is supported by several pericentric inversions, suggesting that genetic differentiation is initiated by suppression of recombination within inverted segments. Furthermore, the accumulation of multiple pericentric inversions reinforced genetic isolation and encompassed the subsequent speciation events (García, 2006; see Chapter 17).

Figure 16.3 shows the ideogram for each *Austrolebias* species group, which allows visualization of alternative rearrangement pathways between the two major clades of *Austrolebias* and yields different karyotype structures that may support a karyotypic orthoselection hypothesis

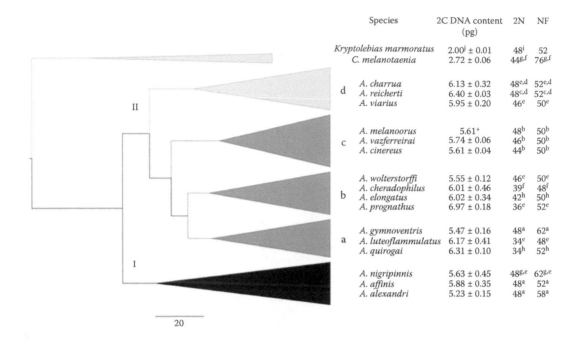

Species	2C DNA content (pg)	2N	NF
Kryptolebias marmoratus	2.00[j] ± 0.01	48[i]	52
C. melanotaenia	2.72 ± 0.06	44[g,f]	76[g,f]
d *A. charrua*	6.13 ± 0.32	48[e,d]	52[e,d]
A. reicherti	6.40 ± 0.03	48[c,d]	52[c,d]
A. viarius	5.95 ± 0.20	46[e]	50[e]
A. melanoorus	5.61[+]	48[b]	50[b]
c *A. vazferreirai*	5.74 ± 0.06	46[b]	50[b]
A. cinereus	5.61 ± 0.04	44[b]	50[b]
A. wolterstorffi	5.55 ± 0.12	46[e]	50[e]
A. cheradophilus	6.01 ± 0.46	39[f]	48[f]
b *A. elongatus*	6.02 ± 0.34	42[h]	50[h]
A. prognathus	6.97 ± 0.18	36[e]	52[e]
A. gymnoventris	5.47 ± 0.16	48[a]	62[a]
a *A. luteoflammulatus*	6.17 ± 0.41	34[e]	48[e]
A. quirogai	6.31 ± 0.10	34[h]	52[h]
A. nigripinnis	5.63 ± 0.45	48[g,e]	62[g,e]
A. affinis	5.88 ± 0.35	48[a]	52[a]
A. alexandri	5.23 ± 0.15	48[a]	58[a]

20

Figure 16.2 Tree topology based on *cyt*-b sequences from *Austrolebias* and the outgroup taxa. Bayesian phylogeographic inference framework generated using the GTR + G model of molecular evolution implemented in BEAST 1.5.4. *Cynopoecilus melanotaenia*, *Kryptolebias marmoratus*, and *Nothobranchius furzeri* were included as outgroup taxa. Average of DNA content (mean ± SE) in each *Austrolebias* species group and *C. melanotaenia* determined by flow cytometry. 2N: chromosome number; NF: chromosome arm number. [a]García et al. (1995), [b]García et al. (2001), [c]Loureiro and García (2004), [d]García (2006), [e]García et al. (1993), [f]Máspoli and García (1988), [g]Scheel (1972), [h]García et al. (2014), [i]Scheel (1972), [j]Kelly et al. (2012). [+]Only one flow cytometry estimation was obtained. (Modified from García, G. et al. 2014. *Genetica* 142: 87–98.)

(*sensu* White, 1973). Extensive variation by means of chromosome rearrangements involving both Robertsonian and non-Robertsonian changes at intra and interspecific level shows that the genome of *Austrolebias* is highly unstable.

Remarkably, the frequency of these different types of chromosomal rearrangements between the two major clades has occurred without statistically significant differences in nuclear DNA content. Further analyses using NGS techniques would clarify whether the different types and/or proportions of TE families are responsible for the aforementioned different chromosomal evolutionary pathways.

Chromosome repatterning could be occurring in other closely related Rivulidae genera, for example, *Trigonectes balsanii* and *Papiliolebias hatinne* (Figure 16.4). *Trigonectes balzanii* has a 2N = 38, NF = 72 and showed a reduction in chromosome number from an ancestral hypothetical ancestor of 48 acrocentric chromosomes and an increase of medium-sized and meta-submetacentric chromosomes (Figure 16.4a). This transient polymorphism could represent a step in the karyotypic orthoselection process (*sensu* White, 1973). However, *Papiliolebias hatinne* (2N = 28, NF = 52) showed an extreme reduction of diploid number with the presence of distinctive extra-large biarmed chromosomes that could be the result of multiple centric fusion events producing extensive karyotypic restructuring (Figure 16.4b). A similar variation was described in other Rivulidae. *Krytolebias marmoratus* exhibits synchronous hermaphroditism with self-fertilization, a remarkable constancy in chromosome number (2N = 48) and morphology that may represent the ancestral condition (Sola et al., 1997), whereas other species have diploid number ranging from 48 to 40 and

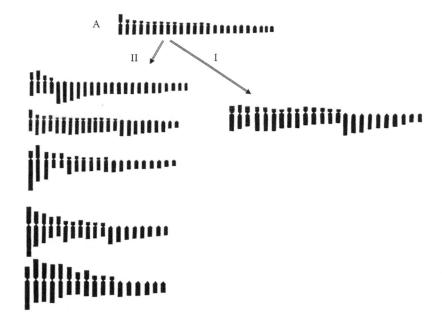

Figure 16.3 Ideogram showing the two different chromosome pathways found in this study. A. Hypothetical ancestral Rivulidae karyotype constituted by 48 acro-subtelocentric chromosomes. I. Ideogram representing species from clade I (Figure 16.2) showing 2N = 48 and different numbers of NF, the existence of extra-large acrocentric and subtelocentric chromosomes as markers, and the occurrence of predominantly pericentric inversions, including heterochromatic addition events. II. Ideogram representing species from clade II (Figure 16.2) showing a reduced chromosome number in the derivative taxa and the occurrence of major karyotype repatterning, probably involving centric fusions and pericentric inversions, including heterochromatic addition events.

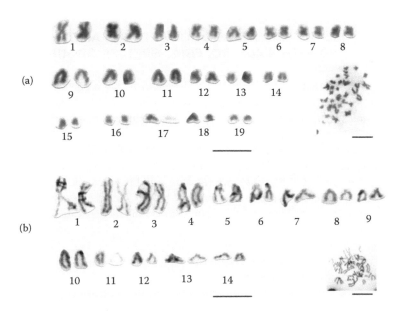

Figure 16.4 Conventional karyograms of somatic cells of two *Rivulidae* species (a) *Trigonectes balzanii*, 2N = 38, NF = 72; (b) *Papiliolebias hatinne*, 2N = 28, NF = 52. Bar = 10 μm.

chromosome arms from 54 to 92 (Scheel, 1972). The genus *Pterolebias* has variation in the diploid number ranging from 42 to 54 chromosomes (Elder et al., 1993).

Chromosome number and morphology are highly variable among Old World aplocheiloid species. This is evident within the genus *Aphyosemion*, where several species show karyotypes differing both in the diploid number and in the number of chromosome arms, for example, *A. rubrolabiale* (Scheel, 1972, 1990). The occurrence of alternating fusions and pericentric inversions may explain karyotype evolution in Old World and New World genera (Scheel, 1972). This author suggested that a reduced number of chromosomes represent a derived condition.

The role played by chromosomal changes in the evolution and cladogenetic events is unclear. The correlation between both factors has led to models that assume that chromosomal variation could cause barriers to gene flow, creating RI mechanisms among populations (Sites and Moritz, 1987; King, 1993) . Studies in Rivulidae could indicate that the occurrence of a high rate of chromosomal evolution is associated with high levels of differentiation in this fish group, supporting the previous hypothesis. This is likely due to the greater chance that chromosomal changes are stochastically fixed in small demes with high rates of local extinction and colonization (Lande, 1984). Since annual life cycle occurs in highly heterogeneous environments and seasonal habitats, several parameters need to be considered associated with this lifestyle: age-dependent mortality, nonoverlapping generations, and inbreeding in a full-sib group in small isolated ponds. In particular, when the individuals are very closely related, the probability of fixation of a chromosomal variant is higher than in a group of unrelated individuals (Hedrick and Levin, 1984). Deterministic and stochastic factors could shape the population structure and chromosomal rearrangements and reach fixation in the homozygous state, as shown in natural populations of *Austrolebias* (García, 2006). New chromosome rearrangements might be fixed in populations after severe bottlenecks, promoting substantial variation in their population sizes, which at the same time might promote rapid fixation of new variants. These scenarios of *Austrolebias* differentiation are concordant with Quaternary environmental changes in several South American basins (García et al., 2001; García, 2006), particularly, in the lowlands of the Patos-Merín drainage (PMD) system, which is a "hot spot" of speciation for *Austrolebias* (Loureiro and García, 2006).

16.3 DIFFERENT MEIOTIC PATTERNS AMONG AND WITHIN *AUSTROLEBIAS* SPECIES GROUP

King (1993) proposed that the final arbiter that decides whether a particular rearrangement can or cannot become involved in the speciation process is the meiotic system of the individual and its population. A complex structural rearrangement may segregate in a balanced fashion in one organism and not in another, and this single event decides whether a rearrangement can ever play a role in speciation. During the formation of eggs and sperm, the chromosomes must pair up in an elaborate sequence of events that result in cells with exactly half the number of chromosomes as the parent cell. A single misstep can cause infertility, miscarriage, and birth defects.

Here we review the meiotic patterns and process in different *Austrolebias* species groups to assess how major chromosomal changes may occur and became incorporated in the meiotic systems encompassing species diversity.

Overall, early prophase I is highly conserved among different *Austrolebias* species groups (García, 1996). Figure 16.5 shows a general pattern of early meiotic stages: leptotene–zygotene stage homologous chromosomes come together to form a synapse (Figure 16.5a); this stabilizing process is promoted by a meiosis-specific organization of chromosomes called the "bouquet configuration" that is initiated by clustering of telomeres on the inner nuclear envelope (Figure 16.5b). The bouquet appears to facilitate homologous recognition and alignment by concentrating chromosomes within a limited region of the nuclear volume. Condensation progresses in the bivalent elements during

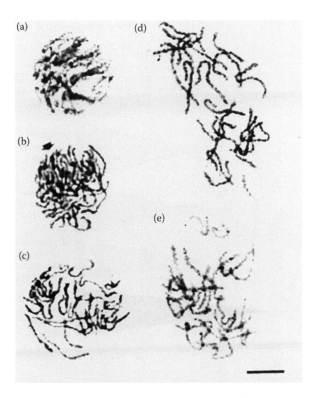

Figure 16.5 Early prophase I events in *A. bellottii* taxon from meiotic studies in male specimens performed following standard protocols: (a) leptotene-zygotene meiotic stage; (b) broad arrowhead indicates bouquet configuration; (c) early pachytene stage; (d) late pachytene stage; (e) diplotene stage. A "diffuse" overlaps the late pachytene and diplotene stages. Bar = 10 μm. (Adapted from Kligerman, A. D. and S. E. Bloom. 1977. *Journal of the Fisheries Research Board of Canada* 34: 266–269; modified from García, G. 1996. Reconstrucción de la filogenia y de los procesos de cladogénesis en el género *Cynolebias* en base a datos de ADN mitocondrial y cromosómicos PhD dissertation, ROU University.)

the pachytene stage, in which the chromosomes are bound tightly together and in perfect alignment by a protein lattice called the synaptonemal complex and by cohesin proteins at the centromere (Figure 16.5c). When the synaptonemal complex disappears, the homologous chromosomes remain attached to each other at the centromere and at chiasmata in the late pachytene (Figure 16.5d) and diplotene (Figure 16.5e) stages. *Austrolebias* species showed a "diffuse" phase overlapping these two last stages, as was previously described for many groups (John, 1976). The chiasmata remain in anaphase I until they separate to opposite ends of the dividing cell.

Conversely to this early conserved series of events, late prophase I and the remaining meiotic stages possess greatly distinctive pathways among *Austrolebias* species groups. Figure 16.6 shows the meiotic patterns in a phylogenetic analysis context. The ancestral pathway is present in the outgroup, assuming a hypothetical karyotype constituted by 48 acrocentric chromosomes, N = 24, and the presence of at least one chiasma per chromosome for proper separation of homologous chromosomes during meiosis. This pattern remains conserved in subclades II.c and II.d and in the basal taxa from the remaining subclades. However, two to three chiasmata per bivalent are observed in clade I, which is characterized by the presence of extra-large acrocentric and subtelocentric chromosomes as a product of pericentric inversions or by the addition/deletion of heterochromatic blocks. Two to three chiasmata are also detected among the derived taxa of the subclades II.a and II.b, corresponding to those species showing reduced chromosome numbers and the presence of

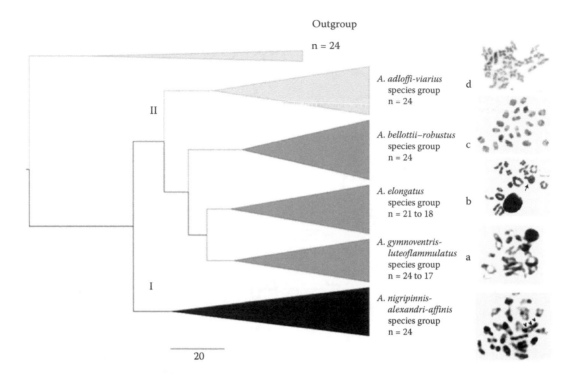

Figure 16.6   Late prophase I–metaphase I patterns in different *Austrolebias* species groups under a phylogenetic context. Tree topology based on *cyt*-b sequences from *Austrolebias* and the outgroup taxa.
Bayesian phylogeographic inference framework generated using the GTR + G model of molecular
evolution implemented in BEAST 1.5.4. *Cynopoecilus melanotaenia*, *Kryptolebias marmoratus*,
and *Nothobranchius furzeri* were included as outgroup taxa. Arrows and arrowhead indicate chiasmata in three bivalent positions: proximal, distal, and interstitial. (Modified from García, G. 1996.
Reconstrucción de la filogenia y de los procesos de cladogénesis en el género *Cynolebias* en base
a datos de ADN mitocondrial y cromosómicos PhD dissertation, ROU University; García, G. et al.
2014. *Genetica* 142: 87–98.)

extra-large biarmed elements. Moreover, as a result of multiple fusion events and/or pericentric
inversion, these taxa showed peculiar bivalent configurations in the diakinesis–metaphase I stages,
according to their chiasmata distribution and resolution in the advanced prophase I–anaphase I. In
conclusion, meiotic processes were consistent with the proposed chromosome evolution hypothesis
encompassing the cladogenetic events in the genus *Austrolebias*.

16.4 MEIOTIC HYBRID PRODUCTS BETWEEN HIGHLY RELATED TAXA

Austrolebias constitutes an excellent model to explore the nature of possible RI, in particular the
existence of postzygotic barriers among taxa belonging to the same species group or among groups.
Since it is possible to obtain laboratory hybrids, these taxa allow testing if they conform to hybrid-
sterility models, reinforcing the role of CRs in the speciation process.

Katz (1982) reported F1 hybrid success from crosses between a male of *A. viarius* and a female
of *A. adloffi*. In these experiments, individuals that presented low viability had a female appearance,
such as caudal black spots similar to *A. adloffi* and dark flank spots similar to *A. viarius*.

Here, we analyze the meiotic patterns of the restricted hybrid progeny from crosses between
allopatric pairs of taxa within the *A. adloffi* species group. *A. adloffi* female and *A. charrua* male

cross produces a single male offspring that has an overall appearance of a normal testis. The reciprocal cross resulted in two morphological male offspring, one of which showed a hypertrophied
white tissue within the abdominal cavity (most probably a nonfunctional testis) and the other offspring died as a juvenile, suggesting the possible occurrence of postzygotic RI mechanisms between
both taxa.

With conventional techniques, meiotic analysis of a specimen with an apparently normal testis showed stages of meiotic processes similar to those found in parental species (Figure 16.7a).
Nevertheless, the F1 viable hybrid studied showed two types of metaphase I (Figure 16.7b and c). In
Figure 16.7b "normal" pairing bivalents display a typical image showing one interstitial chiasma as
in parental species, whereas a lower number of mismatched pairing meiotic bivalents are observed
in others in metaphase I, yielding to loose meiotic products and suggesting that F1 hybrids were
not completely fertile (Figure 16.7c). This could be due to the existence of different types of CR,
like pericentric inversions, in both parental genomes that could be acting as a postzygotic barrier
between those taxa (García, 2006).

The comparative analysis of early meiotic stages (diakinesis) between parental populations and
F1 hybrid progeny revealed interesting differences in the number and localization of chiasmata. F1
hybrid cells showed a significant reduction (Mann-Whitney U test, $p < 0.01$) in the mean chiasmata number per cell (mean = 19.7, S.D. = 2.28) compared with the parental populations (*A. adloffi*:
mean = 23.3, S.D. = 1.15; *A. charrua*: mean = 22.6, S.D. = 1.57). A Mann–Whitney U test showed

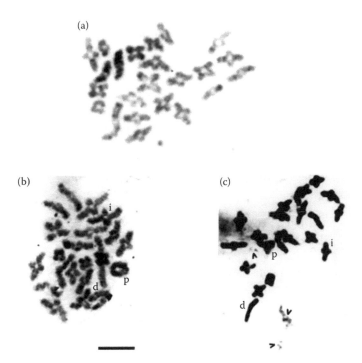

Figure 16.7 Meiotic studies in male individuals of parental species and the interspecific F1 hybrids were performed following standard protocols. Early meiotic stages in parental populations: (a) diakinesis
in *A. adloffi*; (b) diakinesis in F1 hybrid progeny showing "normal" pairing bivalents; (c) metaphase
I in F1 hybrid progeny. Arrowheads indicate mismatched pairing meiotic bivalents. Three bivalent
chiasmata positions: p, proximal; d, distal; v, mismatched pairing meiotic bivalents; and i, interstitial. Bar = 10 μm. (Adapted from Kligerman, A. D. and S. E. Bloom. 1977. *Journal of the Fisheries
Research Board of Canada* 34: 266–269; modified from García, G. 1996. Reconstrucción de la
filogenia y de los procesos de cladogénesis en el género *Cynolebias* en base a datos de ADN
mitocondrial y cromosómicos PhD dissertation, ROU University.)

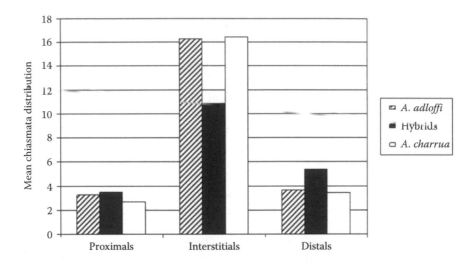

Figure 16.8 Statistical comparisons of the chiasmata location and frequency in parental species (*A. adloffi*, *A. charrua*) and F1 hybrid progeny. Thirty-five F1 hybrid cells and 20 cells of each parental population were examined. Mann-Whitney *U* tests were performed using STATISTICA (Statsoft, Inc., 1999) to detect significant differences in the mean of chiasmata number per cell and its position.

there are not significant differences between parental meiotic cells considering the total number and distribution of chiasmata. These chiasmata were localized in all possible positions: proximal, distal, and interstitial, but mainly interstitial (Figure 16.8). However, hybrid specimens showed that chiasmata frequency is significantly reduced, and a significant increase in distal and proximal chiasmata positions ($p < 0.01$), with a significant reduction in the interstitial ones ($p < 0.01$) with respect to parental meiotic patterns, was observed (Figure 16.8).

These results were concordant with the idea that interspecific hybrids and their offspring have reduced chiasmata frequency and showed unrestricted chiasmata distribution (Sybenga, 1975). Furthermore, these data about both reciprocal interspecific crosses could suggest that introgression of the F1 hybrids toward the *A. adloffi* genome mediated by females of this taxon could be possible.

16.5 CHROMOSOMAL REARRANGEMENTS, REPRODUCTIVE ISOLATION, AND SPECIATION

Recently, large amounts of data have become available and new theoretical models have been developed explaining how rearrangements facilitate speciation in the face of gene flow (Faria and Navarro, 2010).

Although chromosomal rearrangements restrict gene flow, whether chromosome number differences can create a complete postzygotic barrier in many groups remains an open question (Kandul et al., 2007; Vila et al., 2010; Lukhtanov et al., 2011).

Kirkpatrick and Barton's (2006) analysis shows that selection can favor inversions that decrease recombination between alleles involved in local adaptation, which in turn favors the fixation of alternative CRs in different subpopulations connected by gene flow. These CRs can become strong genetic barriers and lead to speciation, for example, by trapping more locally adapted genes. CRs are known to suppress recombination in heterokaryotypes in two ways: (1) mechanical pairing problems within CRs, which inhibit crossover during meiosis, as was observed in the viable F1 hybrid from the interspecific cross previously mentioned and (2) unbalanced gametes produced by crossover events within the rearranged regions of heterokaryotypes (Spirito, 1998).

Evidence that CRs can induce RI mainly conforms to hybrid-sterility models. One of the most simple and convincing results was obtained by crosses between two yeast species, *Saccharomyces cerevisiae* and *Saccharomyces mikatae* (Greig, 2009). The resulting hybrids show markedly reduced fertility due to reciprocal translocations. Another recent study showed that a single translocation is the basis for RI between certain *Drosophila melanogaster* and *Drosophila simulans* hybrid strains (Masly et al., 2006). However, in animals, most reported cases of hybridization resulting in viable offspring involve taxa with small differences in chromosome number, and a cumulative effect generally has been observed (i.e., fertility decreases proportionally with the level of chromosomal differences; King, 1993; Lyapunova et al., 2010). In the present study, we report cases of introgression occurring between taxa that share the same chromosome numbers, but showing different chromosome arm numbers as a product of pericentric inversion events among taxa of the *A. adloffi* species group (García, 2006). The analyses of the construction of linkage groups using laboratory hybrid progeny between highly related taxa belonging to this species group will be more deeply and extensively considered in the next chapter (see Chapter 17).

16.6 GENERAL CONCLUSIONS

Neotropical fish belonging to the family Rivulidae represent an exciting model to access the mechanisms and processes driving chromosomal evolution and karyotypic orthoselection. These mechanisms, consistent with clade divergence, have enhanced the great diversity detected at different levels in this family. At the same time, the possible occurrence of viable and fertile hybrid progenies between taxa within a species group allow us to dissect and test the weight of potential RI barriers between allopatrically or parapatrically distributed sister taxa. Additionally, meiotic process analysis in both parental species and the hybrid progeny could clarify the intrinsic incompatibilities between several genomic regions that share structural differences between the parental sister taxa.

ACKNOWLEDGMENTS

We thank L. Malabarba, who kindly provided in 1991 *A. adloffi* fish specimens to implement laboratory interspecific crosses. G.G. acknowledges the research support of SNI (ANII, Uruguay), PEDECIBA (Programa de Desarrollo de Ciencias Básicas), and her DT research CSIC_UdelaR Project.

REFERENCES

Azpelicueta, M. M., M. C. Butí, and G. García. 2009. *Papiliolebias hatinne,* a new annual fish species (Cyprinodontiformes: Rivulidae) from Salta, Argentina. *Revue Suisse de Zoologie* 6: 313–323.

Böhne, A., F. Brunet, D. Galiana-Arnoux, C. Schultheis, and J. N. Volff. 2008. Transposable elements as drivers of genomic and biological diversity in vertebrates. *Chromosome Research* 16: 203–215. doi: 10.1007/s10577-007-1202-6

Costa, W. J. E. M. 1998. Phylogeny and classification of Rivulidae revisited: Origin and evolution of annualism and miniaturization in rivulid fishes (Cyprinodontiformes: Aplocheiloidei). *Journal of Comparative Biology* 3: 33–94.

Costa, W. E. J. M. 2006. The South American annual killifish genus *Austrolebias* (Teleostei: Cyprinodontiformes: Rivulidae): Phylogenetic relationships, descriptive morphology and taxonomic revision. *Zootaxa* 1213: 1–162. http://dx.doi.org/

Elder, J. F., B. J. Turner, J. E. Thomerson, and D. C. Taphorn. 1993. Karyotypes of nine Venezuelan annual killifishes (Cyprinodontidae), with comments on karyotype differentiation in annual killifishes. *Ichthyological Exploration Freshwater* 4: 261–268.

Faria, R. and A. Navarro. 2010. Chromosomal speciation revisited: Rearranging theory with pieces of evidence. *Trends in Ecology and Evolution* 25: 660–669. doi: 10.1016/j.tree.2010.07.008

García, G. 1996. Reconstrucción de la filogenia y de los procesos de cladogénesis en el género *Cynolebias* en base a datos de ADN mitocondrial y cromosómicos PhD dissertation, ROU University.

García, G. 2006. Multiple simultaneous speciation in killifishes of the *Cynolebias adloffi* species complex (Cyprinodontiformes, Rivulidae) from phylogeography and chromosome data. *Journal of Zoological Systematics and Evolutionary Research* 44: 75–87. doi: 10.1111/j.1439-0469.2005.00346.x

García, G., V. Gutiérrez, N. Ríos et al. 2014. Burst speciation processes and genomic expansion in the neotropical annual killifish genus *Austrolebias* (Cyprinodontiformes, Rivulidae). *Genetica* 142: 87–98. doi: 10.1007/s10709-014-9756-7

García, G., A. I. Lalanne, G. Aguirre, and M. Cappetta. 2001. Chromosome evolution in annual killifish genus *Cynolebias* and mitochondrial phylogenetic analysis. *Chromosome Research* 9: 93–100. doi: 10.1023/A:1011664009509

García, G., N. Ríos, and V. Gutiérrez. 2015. Next-generation sequencing detects repetitive elements expansion in giant genomes of annual killifish genus *Austrolebias* (Cyprinodontiformes, Rivulidae). *Genetica* 143: 353–360. doi: 10.1007/s10709-015-9834-5

García, G., E. Scvortzoff, and A. Hernández. 1995. Karyotypic heterogeneity in South American annual killifishes of the genus *Cynolebias* (Pisces, Cyprinodontiformes, Rivulidae). *Cytologia* 60: 103–110. doi: 10.1508/cytologia.60.103

García, G., E. Scvortzoff, M. C. Máspoli, and R. Vaz-ferreira. 1993. Analysis of karyotypic evolution in natural populations of *Cynolebias* (Pisces, Cyprinodontiformes, Rivulidae) using banding techniques. *Cytologia* 58: 85–94. http://dx.doi.org/10.1508/cytologia.58.85

Greig, D. 2009. Reproductive isolation in *Saccharomyces*. *Heredity* 102: 39–44. doi: 10.1038/hdy.2008.73

Hedrick, P. W. and D. A. Levin. 1984. Kin founding and the fixation of chromosomal variants. *Evolution* 35: 322–332.

John, B. 1976. Myths and mechanisms of meiosis. *Chromosoma* 54: 295–325.

Kandul, N. P., V. A. Lukhtanov, and N. P. Pierce. 2007. Karyotypic diversity and speciation in *Agrodiaetus* butterflies. *Evolution* 61: 546–558. doi: 10.1111/j.1558-5646.2007.00046.x

Katz, D. 1982. Notes on *Cynolebias adloffi* Ahl. *Journal of the American Killifish Association* 15: 237–241.

Kirkpatrick, M. and N. Barton. 2006. Chromosome inversions, local adaptation and speciation. *Genetics* 173: 419–434. doi: 10.1534/genetics.105.047985

King, M. 1993. *Species Evolution: The Role of Chromosome Change*. Cambridge University Press, New York.

Kligerman, A. D. and S. E. Bloom. 1977. Rapid chromosome preparations from solid tissues of fishes. *Journal of the Fisheries Research Board of Canada* 34: 266–269.

Kornfield, I. L. 1984. Descriptive genetics of Cichlid fishes. In *Evolutionary Genetics of Fishes*, ed. B. J. Turner, 591–616. New York: Plenum Press.

Lande, R. 1984. The expected fixation rate of chromosomal inversions. *Evolution* 38: 743–752.

Loureiro, M. and R. O. de Sá. 1998. Osteological analysis of the killifish genus Cynolebias (Cyprinodontiformes: Rivulidae). *Journal of Morphology* 238: 109–262. doi: 10.1002/(SICI)1097-4687 (199811)238:2<245::AID-JMOR5>3.0.CO;2-F

Loureiro, M. and G. García. 2004. *Cynolebias reicherti* a new annual fish (Rivulidae: Cynolebiatinae) from southern Laguna Merim basin. *Acta Zoológica Lilloana* 48: 13–25.

Loureiro, M. and G. García. 2006. Transgresiones y regresiones marinas en la costa Atlántica y lagunas costeras de Uruguay: efectos sobre los peces continentales. In *Bases para la conservación y el manejo de la costa uruguaya*, eds. R. Menafra, L. Rodríguez-Gallego, F. Scarabino, and D. Conde, 545–556. Uruguay: Vida Silvestre.

Lukhtanov, V. A., V. Dinca, G. Talavera, and R. Vila. 2011. Unprecedented within-species chromosome number cline in the Wood White butterfly *Leptidea sinapis* and its significance for karyotype evolution and speciation. *BMC Evolutionary Biology* 11: 109. doi: 10.1186/1471-2148-11-109

Lyapunova, E. A., I. Y. Bakloushinskaya, A. S. Saidov, and K. K. Saidov. 2010. Dynamics of chromosome variation in mole voles *Ellobius tancrei* (Mammalia, Rodentia) in Pamiro-Alai in the period from 1982 to 2008. *Russian Journal of Genetics* 46: 566–571. doi: 10.1134/S1022795410050091

Mank, J. E. and J. C. Avise. 2006. Cladogenetic correlates of genomic expansion in the recent evolution of actinopterygiian fishes. *Proceedings of the Royal Society B* 273: 33–38. doi: 10.1098/rspb.2005.3295

Masly, J. P., C. D. Jones, M. A. F. Noor et al. 2006. Gene transposition as a novel cause of hybrid male sterility. *Science* 313: 1448–1450.

Rebollo, R., B. Horard, B. Hubert, and C. Vieira. 2010. Jumping genes and epigenetics: Towards new species. *Gene* 454: 1–7. doi: 10.1016/j.gene.2010.01.003

Scheel, J. J. 1972. Rivulinae karyotypes and their evolution (Rivulinae, Cyprinodontidae, Pisces). *Sonderdruck aus Z. F. Zool. Systematik u Evolutions forschung* 10: 108–209.

Scheel, J. J. 1990. *Atlas of Killifishes of the Old World*. T. F. H. Publications, Neptune City, NJ.

Sites, J. W. and C. Moritz. 1987. Chromosome evolution and speciation revisited. *Systematic Zoology* 36: 153–174.

Sola, L., M. Marzovillo, A. R. Rossi, S. Bressanello, and B. J. Turner. 1997. Cytogenetic analysis of a self-fertilizing fish, *Rivulus marmoratus*: Remarkable chromosomal constancy over a vast geographic range. *Genome* 40: 945–949. doi: 10.1139/g97-121

Spirito, F. 1998. The role of chromosomal rearrangements in speciation. In *Endless Forms*, eds. D. J. Howard and S. H. Berlocher, 320–329. Oxford University Press, Oxford.

Sybenga, J. 1975. *Meiotic Configurations*. Berlin: Springer-Verlag.

Talavera, G., V. A. Lukhtanov, L. Rieppel, N. P. Pierce, and R. Vila. 2013. In the shadow of phylogenetic uncertainty: The recent diversification of Lysandra butterflies through chromosomal change. *Molecular Phylogenetics and Evolution* 69: 469–478. doi: 10.1016/j.ympev.2013.08.004

Vaz-Ferreira, R. and A. R. Melgarejo. 1984. La distribución de las especies del género *Cynolebias* Steindachner 1876, en el Uruguay, con notas sobre *C. alexandri* Castello y López, 1974. *Boletín de la Sociedad Zoológica*, Uruguay 2: 41–46.

Vila, R., V. A. Lukhtanov, G. Talavera, T. F. Gil, and N. E. Pierce. 2010. How common are dot-like distribution ranges? Taxonomical oversplitting in Western European *Agrodiaetus* (Lepidoptera, Lycaenidae) revealed by chromosomal and molecular markers. *Biological Journal of the Linnean Society* 101: 130–154. doi: 10.1111/j.1095-8312.2010.01481.x

Volff, J. N., C. Korting, A. Froschauer, K. Sweeney, and M. Schartl. 2001. Non-LTR retrotransposons encoding a restriction enzyme-like endonuclease in vertebrates. *Journal of Molecular Evolution* 52: 351–360.

White, M. J. D. 1973. *Animal Cytology and Evolution*, 3rd Edition, Cambridge University Press, London.

CHAPTER 17

Genomic Isolated Regions
Linkage Groups in Parental and Laboratory Hybrids between Austrolebias adloffi *Species Group*

Sebastián Oviedo, Mareney Rovira, and Graciela García

CONTENTS

17.1 INTRODUCTION

New insights have emerged in the recent years about the role of the structural chromosome rearrangements (CRs) in genomic speciation models, mostly due to advances using next-generation sequencing (NGS) (Feder et al., 2012; Feder and Nosil, 2012; Via, 2012). A decade ago, new empirical evidence showed a new framework for chromosomal speciation beyond conventional views: the suppressed-recombination models (Faria and Navarro, 2010). These models propose that the reduction in recombination by means of divergent selection in several genomic regions and/or by means of chromosomes carrying different structural rearrangements could represent determining factors in speciation processes (Noor et al., 2001; Rieseberg, 2001; Navarro and Barton, 2003; Kulathinal et al., 2009). In other words, CRs allow genes located in these regions to differentiate, in contrast to genes in freely recombining collinear regions, and therefore it is possible that divergent genomic regions that promote speciation *with* gene flow exist (Faria and Navarro, 2010; Feder and Nosil, 2012; Via, 2012).

Chromosomal rearrangements such as inversions (i.e., an orientation reversal of a piece of DNA within a single chromosome) impact species divergence by reducing recombination in heterokaryotypes (Kulathinal et al., 2009; Stevison et al., 2011). Chromosomal regions that experience reduced recombination in hybrids, such as within inversions, have been hypothesized to contribute to the maintenance of species integrity. Therefore, inversions can facilitate speciation because they are potent recombination modifiers (Noor et al., 2001; Ortiz-Barrientos et al., 2002). Recombination in heterokaryotypes is severely reduced, causing associations between sets of alleles within the inversion region (Navarro and Barton, 2003). Consequently, alternative chromosomal arrangements can play a key role in speciation. In diverging populations with alternative arrangements, they may protect the inverted region from introgression and allow the accumulation of alleles that contribute to reproductive isolation (Noor et al., 2001; Reiseberg, 2001; Feder and Nosil, 2009).

Among the Neotropical fish fauna, annual fishes of the *Austrolebias adloffi* species group constitute an excellent model to test what structural features could explain the existence of partial or total isolated genomic regions between recently differentiated sister parapatric taxa (García, 2006; García et al., 2009; see Chapter 15). Different karyotypic structure among taxa in this species group could have triggered cladogenetic events from a highly polymorphic ancestral population. Resulting speciation could have included the fixation of CRs, such as pericentric inversions in isolated populations, during the major Pleistocene events in such a region (García, 2006).

The use of partial genome sequences to reconstruct the evolutionary process could recover genomics regions of low recombination that could provide a means to create "islands of differentiation" between species (Kulathinal et al., 2009).

Genetic linkage maps constitute a useful tool for the elucidation of genomic evolution and organization (Danzmann and Gharbi, 2001) and for identification of the possible existence of "islands of differentiation" and for localization and identification of associated loci in polygenic traits (Morizot et al., 2001; Khoo et al., 2003). Genetic maps constructed by the random amplified polymorphic DNA (RAPD) method (Welsh and McClelland, 1990; Williams et al., 1990), which requires no previous DNA sequence information, were reported for fishes, for example, *Poecilia reticulata* (Khoo et al., 2003), *Astyanax mexicanus* (Borowsky and Wilkens, 2002), and *Phalloceros cuadimaculatus* (Gutiérrez and García, 2011). Since chromosomal rearrangements have been associated with both among-clade (García et al., 2002) and within-clade (García, 2006) differentiation processes, the question is what genomic regions remain isolated between parapatric sister taxa of the *A. adloffi* species group. We implement the construction of linkage group (LG) maps using RAPD molecular markers applied to a laboratory hybrid between *A. viarius* and *A. charrua* and between *A. charrua* and *A. reicherti* taxa to study the effects of chromosomal structural differences on the introgression in backcrossed laboratory progeny from these closely related taxa.

17.2 MATERIAL AND METHODS

17.2.1 Experimental Crosses and Mapping Population

Interspecific hybrids from taxa belonging to the *A. adloffi* species group were used to carry out the analyses. The experiment consisted of a 3 × 3 diallele cross as a mating scheme, including males and females of three parapatric species, *A. viarius*, *A. charrua*, and *A. reicherti*, with their corresponding three replicates. All fishes were maintained by sex in freshwater aquaria under environmental conditions of temperature (17–19°C on natural photoperiod). Fish were fed with *Artemia salina* and *Tubifex* spp. ad libitum. For each breeding experiment, a male and a female were placed in a 220-l aquarium under the same laboratory conditions. All sampling protocols for this scientific study were approved by CNEA (Comisión Nacional de Experimentación Animal) of Uruguay.

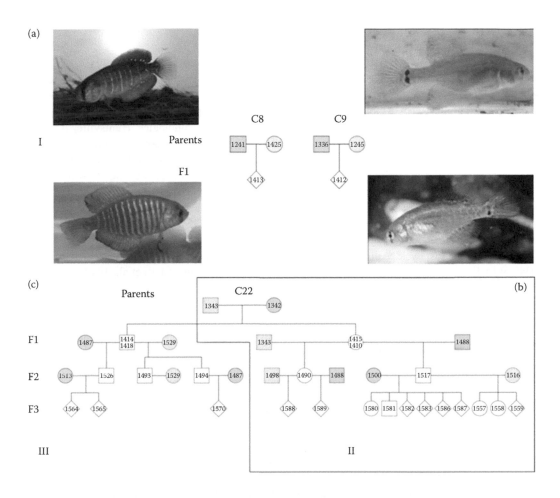

Figure 17.1 Interspecific crosses and laboratory hybrids for linkage-mapping analysis. In laboratory pedi-grees: square figures, males; circle figures, females; rhombus, indeterminate sex of juvenile indi-viduals. (a) Two replicates (C8, C9) from interspecific cross between ♂ *A. viarius* (left image, green squares) and ♀ of *A. charrua* (right image, yellow circles) and embryo individuals of two unknown sex (white rhombus) for mapping design. (b) Partial genealogy of 22 individuals from ♂ *A. charrua* (yellow) × ♀ *A. reicherti* (light blue) cross (C22) and the successive backcrosses of F1 hybrid females, as well as from the F2-F3 male and female hybrid progeny with the two parental species. (c) Total genealogy, including 34 Individuals from the interspecific crosses between ♂ *A. charrua* × ♀ *A. reicherti* cross (C22) and the successive backcrosses (F1 to F3 hybrids progeny [white circles, squares, and rhomboids]) with the two parental species.

Although large numbers of viable interspecific hybrid offspring were obtained in all crosses, only two nonsexually mature juvenile hybrids from two different crosses between ♂ *A. viarius* and ♀ *A. charrua* were available for mapping purposes (Figure 17.1; Appendix 17.1). All remaining individuals from this progeny died at different development stages. Among all crosses involving males and females of *A. charrua* and *A. reicherti* taxa, numerous viable interspecific hybrid off-spring were produced. Nevertheless only four sexually mature adult F1 hybrids (two males and two females) were obtained from the ♂ *A. charrua* and ♀ *A. reicherti* cross to continue the experimental backcross design. These adults were used to obtain a hybrid population during three consecutive backcross generations using males and females of *A. charrua* and *A. reicherti*, respectively, from wild populations. Therefore, numerous and viable F1 to F3 hybrid progeny were obtained from ♂ *A. charrua* × ♀ *A. reicherti* and its successive backcrosses (Figure 17.1; Appendix 17.1).

17.2.2 DNA Extraction, Amplification of Genomic DNA, and Visualization of Amplified Fragments

DNA was extracted from muscle, fin, and total body of preserved juveniles using proteinase K digestion, protein precipitation through sodium chloride, and DNA total precipitation with ethanol (modified from Medrano et al., 1990). Among several RAPD analyzed, the present work included a total of five 10-mer primers (OPA10, 14, 15, 18, and 19) from Operon Life Technologies, Huntsville, AL, USA. Among 10 primers tested, these five were selected after a population polymorphism screening.

RAPD-PCR reactions were assayed under the following conditions: total reaction volumes of 10 µL consisting of 1.5 mM $MgCl_2$, 1.25 µM primer, 0.2 mM each dNTP, 0.5 U Taq DNA polymerase (Invitrogen), 1 µL of 10× reaction buffer, and 1 µL of DNA sample. Amplifications were performed using a PTC-100 (MJ Research, Waltham, MA, USA) thermocycler with the following experimental parameters: an initial denaturation of 5 min at 95°C, followed by 35 cycles of 1 min at 95°C, annealing for 1 min at 36°C, and elongation of 1 min at 72°C. An additional elongation period of 5 min at 72°C followed the last cycle. Amplified RAPD products were electrophoresed in 6% acrylamide gels at 140 V in 1× TBE buffer for 5 h and stained with silver nitrate (Sanguinetti et al., 1994).

17.2.3 Band Analysis

Band sizes were estimated against 1 kb DNA Extension Ladder (Invitrogen) and scored between 200 to 1018 bp for detection of presence or absence of polymorphic markers. Allele markers of different loci show different molecular weights and do not comigrate equally in the gel. Loci were named according to their size and the primer used. Only reproducible and polymorphic loci were selected to construct a corrected reproducibility band matrix that was used to perform linkage analysis.

17.2.4 Linkage Analysis Strategy

In linkage analysis I six individuals (four parents and two juvenile hybrids) from ♂ A. *viarius* and ♀ of A. *charrua* crosses have been included in the mapping linkage analysis (Figure 17.1). In linkage analysis II, a mapping population of 22 individuals was obtained from F1 ♀ hybrid females (1415 and 1419) and the successive backcrosses (F2 to F3) with ♂ and ♀ of A. *charrua* and ♂ and ♀ of A. *reicherti*, respectively (Figure 17.1). In linkage analysis III, a total mapping population of 34 individuals was used for the analysis, including both lines of successive backcrosses: ♂ A. *charrua* and ♂ A. *reicherti* with ♀ hybrid (1415 and 1419) and ♀ A. *charrua* and ♀ A. *reicherti* with ♂ hybrid (1414 and 1418) and their successive backcrosses F2 to F3 with parental species (Figure 17.1).

17.2.5 Genetic Linkage Mapping

Hundreds of specific polymorphic molecular markers of the RAPD technique were obtained for each taxon and their respective hybrid progeny to construct a high-density genomic map. Raw genotypic and phenotypic data of RAPD-scored bands were recorded, with no previous knowledge of phase relationships or inheritance, and were converted into an "ancestry known" format that resembled an inbred pedigree structure. Alleles for each progeny and parent genotype were initially oriented to reflect their inheritance from the previous generation. If ancestry was inconclusive owing to missing grandparental data, the data were considered as "ancestry unknown" and not included during mapping.

The reconstruction of the LGs with the markers and linkage analysis were done using the MapManager QTX v.3.0 software package (Manly et al., 2001). Linkage analyses were oriented to reflect the ancestry of each parental line. For all loci scored, the approach included only markers for which one parent was scored as band-present and band-absent in the other parent and also present in at least one offspring.

Different alternative maps were constructed by using different sets of data: excluding and including markers and using different stringency between marker associations (data not shown). The final linkage analyses were constructed using the arbitrary crosses function implemented by the MapManager QTX v. 3.0, in which populations derived by crossing any parents—at least one of which is heterozygous at some loci—are used. The minimum logarithm of the likelihood of odds (LOD) score for determining linkage was taken as 3.0, while a significant level to provide evidence of linkage of $P = 0.001$ was used. Map units used were Kosambi centimorgans (cMκ; Kosambi, 1944).

17.3 RESULTS

17.3.1 Linkage-Mapping Analysis I

The first linkage map was obtained using viable F1 interspecific hybrids from ♂ *C. viarius* × ♀ *C. charrua* crosses (Figure 17.1). Based on 232 markers, 29 LGs were constructed (Figure 17.2). Remarkably, these LGs included markers from one or the other species. Thus, eight LGs (LG II, IV, V, X, XIII, XV, XIX, and XXIX) incorporated only *A. viarius* markers and seven LGs (LG I, VI, VII, VIII, XI, XXVI, and XXVII) displayed only *A. charrua* markers. Furthermore, 14 LGs (LG III, IX, XII, XIV, XVI, XVII, XIX, XX, XXI, XXII, XXIII, XXIV, XXV, and XXVIII) resulted in a new marker combination in the hybrid genome. Neither LG contained mixed markers from both parental taxa. This could be validating the absence of recombination events between the two parental genomes present in the viable hybrid progeny.

LG II, IV, and XI contained larger number of markers, whereas LG XIX, XX, XXV, XXVI, and XXVII represented only minor ones, including two linked markers. In this linkage analysis, due to the scarcity of F1 viable hybrid progeny, it was not possible to estimate a total map distance.

17.3.2 Linkage-Mapping Analysis II

We report the first linkage map using viable and fertile interspecific hybrids from ♂ *A. charrua* × ♀ *A. reicherti* cross and the successive backcrosses of F1 hybrid females, as well as from the F2-F3 male and female hybrid progeny with the two parental species (Figure 17.1b). Based on 197 RAPD markers and 22 individuals, 15 LGs were obtained (Figure 17.3): nine LGs (LG I, V, VI, VII, VIII, X, XI, XII, and XIII) shared markers with *A. reicherti* and one (LG IV) with *A. charrua*. Remarkably, four LGs (LG II, III, V, and XIV) represented a mixed-marker combination from two parental species, suggesting the occurrence of recombination events starting from F1 hybrid female genomes.

17.3.3 Linkage-Mapping Analysis III

The present approach included all 34 individuals of the genealogy (Figure 17.1) from the interspecific crosses between ♂ *A. charrua* × ♀ *A. reicherti* and the successive backcrosses from F1 to F3 in the linkage analysis map. Based on 197 RAPD markers, 17 LGs were obtained: 10 LGs (LG I, II, VI, VII, VIII, IX, X, XI, XII, and XVI) constituted by *A. reicherti* genome markers and 7 (LG III, IV, V, XIII, XIV, XV, and XVII) conformed only with *A. charrua* genome markers (Figure 17.4). Again, in this approach, neither of the GLs resulted in a mix between markers from both parental genomes. The absence of mixed GLs between both taxa was concordant with those results obtained when F1 hybrid males were backcrossed with both parental taxa (data not shown). The present approach allowed us to estimate a genetic linkage map coverage of 707.5 cM using viable and fertile interspecific hybrid progeny.

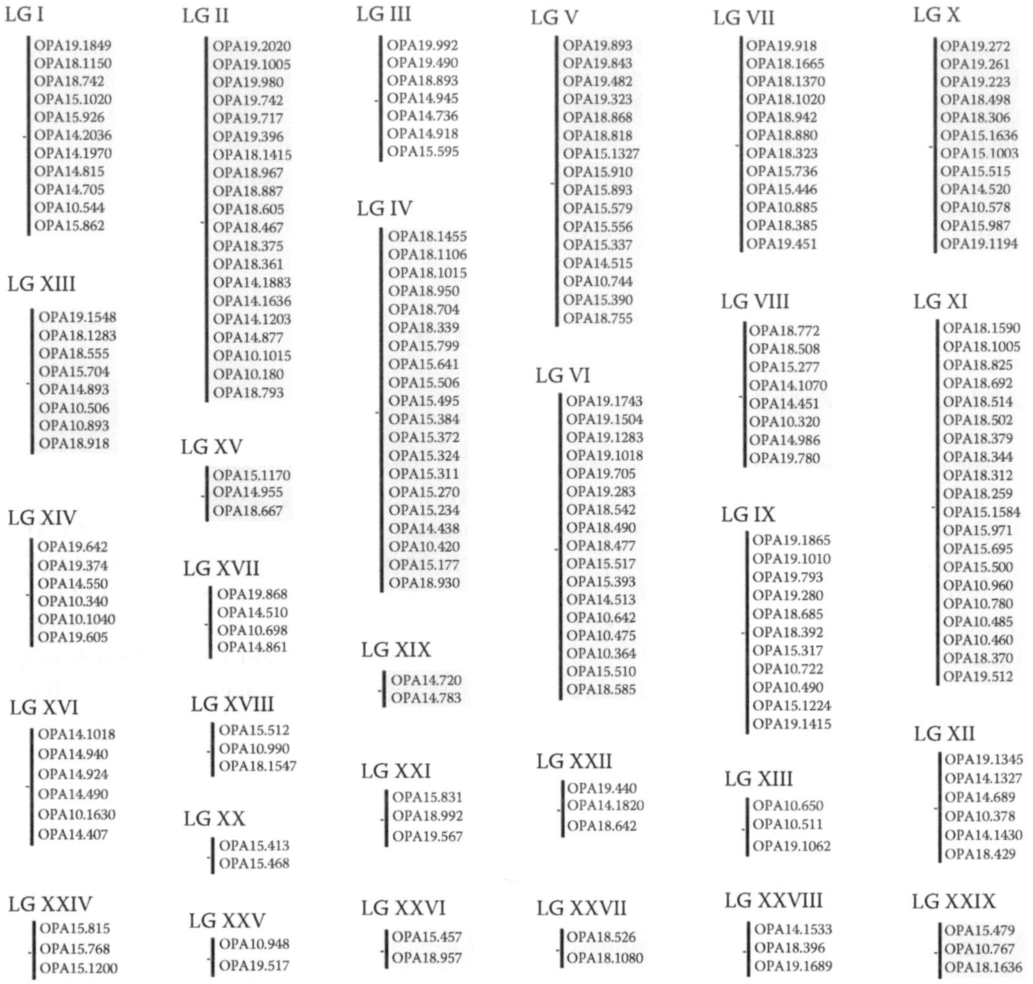

LG I
OPA19.1849
OPA18.1150
OPA18.742
OPA15.1020
OPA15.926
OPA14.2036
OPA14.1970
OPA14.815
OPA14.705
OPA10.544
OPA15.862

LG XIII
OPA19.1548
OPA18.1283
OPA18.555
OPA15.704
OPA14.893
OPA10.506
OPA10.893
OPA18.918

LG XIV
OPA19.642
OPA19.374
OPA14.550
OPA10.340
OPA10.1040
OPA19.605

LG XVI
OPA14.1018
OPA14.940
OPA14.924
OPA14.490
OPA10.1630
OPA14.407

LG XXIV
OPA15.815
OPA15.768
OPA15.1200

LG II
OPA19.2020
OPA19.1005
OPA19.980
OPA19.742
OPA19.717
OPA19.396
OPA18.1415
OPA18.967
OPA18.887
OPA18.605
OPA18.467
OPA18.375
OPA18.361
OPA14.1883
OPA14.1636
OPA14.1203
OPA14.877
OPA10.1015
OPA10.180
OPA18.793

LG XV
OPA15.1170
OPA14.955
OPA18.667

LG XVII
OPA19.868
OPA14.510
OPA10.698
OPA14.861

LG XVIII
OPA15.512
OPA10.990
OPA18.1547

LG XX
OPA15.413
OPA15.468

LG XXV
OPA10.948
OPA19.517

LG III
OPA19.992
OPA19.490
OPA18.893
OPA14.945
OPA14.736
OPA14.918
OPA15.595

LG IV
OPA18.1455
OPA18.1106
OPA18.1015
OPA18.950
OPA18.704
OPA18.339
OPA15.799
OPA15.641
OPA15.506
OPA15.495
OPA15.384
OPA15.372
OPA15.324
OPA15.311
OPA15.270
OPA15.234
OPA14.438
OPA10.420
OPA15.177
OPA18.930

LG XIX
OPA14.720
OPA14.783

LG XXI
OPA15.831
OPA18.992
OPA19.567

LG XXVI
OPA15.457
OPA18.957

LG V
OPA19.893
OPA19.843
OPA19.482
OPA19.323
OPA18.868
OPA18.818
OPA15.1327
OPA15.910
OPA15.893
OPA15.579
OPA15.556
OPA15.337
OPA14.515
OPA10.744
OPA15.390
OPA18.755

LG VI
OPA19.1743
OPA19.1504
OPA19.1283
OPA19.1018
OPA19.705
OPA19.283
OPA18.542
OPA18.490
OPA18.477
OPA15.517
OPA15.393
OPA14.513
OPA10.642
OPA10.475
OPA10.364
OPA15.510
OPA18.585

LG XXII
OPA19.440
OPA14.1820
OPA18.642

LG XXVII
OPA18.526
OPA18.1080

LG VII
OPA19.918
OPA18.1665
OPA18.1370
OPA18.1020
OPA18.942
OPA18.880
OPA18.323
OPA15.736
OPA15.446
OPA10.885
OPA18.385
OPA19.451

LG VIII
OPA18.772
OPA18.508
OPA15.277
OPA14.1070
OPA14.451
OPA10.320
OPA14.986
OPA19.780

LG IX
OPA19.1865
OPA19.1010
OPA19.793
OPA19.280
OPA18.685
OPA18.392
OPA15.317
OPA10.722
OPA10.490
OPA15.1224
OPA19.1415

LG XIII
OPA10.650
OPA10.511
OPA19.1062

LG XXVIII
OPA14.1533
OPA18.396
OPA19.1689

LG X
OPA19.272
OPA19.261
OPA19.223
OPA18.498
OPA18.306
OPA15.1636
OPA15.1003
OPA15.515
OPA14.520
OPA10.578
OPA15.987
OPA19.1194

LG XI
OPA18.1590
OPA18.1005
OPA18.825
OPA18.692
OPA18.514
OPA18.502
OPA18.379
OPA18.344
OPA18.312
OPA18.259
OPA15.1584
OPA15.971
OPA15.695
OPA15.500
OPA10.960
OPA10.780
OPA10.485
OPA10.460
OPA18.370
OPA19.512

LG XII
OPA19.1345
OPA14.1327
OPA14.689
OPA10.378
OPA14.1430
OPA18.429

LG XXIX
OPA15.479
OPA10.767
OPA18.1636

Figure 17.2 Preliminary genetic linkage map using viable F1 interspecific hybrids from ♂ *C. viarius* × ♀ *C. charrua* cross based on RAPD markers. Assignment of linkage groups, positioning of markers, and application of mapping function were carried out with MapManager QTX. All markers were linked at LOD > 3.0, while those that did not were excluded. RAPD markers are identified according to the following nomenclature: i.e., in GL 1, OPA19.1849, where the number on the right is the size in base pairs and the remaining characters are the primer designation. Eight LGs (green marker blocks) incorporated only *A. viarius* markers. Seven LGs (yellow marker blocks) displayed only *A. charrua* markers. Fourteen LGs (white marker blocks) resulted in a new marker combination in the hybrid genome.

17.4 DISCUSSION

We report, for the first time, the construction of linkage maps using viable interspecific F1 hybrids between ♂ *A. viarius* × ♀ *A. charrua* cross and between the sister taxa ♂ *A. charrua* × ♀ *A. reicherti* and their viable and fertile interspecific hybrids through three backcross generations.

Only two nonsexually mature juvenile hybrids between ♂ *A. viarius* × ♀ *A. charrua* crosses were available for mapping purposes, since most of the progeny died during different developmental stages. Similarly, only few F1 hybrid viable and fertile males in interspecific crosses between *A. adloffi* × *A. charrua* taxa were previously reported (García et al., 2002; see Chapter 16). This could indicate that major genomic incompatibilities are present in the hybrid progeny between

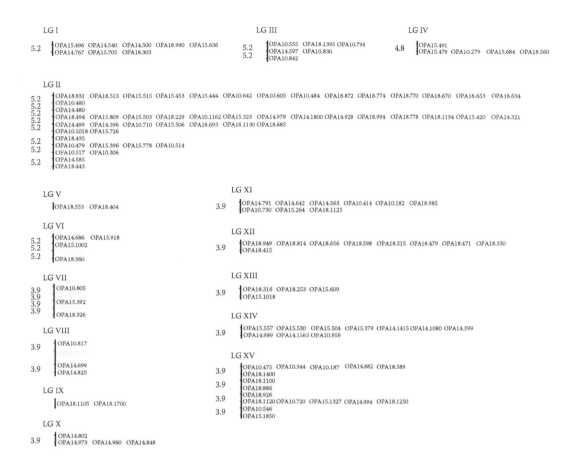

Figure 17.3 Preliminary genetic linkage map based on 197 RAPD markers and using 22 interspecific viable hybrids from ♂ *A. charrua* and ♂ *A. reicherti* with ♀ F1 hybrids (1415 and 1419) and their successive backcrosses of F1 to F3 with parental species. Assignment of linkage groups, positioning of markers, and application of mapping function were carried out with MapManager QTX. All markers were linked at LOD > 3.0, while those that did not were excluded. RAPD markers are identified according to the following: i.e., in GL 1, OPA15.496, where the number on the right is the size in base pairs and the remaining characters are the primer designation. Nine LGs (sky blue marker blocks) shared markers with *A. reicherti*, and one (yellow marker block) with *A. charrua*. Remarkably, four LGs (sky blue and yellow marker blocks) represented a mixed-marker combination from two parental species. Map intervals in cM_K and LOD score are on the left of each LG.

both *A. viarius* and *A. charrua* parental genomes. Linkage analysis I showed that hybrid genomes receive complete blocks of markers from one or another parental species, confirming separate LGs. Therefore, LGs from both mixed genomes have not been detected, perhaps suggesting that parental genome regions could be protected from recombination by unique structural differences. A previous study reported an $N = 24$ and $N = 23$ for *A. viarius* and *A. charrua*, respectively (García, 2006). Therefore, by increasing the number of molecular markers, it would be possible to link several separated minor LGs obtained in the present map approach to achieve the concordance with the expected haploid number of chromosomes from both taxa.

Conversely, most extensive viable and fertile interspecific F1 to F3 hybrid progeny resulted from the ♂ *A. charrua* × ♀ *A. reicherti* cross and their respective backcrosses. Linkage analyses II and III recovered 15 major LGs and 17 LGs, respectively, representing a preliminary linkage map approach in relation to the expected haploid number ($N = 24$). Genetic linkage maps are indirect measures of the number of chromosomes and should be consistent with the haploid

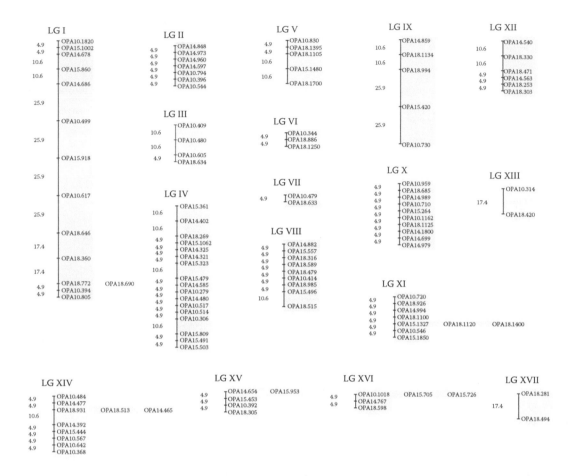

Figure 17.4 Preliminary genetic linkage map based on 197 RAPD markers and the total genealogy including 34 individuals. The genealogy incorporated individuals from the initial ♂ *A. charrua* and ♂ *A. reicherti* cross (C22) and the successive backcrosses, F1 to F3 viable and fertile males and females with parental species. Assignment of linkage groups, positioning of markers, and application of mapping function were carried out with MapManager QTX. All markers were linked at LOD > 3.0, while those that did not were excluded. RAPD markers are identified according to the following: i.e., in GL 1, OPA10.1820, where the number on the right is the size in base pairs and the remaining characters are the primer designation. Ten LGs (sky-blue marker blocks) incorporated only *A. reicherti* markers. Seven LGs (yellow marker blocks) displayed only *A. charrua* markers. Map intervals in cM$_K$ and LOD score are on the left of each LG.

chromosome number of the genome when sufficient coverage is achieved. There are a few minor LGs in linkage analyses II and III, indicating that the maps were not saturated in the hybrid genomes (Figures 17.3 and 17.4). The discrepancy between the haploid number and the expected number of LGs was previously reported in other fish studies and could be due to large gaps among markers because of their low density in the genetic linkage map analysis (Gutiérrez and García, 2011; Keong et al., 2014). Additional markers are needed to allow the saturation of the expected 24 LGs from major distinctly genomic regions between both taxa. In the present analyses, the 17 LG spanned a map size of 707.5 cM$_K$, which represents a good approximation in a preliminary linkage study.

Linkage analysis II detected interesting features in the interspecific F1 hybrid females and a F2 hybrid male in relation to their recombination performance in the backcrosses with parental species, in particular with *A. reicherti*. In fact, in this analysis, four LGs represented a mix of both rearranged markers, a mosaic between two parental genomes, as a result of recombination events. Conversely, neither recombination event was detected in backcrosses using interspecific F1 hybrid males and a F2 hybrid female (data not shown). Two major issues emerge from these results: (1) introgression of the interspecific hybrids toward the *A. reicherti* genome could be possible and could be mediated mostly by F1 hybrid females and their hybrid male progeny; (2) the possible existence of sex-bias differences in the recombination rate. In several fish species, for example, catfish (Waldbieser et al., 2001) and zebrafish (Knapik et al., 1998), sex-bias recombination rate differences were reported. According to the Haldane (1922) rule, the heterogametic sex showed decreasing recombination events. Moreover, Haldane (1922) and Huxley (1928) proposed that in interspecific hybridization events when one sex is reduced, absent, or sterile, this constitutes the heterogametic sex. In our cross and backcross designs, the total number of hybrid males was higher than that of the hybrid females. Similar results were obtained in interspecific crosses between males and females of *A. charrua* and *A. adloffi*, from which only a scarce number of F1 hybrid male progeny was obtained (García et al., 2002). Therefore, the present data suggest that in these *Austrolebias* species, the heterogametic sex could be the female. Other experimental approaches found a partial cDNA fragment in *A. charrua* with a sexually dimorphic pattern from early developmental stages. This sequence shares highly homology with the *dsx D. melanogaster* gene, and their early expression pattern suggests that this region might be located near the top of the sex determination cascade in this species (Arezo et al., 2014; Chapter 5). Further analyses, including a massive set of new molecular markers, will allow for the detection of highly sex-determining linked loci in absence of heteromorphic sex chromosomes (García, 2006).

All linkage analysis results using interspecific hybrid progeny were as expected, considering the estimate of pairwise genetic distances between these taxa: *A. viarius* and *A. charrua* (10.2%) and *A. charrua* and *A. reicherti* (8.4%), respectively (see Chapter 15). Moreover, based on available molecular phylogenies, the monophyly for species in the *A. adloffi* species group is well supported (García, 2006; García et al., 2009; Chapter 15). According to these results, we would expect larger genomic compatibility between the sister species *A. charrua–A. reicherti* than in the pair *A. viarius–A. charrua*. Additionally, previous chromosomal studies support the great taxa affinity in this *Austrolebias* species group (García et al., 1993, 1995). This affinity is given by the following diploid numbers 46–48, NF = 48–50, the predominance of subtelocentric–acrocentric chromosomes, and the presence of only one or two pairs of medium-sized and biarmed chromosomes (García et al., 1993, 1995). Pericentric inversions were likely the proposed rearrangements involved in the shift from acrocentric to metacentric chromosomes to explain chromosomal evolution in a wild population of the *A. adloffi* species group (García, 2006). This type of CR and its fixation in a homozygous state has already been described in several species of *Austrolebias* in particular among the *A. adloffi* species group (García et al., 1995, 2001; García, 2006).

We cannot disregard the possibility that the fixation of different structural genomic rearrangements by means of pericentric inversions and the accumulation of genetic divergence in these regions could be responsible for the existence of genomic hybrid incompatibility between *A. viarius* and *A. charrua* and, therefore, for the absence of recombination events. Inversions could create LGs that cause sterility to persist between hybridizing taxa (Noor et al., 2001). The maintenance of this sterility allows the species to persist with ongoing gene flow for longer periods than without such inversions; if so, natural selection could favor a decrease in the frequency of interspecies matings. Moreover, Buerkle and Rieseberg (2001) have proposed that barriers to gene flow between species result from selection against foreign linkage blocks in hybrids.

Several models support that the recombination-suppressing effects of inversions facilitate the maintenance of differences between interbreeding populations in genes affecting adaptive divergence and reproductive isolation (Feder and Nosil, 2009). These models predict that such genes will disproportionately reside within inversions rather than in collinear region. These authors concluded that strong selection facilitates maintenance of divergence in a manner analogous to inversions facilitating the maintenance of species differences under some conditions.

Our data emphasize the role of the genomic structural changes as possible postzygotic reproductive barriers among sister taxa of the *A. adloffi* species group; at the same time, they yielded support to the existence of genomic permeable regions among these recently diverged sister taxa. In fact, it has been proposed that *A. charrua* and *A. reicherti* are involved in a hybrid zone as a subproduct of secondary contact between populations from the lowlands of the Patos-Merín drainage (PMD) system (García et al., 2009). Present linkage analysis suggests the possible introgression between both taxa, through F1 hybrid females in the backcross with males of *A. reicherti*. These results are consistent with the estimate of asymmetric levels of migration from *A. charrua* toward *A. reicherti*, inferred by means of matrilineal genomes (Chapter 15).

Genome-wide analyses of introgression in organisms show that a substantial fraction of their genomes are permeable to allele exchange from related species. Hybridization can lead to rapid genomic changes, including chromosomal rearrangements, genome expansion, differential gene expression, and gene silencing, some of which are mediated by transposable elements (Baack and Reiseberg, 2007). All these processes have been detected in the genus *Austrolebias* (García et al., 2014; Chapter 14).

The dramatic increases in the availability of genomic tools will produce new insights about the naturally occurring recombinants found in hybrid zones, which will allow genetic mapping of species differences and reproductive barriers in nonmodel organisms (Baack and Reiseberg, 2007).

17.5 CONCLUSIONS AND FUTURE WORK

The existence of LGs formed by specific markers belonging to one parental taxon could suggest that parental genome regions could be protected from recombination events by structural features such as pericentric inversions, reinforcing the existence of "islands of differentiation" that possibly represent regions of reproductive barriers between the highly related taxa of the *A. adloffi* species group. Conversely, those LG groups representing a mosaic of markers between two parental genomes could indicate the existence of genomic regions permeable to the genetic exchange between these taxa. Construction of a genetic linkage map, especially a high-density genetic linkage map, requires not only an appropriate mapping population but also a large number of molecular markers. Therefore, application of massive molecular markers (SNPs) by means of NGS techniques will need to be obtained to increase LG resolution to allow high-density linkage analysis to map those regions sharing structural genomic differences among sister genomes. As implemented by other authors (Kulathinal et al., 2009; McGaugh and Noor, 2012), these approaches will allow analysis of the level of molecular divergence inside and outside the inversion in relation to collinear region and access to candidate genomic regions associated with possible reproductive barriers and their dynamics maintenance in a hybrid zone.

ACKNOWLEDGMENTS

We would like to express our gratitude to V. Gutiérrez and W.S. Serra, who furnished images of adult specimens of female *A. charrua* and male *A. viarius*, respectively.

We thank L. Malabarba, who kindly provided *A. adloffi* fish specimens during 1991 for implementation of laboratory interspecific crosses. The authors are also grateful to the Japanese government for the donation of equipment. G.G. acknowledges the research support of SNI (ANII, Uruguay) and her DT research CSIC_UdelaR Project.

Appendix 17.1 List of Specimens Included in the Experimental Laboratory Cross Design

Catalog N	Species	Pond	Cross	Sex
1241	*A. viarius*	3		M
1336	*A. viarius*	3		M
1245	*A. charrua*	32		F
1343	*A. charrua*	32		M
1425	*A. charrua*	32		F
1496	*A. charrua*	32		M
1498	*A. charrua*	32		M
1511	*A. charrua*	32		M
1516	*A. charrua*	32		F
1529	*A. charrua*	32		F
1533	*A. charrua*	32		F
1537	*A. charrua*	32		M
1538	*A. charrua*	32		F
1342	*A. reicherti*	56		F
1487	*A. reicherti*	56		F
1488	*A. reicherti*	56		M
1499	*A. reicherti*	56		F
1503	*A. reicherti*	56		F
1512	*A. reicherti*	56		F
1513	*A. reicherti*	56		F
1546	*A. reicherti*	56		F
1547	*A. reicherti*	56		F
1412	F1 Hybrid		C8	I
1413	F1 Hybrid		C9	I
1414	F1 Hybrid		C22	M
1415	F1 Hybrid		C22	F
1418	F1 Hybrid		C22	M
1419	F1 Hybrid		C22	F
1493	F2 Hybrid		D4	M
1494	F2 Hybrid		D5	M
1490	F2 Hybrid		D6	F
1492	F2 Hybrid		D6	F
1517	F2 Hybrid		D12	M
1526	F2 Hybrid		D20	M
1568	F3 Hybrid		E14	I
1569	F3 Hybrid		E10	I
1570	F3 Hybrid		E15	I
1560	F3 Hybrid		E23	F
1561	F3 Hybrid		E23	M
1562	F3 Hybrid		E23	I
1563	F3 Hybrid		E23	I
1500	F3 Hybrid		E23	I
1567	F3 Hybrid		E23	I
1564	F3 Hybrid		E27	I
1565	F3 Hybrid		E27	I
1557	F3 Hybrid		E30	F
1558	F3 Hybrid		E30	F
1559	F3 Hybrid		E30	I
1569	F3 Hybrid		E10	I

Note: Catalog Number per individual from Sección Genética Evolutiva, Facultad de Ciencias Montevideo, Uruguay; species or laboratory hybrid, cross number and sex of specimens: M-male, F-female, I-indeterminate-sex.

REFERENCES

Arezo, M.J., N. Papa, V. Gutiérrez, G. García, and N. Berois. 2014. Sex determination in annual fishes: Searching for the master sex-determining gene in *Austrolebias charrua* (Cyprinodontiformes, Rivulidae). *Genetics and Molecular Biology* 37: 364–374. doi:10.1590/s1415-47572014005000009

Baack, E.J. and L.H. Rieseberg. 2007. A genomic view of introgression and hybrid speciation. *Current Opinion in Genetics and Development* 17: 513–518. doi: 10.1016/j.gde.2007.09.001

Borowsky, R. and H. Wilkens. 2002. Mapping a cave fish genome: Polygenic systems and regressive evolution. *Journal of Heredity* 93: 19–21. doi: 10.1093/jhered/93.1.19

Buerkle, C.A. and L.H. Rieseberg. 2001. Low intraspecific variation for genomic isolation between hybridizing sunflower species. *Evolution* 55: 684–691.

Danzmann, R.G. and K. Gharbi. 2001. Gene mapping in fishes: A means to an end. *Genetica* 111: 3–23.

Faria, R. and A. Navarro. 2010. Chromosomal speciation revisited: Rearranging theory with pieces of evidence. *Trends in Ecology and Evolution* 25: 660–669. doi:10.1016/j.tree.2010.07.008

Feder, J.L. and P. Nosil. 2009. Chromosomal inversions and species differences: When are genes affecting adaptive divergence and reproductive isolation expected to reside within inversions? *Evolution* 63: 3061–3075. doi:10.1111/j.1558-5646.2009.00786.x

Feder, J.L., S.P. Egan, and P. Nosil. 2012. The genetics of speciation with gene flow. *Trends in Genetics* 28: 342–350. doi:10.1016/j.tig.2012.03.009

García, G. 2006. Multiple simultaneous speciation in killifishes of the *Cynolebias adloffi* species complex (Cyprinodontiformes, Rivulidae) from phylogeography and chromosome data. *Journal of Zoological Systematics and Evolutionary Research* 44: 75–87. doi: 10.1111/j.1439-0469.2005.00346.x

García, G., V. Gutiérrez, N. Ríos et al. 2014. Burst speciation processes and genomic expansion in the neotropical annual killifish genus *Austrolebias* (Cyprinodontiformes, Rivulidae). *Genetica* 142: 87–98. doi: 10.1007/s10709-014-9756-7

García, G., M. Loureiro, N. Berois et al. 2009. Pattern of differentiation in the annual killifish genus *Austrolebias* (Cyprinodontiformes: Rivulidae) from a biosphere reserve site in South America: A multidisciplinary approach. *Biological Journal of the Linnean Society of London* 98: 620–635. doi: 10.1111/j.1095-8312.2009.01303.x

García, G., F. Alvarez-Valin, and N. Gómez. 2002. Mitochondrial genes: Signals and noise in phylogenetic reconstruction within killifish genus *Cynolebias* (Cyprinodontiformes, Rivulidae). *Biological Journal of the Linnean Society of London* 76: 49–59. doi: 10.1111/j.1095-8312.2002.tb01713.x

García, G., A.I. Lalanne, G. Aguirre, and M. Cappetta. 2001. Chromosome evolution in annual killifish genus *Cynolebias* and mitochondrial phylogenetic analysis. *Chromosome Research* 9: 93–100. doi: 10.1023/A:1011664009509

García, G., E. Scvortzoff, and A. Hernández. 1995. Karyotypic heterogeneity in South American annual killifishes of the genus *Cynolebias* (Pisces, Cyprinodontiformes, Rivulidae). *Cytologia* 60: 103–110. doi: 10.1508/cytologia.60.103

García, G., E. Scvortzoff, M.C. Máspoli, and R. Vaz-Ferreira. 1993. Analysis of karyotypic evolution in natural populations of *Cynolebias* (Pisces, Cyprinodontiformes, Rivulidae) using banding techniques. *Cytologia* 58: 85–94. http://dx.doi.org/10.1508/cytologia.58.85

Gutiérrez, L. and G. García. 2011. A preliminary linkage map using the spotted melanic laboratory strains of the live bearing *Phalloceros caudimaculatus* var. *reticulata* (Cyprinodontiformes: Poeciliidae. *Journal of Genetics*, Online Resources: 90/e66).

Haldane, J.B.S. 1922. Sex ratio and unisexual sterility in hybrid animals. *Journal of Genetics* 12: 101–109.

Huxley, J. 1928. Sexual difference of linkage in *Gammarus chevreuxi*. *Journal of Genetics* 20: 145–156.

Keong B.P., S.S. Siraj, S.K. Daud, J.M. Panandam, and A.N.A. Rahman. 2014. Identification of quantitative trait locus (QTL) linked to dorsal fin length from preliminary linkage map of molly fish. Poecilia sp. *Gene* 536: 114–11. doi: 10.1016/j.gene.2013.11.068

Khoo, G., M.H. Lim, H. Suresh et al. 2003. Genetic linkage maps of the guppy (*Poecilia reticulata*): Assignment of RAPD markers to multipoint linkage groups. *Marine Biotechnology* 5: 279–293. doi: 10.1007/s10126-002-0072-3

Knapik, E.W., A. Goodman, M. Ekker et al. 1998. A microsatellite genetic linkage map for zebrafish (*Danio rerio*). *Nature Genetics* 18: 338–343. doi:10.1038/ng0498-338

Kosambi, D.D. 1944. The estimation of map distance from recombination values. *Annals of Eugenics* 12: 172–175.

Kulathinal, R.J., L.S. Stevison, and M.A.F. Noor. 2009. The genomics of speciation in *drosophila*: Diversity, divergence, and introgression estimated using low-coverage genome sequencing. *PLoS Genetics* 5:e1000550.

Manly, K.F., Jr. R.H. Cudmore, and J.M. Meer. 2001. MapManager QtX, cross platform software for genetic mapping. *Mammalian Genome* 12: 930–932. doi: 10.1007/s00335-001-1016-3

McGaugh, S.E. and M.A.F. Noor. 2012. Genomic impacts of chromosomal inversions in parapatric *Drosophila* species. *Philosophical Transactions of the Royal Society B, Biological Sciences* 367: 422–429. doi:10.1098/rstb.2011.0250

Medrano, J.F., E. Aasen, and L. Sharrow. 1990. DNA extraction from nucleated red blood cells. *Biotechniques* 8: 43.

Morizot, D.C., R.S. Nairin, P. Simhambhatla et al. 2001. *Xiphophorus* genetic linkage map: Beginnings of comparative gene mapping in fishes. *Marine Biotechnology* 3: S153–S161. doi: 10.1007/s10126001-0037-y

Navarro, A. and N.H. Barton. 2003. Accumulating postzygotic isolation genes in parapatry: A new twist on chromosomal speciation. *Evolution* 57: 447–459. doi: 10.1111/j.0014-3820.2003.tb01537.x

Noor, M.A.F., K.L. Grams, L.A. Bertucci, and J. Reiland. 2001. Chromosomal inversions and the reproductive isolation of species. *Proceedings of the National Academy of Sciences* USA 98: 12 084–12 088.

Ortiz-Barrientos, D., J. Reiland, J. Hey, and M.A. Noor. 2002. Recombination and the divergence of hybridizing species. *Genetica* 116: 167–178. doi:10.1023/A:1021296829109

Rieseberg, L.H. 2001. Chromosomal rearrangements and speciation. *Trends in Ecology and Evolution* 16: 351–358. doi: http://dx.doi.org/10.1016/s0169-5347(01)02187-5

Sanguinetti, C., F. Díaz Neto, and A.J. Simson. 1994. Rapid silver staining and recovery of PCR products separated on acrylamide gels. *Biotechniques* 17: 915–918

Stevison, L.S., K. Hoehn, and M.A.F. Noor. 2011. The effect of chromosomal inversions on recombination in hybrid between *Drosophila pseudoobscura* and *D. persimilis*. *Genome Biology and Evolution* 3: 830–841. doi:10.1093/gbe/evr081

Via, S. 2012. Review article: Divergence hitchhiking and the spread of genomic isolation during ecological speciation with gene flow. *Philosophical Transactions of the Royal Society B, Biological Sciences* 367: 451–460. doi: 10.1098/rstb.2011.0260

Waldbieser, G.C., B.G. Bosworth, D.J. Nonneman, and W.R. Wolters. 2001. A microsatellite based genetic linkage map for channel catfish. *Ictalurus punctatus Genetics* 158: 727–734.

Welsh J. and M. McClelland. 1990. Fingerprinting genomes using PCR with arbitrary primers. *Nucleic Acids Research* 24: 7213–7218. doi: 10.1093/nar/18.24.7213

Williams J.G.K., A.R. Kubelik, K.J. Livak, J.A. Rafalski, and S.V. Tingey. 1990. DNA polymorphisms amplified by arbitrary primers are useful as genetic markers. *Nucleic Acids Research* 18: 6531–6535. doi: 10.1093/nar/18.22.6531

CHAPTER 18

Overview, Future Challenges, and Evolution of Annualism

Rafael O. de Sá, Nibia Berois, and Graciela García

CONTENTS

18.1 INTRODUCTION

The previous chapters have summarized, reviewed, and presented published and original data on the work being done across various disciplines to understand the overall biology and evolution of annual fishes. The term "annual fishes" was coined to apply to fish species living in temporary freshwater biotopes, which dry seasonally (Myers, 1952). This term also describes the atypical, among vertebrates, short lifespan of this group of fishes that would have evolved as an adaptation to inhabit ephemeral pools of water that vanish during the dry season. Adults of Neotropical annual fishes usually live in temporal ponds less than 1 year (e.g., *Austrolebias*), whereas in some species of African annual fishes lifespan may be as low as 17 weeks (e.g., *Nothobranchius furzeri*). The annual condition was described by two characteristics: (1) a unique stage during early development, the dispersion–reaggregation of deep blastomeres that separates epiboly to embryonic axis formation, and (2) the possibility of the embryo to enter in developmental arrests or diapauses before hatching (Myers, 1952; Wourms, 1972a,b,c).

18.2 DISTRIBUTION, DIVERSITY, AND CONFLICTING TAXONOMY

Annual fishes are not a monophyletic group but are distributed within two sister clades of freshwater fishes: the African Nothobranchiidae and the Neotropical Rivulidae. They exhibit a great diversity of morphologies, and the content and relationships among genera and other taxonomic categories have been extensively discussed in the literature. Within these clades we found annual, semiannual, facultative annual, and nonannual species. Although several phylogenies based on non-molecular and molecular data and combined analyses are available, most have lacked an intensive taxon sampling that can account for the different resulting topologies. Also, within and among genera, relationships are missing or unresolved for several groups (see Chapter 1). Additional phylogenetic work is needed and will not only contribute to our understanding of relationships among species but will also provide a framework to assess the evolution of many very interesting traits.

18.3 PHENOTYPIC PLASTICITY IN HETEROGENEOUS ENVIRONMENTS

The short lifespan of annual fishes also resulted in various other ecological, behavioral, reproductive, and developmental adaptations (Wourms, 1972a,b,c; Berois et al., 2012, 2014; Blazek et al., 2013). These adaptations include (1) complex courtship behavior and sexual selection that makes annual fishes a good model to understand the roles of evolution and speciation beyond natural selection (see Chapter 12); (2) a prolonged embryonic stage, sometimes lasting longer than the adult stage, that is able to withstand desiccation as well as other environmental stress factors (see Chapter 10); and (3) developmental arrests (diapauses I–III) before hatching (see Chapter 2). Developmental processes and patterns are complex, particularly the presence of embryos that can skip one or more diapauses, "escape eggs" (Wourms, 1972c). A number of examples have shown that developmental switches of a single genotype triggered by environmental conditions can result in alternative phenotypes related to changing environmental conditions (Stearns, 1989). Several studies have demonstrated that under certain conditions, these developmental arrests show some flexibility and embryos may follow alternative pathways and shorten or escape diapause II (Markofsky and Matias, 1977; see Chapter 4).

Under laboratory conditions, the lifespan of Neotropical annual fishes can be extended to about 15 months (60 weeks) in *Austrolebias* (N. Papa, personal communication), but the individuals start to show signs of aging (e.g., decreased coloration, a curved spine, infections, etc.). In African annual fishes, lifespan under captive conditions has been reported to extend from 17.5 to 29 weeks in *Nothobranchius furzeri* and 42–47 weeks in *N. kuhntae* (Terzibasi Tozzini et al., 2013). Overall, these short lifespans make annual fishes a unique vertebrate model for studying senescence. Most of the senescence studies available have been done using the African annual genus *Nothobranchius*, which has been suggested as a model system for aging research (Herrera and Jagadeeswaran, 2004; Genade et al., 2005; Cellerino et al., 2015; Chapter 6). Research on senescence in the Neotropical genus *Austrolebias* is currently underway. This research, as well as research in a variety of other areas, is facilitated by an increasing knowledge of the requirements and maintenance of adults and embryos under laboratory conditions (see Chapter 7).

A sex chromosome-determining system was originally reported for the African *Nothobranchius guentheri* (Ewulonu et al., 1985) and for the Neotropical genus *Pterolebias* (Elder et al., 1991). Sex determination and differentiation has been explored in both Neotropical (Arezo et al., 2007, 2014; see Chapter 5) and African annual fishes (Valenzano et al., 2009). Although gamete characteristics and other reproductive features can serve as hybridization barriers (García et al., 2009; Berois et al., 2014), little is known about gamete interactions in annual fishes (Berois et al., 2011; see Chapter 3). This is an area that needs further research, particularly considering that several species of annual fishes cohabit the same ephemeral ponds and in instances where other isolating mechanisms fail, gamete interaction may represent an important prezygotic barrier.

Morphological studies have focused on the skeletal system (Parenti, 1981; Costa, 1998, 2001, 2006, 2009a; Loureiro and de Sá, 1998), given that it is used in phylogeny and systematic classification. However, little is known about the diversity of soft anatomy systems, an area that could provide many interesting research and student projects. Brain anatomy has only been reported for four genera of the large Cyprinodontiformes clades, including two annual fishes, *Austrolebias* (Fernández et al., 2011) and *Nothobranchius furzeri* (D'Angelo, 2013); active proliferative zones of the adult brain were also reported for *Austrolebias* and *Nothobranchius* (see Chapter 13).

18.4 ANNUALS AS MODELS IN EVOLUTION AND COMMUNITY ASSEMBLAGES

Because of annual fishes' unique reproductive and developmental adaptations coupled with their nonoverlapping generation in time and space, they are excellent models for analysis of evolution, ecology, and community assemblages. In addition to the previous characteristics, the patchy distribution of annual fish species provides ideal natural conditions to study underlying evolutionary mechanisms (e.g., genetic drift, gene flow, selection, female choice, etc.) that may result in rapid divergence with or without morphological differentiation.

A number of studies have addressed spatial distribution patterns, ecology, and diversity of communities where annual fishes occur. Furthermore, the ephemeral environments occupied by annual fishes and the cyclical collapse of isolated ponds results in unique local population dynamics that facilitate the persistence of metapopulations (Chapters 8 and 9). These studies highlight the unique evolutionary and ecological place of annual fishes; however, basic population data is missing for most known species (e.g., Krause Lanés et al., 2014). Furthermore, long-term survival of small and isolated populations, and consequent long-term or not-so-long-term loss of species diversity due to overall habitat degradation and loss is a high concern. Considering that many species inhabit coastal environments and that their annual life cycle is dependent on annual rains and refilling of ephemeral ponds, the current climate change with the alteration of hydrological environmental conditions, extended droughts, and rising sea levels is a major threat to long-term survival of many annual fishes (Chapter 11). Data on climatic change in relation to its impact on annual fishes is highly deficient. Current and future undergraduate and graduate students, particularly individuals who enjoy field and ecological work, should consider pursuing their research on annual fishes.

18.5 THE MOLECULAR BASIS OF PHENOTYPIC PLASTICITY IN ANNUAL FISHES

Recent technological advances in genomics, transcriptomics, proteomics, and metabolomics will facilitate deciphering the molecular basis of the environmental/genetic networks participating in adaptive responses of alternative development pathways.

Nothobranchius furzeri is currently the most intensively studied species, and its genome was recently sequenced, assembled, annotated, and deposited and serves as an available reference for future genomic studies (Harel et al., 2015). The genus *Nothobranchius* has been considered an experimental model to address different aspects of annual fishes biology (Chapter 6). The recent use of clustered regularly interspaced short palindromic repeat (CRISPR) loci, in conjunction with CRISPR-associated genes (*Cas*), particularly *Cas9*, has facilitated targeted genome editing (Pennisi, 2013). This procedure generates knockout lines and also the insertion of human mutations into the *N. furzeri* endogenous loci, inducing phenotypes that closely resemble human pathologies (Harel et al., 2015).

The Neotropical *Austrofundulus* and *Austrolebias* have emerged as alternative annual fish models to understand the biology and the genetic and molecular basis linked to developmental pathways in annual fishes (see Chapters 2 and 4; Berois et al., 2014). The unusually giant genome size described for 16 species of *Austrolebias* (García et al., 2014) represents a trait not reported for any other annual or nonannual Aplocheiloidei (Reichwald et al., 2009). It may also serve as a model to analyze genomic evolutionary processes underlying speciation events. These genomes are larger than those of nearly all other diploid, that is, non-(paleo) polyploid species of actinopterygian fishes, so far reported (Mank and Avise, 2006; Mable et al., 2011). In all other aspects of their genomes,

Austrolebias species are typical diploids (see Chapters 14 and 16; García et al., 2014). An NGS analysis found that the genomic proportion of moderately repetitive DNA in the genome of *A. charrua* represents approximately twice (45%) the repetitive DNA of the closely related genus *Cynopoecilus* (García et al., 2015).

Also, NGS studies of the *Nothobranchius* genome showed different repeat families (TEs) that could proliferate in different sublineages, probably explaining the great differences in genome sizes and their highly reported instability (see Chapter 14; García et al., 2015). A recent analysis of the genome of 256 species (i.e., animals, plants, fungi, and protists) found that the diversity of TEs, above 500 Mbp, did not increase with genome size (Elliott and Gregory, 2015). Furthermore, they reported no correlation across animals or within vertebrates. Currently, many questions emerge from the giant genomes, with plenty TEs, of *Austrolebias* in relation to their regulatory potential and/or reversible functions and relationships with adaptive evolutionary acquisitions. Further research including extensive genome analysis and higher coverage in many *Austrolebias* species, and other genera from New and Old World annual fishes, could clarify the mechanisms and role of genome size in annual fishes.

18.6 EVOLUTION OF ANNUALISM

Among the diversity of Aplocheiloidei fishes, the most outstanding trait is precisely the evolution of annualism. This trait has sparked much research due to uncommon characteristics among fishes and vertebrates in general: (1) the annual death of the entire population during the dry season (i.e., extrinsic mortality) and (2) the specialized eggs/embryos (i.e., diapausing embryos) that withstand the harsh environment and are key to the survival of the species (Berois et al., 2014).

The single or multiple origins of annualism have been the focus of much debate. Early phylogenetic work did not support the single origin of annualism, and Parenti (1981) considered annualism as independently derived in the New World Rivulidae and the Old World Aplocheilidae: "The definition of the Old World Aplocheiloids as a monophyletic group again supports the contention that annualism is a lifestyle which has either arisen at least twice within cyprinodontiform fishes, or is a characteristic that in some sense is basic to all members" (Parenti, 1981:386). Subsequently, annualism was considered plesiomorphic in the Rivulidae (Costa, 1990a,b). Later, a molecular phylogeny could not unambiguously argue for a single or multiple origins (Murphy and Collier, 1997). However, by weighting the gain of the life history trait of "annualism" over its loss, these authors argued for an early and single origin under arid or seasonally dry conditions during the Late Cretaceous, although considering annualism the plesiomorphic state for the group required an additional step. This was based on the fact that the alternative scenario would require the convergent/parallel evolution of the following traits: (1) thickening of the chorion, (2) diapausing embryo, (3) bottom spawning, and (4) rapid growth to reach sexual maturity. The next important contribution (Costa, 1998) contrasts the single origin of annualism in aplocheilids (Murphy and Collier, 1997) with an analysis (at the time in press) that suggested the independent origin of annualism in Cynolebiatinae and Rivulinae (Rivulidae) (Hrbek and Larson, 1999). This study resulted in two equally parsimonious interpretations on the evolution of annualism in Rivulidae (Costa, 1998:84). Costa's (1998) work is exclusively based on a large and diverse suite of nonmolecular characters, and the author points out that developmental, ecological, and behavioral traits are intrinsically related and coadapted.

An analysis of relationships of the Neotropical Rivulidae adheres to the single origin of annualism, indicating that the annual history trait "has clearly been lost several times in the course of rivulids evolution" (Murphy et al., 1999a:289). The most interesting finding reported in this study was a significant increase of substitution rates in three different annual lineages *Callopanchax* (West African), *Nothobranchius* (East African), and Neotropical Rivulids relative to their nonannual sister species. This was recently confirmed in a phylogenetic analysis of *Nothobranchius*, in which the mitochondrial coding *COI* gene was saturated (Dorn et al., 2014). However, in two other papers published simultaneously (in 1999) that assessed phylogenetic relationships of African genera, the authors presented alternative conclusions. The first paper (Murphy et al., 1999b) assessed relationships among West African aplocheiloids and recovered three monophyletic genera with *Archiaphyosemion* (nonannual) basal to the sister pair *Callopanchax* (annual) and *Scriptaphyosemion* (nonannual), concluding that annualism was reacquired in *Callopanchax* (p. 343), having been previously lost in the West African ancestor (p. 349); the paper also indicated that "a single origin of annualism within the suborder still seems plausible" (p. 349). The second paper (Murphy and Collier, 1999) assessed relationships among the genera *Aphyosemion* (then a nonannual) and *Fundulopanchax* (annual) and transferred the *Fundulopanchax batesi* group (annual) to *Aphyosemion* to render the genera monophyletic, making *Aphyosemion* a genus of annual and nonannual species. Both papers concluded that traits associated with annualism are evolutionarily more plastic than once thought (Murphy et al., 1999b; Murphy and Collier, 1999). A recent paper also supports the origin of annualism in *Callopanchax* independent from that of other African annual genera (Costa, 2015). The next molecular analyses supported two independent origins of the diapausing egg, and hence of annualism, in the Rivulidae, suggesting convergent evolution to harsh environmental conditions (Hrbek and Larson, 1999). Furthermore, based on previous work (Murphy and Collier, 1997; Hrbek, 1999), these authors suggested that parallel evolution of annualism occurred twice in African groups and concluded that annualism was independently gained four times in the Aplocheiloidei. Moreover, they suggested that the prolonged development of cyprinodontiform eggs may have favored the evolution of annualism in this group (Hberk and Larson, 1999). An osteological analysis of the African genus *Callopanchax* and related genera compared the resulting topology with previous topology (Murphy and Collier, 1997; Murphy et al., 1999b) and concluded that the osteological phylogeny supported a triple origin of annualism (Costa, 2009b).

It is precisely the difficulty to envision the repeated convergent evolution of a complex diapausing embryo (as well as all other traits associated with annualism, such as rapid growth to maturity, senescence, etc.) without accepting multiple origins of annualism. However, a recent article suggested that key processes of the diapausing embryo, for example, cell dispersion stage and reaggregation phase during gastrulation in annual fishes, may be extended stages or a derived condition (Berois et al., 2014) of processes already in place and long reported in other teleosts (e.g., *Salmo*, *Salvelinus*: Yokoya, 1966; *Oryzias latipes*: Ballard and Dodes, 1968). The most recent analysis does support convergent evolution of the diapausing embryos with four independent evolutions in the New World and at least two times among the Old World annual fishes (Furness et al., 2015).

The analyses of annual fishes over the last 50 years have accumulated data to conclude that annualism has evolved more than once among annual fishes. This statement opens new opportunities for future students and researchers to explore and test the underlying mechanisms of evolution at multiple levels in annual fishes. Parallel evolution of other traits had been documented for African and South American annuals, for example, icthyophagy (Costa, 2011).

In addition to the defined developmental characteristics, a short lifespan, with an early onset of senescence, also characterizes annual fishes. But which of these characteristics was *key* in the evolution of annualism and how does it relate to senescence?

During the development of annual species, the embryo undergoes three facultative or obligate developmental arrests (diapauses I, II, and III). These developmental arrests do not represent "just a pause in development but an alternative developmental pathway that differs morphologically, physiologically, and biochemically from development" in the absence of arrest (see Podrabsky et al., 2010a:3285). A required step for annualism is that at least one of these arrests would be an obligate arrest. For example, in the Neotropical (e.g., *Austrolebias*, *Austrofundulus*) and African (e.g., *Nothobranchius*) genera, diapauses I and II are facultative, and embryos that bypass diapause II are called "escape" or "direct-developing embryos," whereas diapause III is obligate (Podrabsky et al.,

2010a; Berois et al., 2014; Furness et al., 2015). Consequently, diapause III is the developmental arrest that carries the embryos through most of the duration of the dry period and allows the embryos to withstand extreme environmental conditions. It is important to note that under laboratory conditions (e.g., control temperature, plenty of food, no predators, etc.), embryos can hatch without undergoing a dry period. For example, in *Nothobranchius* species under laboratory conditions in which embryos are maintained in aqueous aerobic media, the embryos develop and hatch (Valenzano et al., 2011; see Chapter 9). In *Austrolebias* species, embryos maintained in laboratory conditions can bypass all three diapauses and hatch (M. Arezo, personal observation). However, some of these hatched alevins, called "belly sliders," swim close to the bottom, probably due to air bladder problems, but others are normal. In the laboratory, belly sliders can reach sexual maturity and reproduce, although they are significantly smaller relative to their siblings (C. Passos, personal communication). Furthermore, in *Austrofundulus limnaeus*, normal and escape embryos show significant morphological and physiological differences (Podrabsky et al., 2010a,b). Currently, there is no available field data about survivorship of "escape" embryos in the wild for any annual species. However, the likelihood of reaching sexual maturity would be severely compromised, since the reduced swimming ability would result in a lower chance of securing food and escaping predation. Furthermore, no laboratory study has compared the performance of "laboratory escape adults" and wild adults in courtship and mating success.

In answering which of the early identified characteristics of annual fishes was key in the evolution of annualism, and considering that (1) a dispersion–reaggregation phase may be a derived state of processes present in other teleosts (Berois et al., 2014) and (2) diapauses I and II could be facultative, then an obligate diapause III would be a defining trait of annualism. Furthermore, the annual cycle results in the massive death of the adults in the populations during the dry season. However, in the laboratory, with ideal environmental conditions, annual fishes still exhibit an early onset of senescence; alternatively, could senescence have driven the evolution of annualism? Taking into account that diapauses I and II are not only facultative but exhibit extensive developmental plasticity (Podrabsky et al., 2010a,b), populations producing diapausing embryos, particularly those that added an obligate diapause III, would have been favored by natural selection under harsh environmental conditions. Under this scenario, senescence would have evolved in parallel or later to the appearance of diapause III, as selection favored shifting individual resources and energy to sustain a longer lifespan to reach rapid sexual maturity. Extrinsic mortality has been linked to senescence in other fishes, for example, salmon and guppies (Hendry et al., 2004; Reznick et al., 2004). Furthermore, a recent study using species of *Nothobranchius* occurring in humid (*N. kuhntae*) and drier regions (*N. furzeri*) of Africa demonstrated a shorter lifespan for *N. furzeri*, concluding that extrinsic mortality shaped the evolution of senescence in *Nothobranchius* (Terzibasi Tozzini et al., 2013). The repeated adaptive evolution of annualism in Neotropical and African killifishes fits the pattern reported by Ord and Summers (2015) where similarity in habitat characteristics was the overwhelming factor (48%) of studies reporting parallel or convergent adaptations. Furthermore, the study also showed that fishes have the highest reported (23%) level of adaptations due to repeated evolution.

Moreover, the terms "semiannual" and "facultative annual" are used interchangeably in the literature in reference to life cycles that are somehow intermediate between those of annual and nonannual species. These facultative or semiannual species inhabit areas that sometimes dry out but may retain water throughout the year, for example, *Fundulopanchax* (Suffia, 2004; Collier, 2010). Moreover, the eggs of several nonannual genera or species may survive exposure to air over variable periods of time. For example, the eggs of *Kryptolebias sepia* kept under laboratory conditions were reported to develop normally and equally in water, wet peat moss, and moist peat moss (Vermeulen and Hrbek, 2005). However, eggs in water hatched after 14–16 days (24°C), whereas eggs in peat moss did not hatch and, unless immersed in water, died within 7 days. The authors concluded that for *K. sepia*: (1) 3 weeks is the maximum length of habitat desiccation and (2) the species is "unlikely to lay embryos capable of undergoing a developmental diapause" (Vermeulen and Hrbek, 2005:11). Future comparative developmental and genomic studies, particularly among annual species and nonannual but air-resistant species may provide new insights into the evolutionary processes and mechanisms of annualism.

ACKNOWLEDGMENTS

R.dS. was supported by award NSF–DEB 1144692. G.G. and N.B. acknowledge full-time support from CSIC, Universidad de la República, PEDECIBA (Programa de Desarrollo de Ciencias Básicas), CSIC (Comisión Sectorial de Investigación Científica, UdelaR), SNI (Sistema Nacional de Investigadores), and ANII (Agencia Nacional de Investigación e Innovación).

REFERENCES

Arezo, M.J., S. D'Alessandro, N.G. Papa, R.O. de Sá, and N. Berois. 2007. Sex differentiation pattern in the annual fish *Austrolebias charrua* (Cyprinodontiformes: Rivulidae). *Tissue Cell*. 39: 89–98. doi:10.1016/j.tice.2007.01.004

Arezo, M.J., N.G. Papa, V. Guttierrez, G. García, and N. Berois. 2014. Sex determination in annual fishes: Searching for the master sex-determining gene in *Austrolebias charrua* (Cyprinodontiformes, Rivulidae). *Genetics and Molecular Biology*, 37(2): 364–374. doi:10.1590/s1415–47572014005000009

Ballard, W.W., and L.M. Dodes. 1968. The morphogenetic movements of the lower surface of the blastodisc in *Salmonid* embryos. *Journal of Experimental Zoology* 168: 76–84. doi: 10.1002/jez.1401680107

Berois, N., M.J. Arezo, and R.O. de Sá. 2014. The neotropical genus *Austrolebias*: An emerging model of annual killifishes. *Cell and Developmental Biology* 3: 136. doi:10.4172/2168–9296.1000136

Berois, N., M.J. Arezo, and N.G. Papa. 2011. Gamete interactions in teleost fish: The egg envelope. Basic studies and perspectives as environmental biomonitor. *Biological Research* 44: 119–124. doi: /S0716–97602011000200002

Berois, N., M.J. Arezo, N.G. Papa, and G.A. Clivio. 2012. Annual fish: Developmental adaptations for an extreme environment. *WIREs Developmental Biology* 1: 595–602. http://dx.doi:10.1002/wdev.39

Blazek, R., M. Polacik, and M. Reichard. 2013. Rapid growth, early maturation and short generation time in African annual fishes. *EvoDevo* 4: 24. doi:10.1186–9139–4–24

Cellerino, A., D.R.Valenzano, and M. Reichard. 2015. From the bush to the bench: The annual *Nothobranchius* fishes as a new model system in biology. *Biological Review*. doi:10.1111/brv.12183

Collier, G.E. 2010. The genus *Fundulopanchax* taxonomic history and molecular phylogeny. *Journal of the American Killifish Association* 43(1): 3–16.

Costa, W.J.E.M. 1990a. Análise filogenética da família Rivulidae (Cyprinodontiformes, Aplochciloidei). *Revista Brasileira de Biologia* 50: 65–82.

Costa, W.J.E.M. 1990b. classificação e distribuição da família Rivulidae (Cyprinodontiformes, Aplocheiloidei). *Revista Brasileira de Biologia* 50: 83–89.

Costa, W.J.E.M. 1998. Phylogeny and classification of Rivulidae revisited: Origin and evolution of annualism and miniaturization in rivulid fishes (Cyprinodontiformes: Aplocheiloidei). *Journal of Comparative Biology* 3(1): 33–94.

Costa, W.E.J.M. 2001. The neotropical annual fish genus *Austrolebias* (Teleostei: Cyprinodontiformes: Rivulidae): Phylogenetic relationships, taxonomic revision and biogeography. *Ichthyological Explorations of Freshwaters* 12: 333–383.

Costa, W.E.J.M. 2006. The South American annual killifish genus *Austrolebias* (Teleostei: Cyprinodontiformes: Rivulidae): Phylogenetic relationships, descriptive morphology and taxonomic revision. *Zootaxa* 1213: 1–162.

Costa, W.E.J.M. 2009a. Trophic radiation in the South American annual killifish genus *Austrolebias* (Cyprinodontiformes: Rivulidae). *Ichthyological Explorations of Freshwaters* 20: 179–191.

Costa, W.J.E.M. 2009b. Osteology of the African annual killifish genus *Callopanchax* (Teleostei: Cyprinodontiformes: Nothobranchiidae) and phylogenetic implications. *Vertebrate Zoology* 59(1): 31–40.

Costa, W.J.E.M. 2011. Parallel evolution in icthyophagous annual killifishes of South American and Africa. *Cymbium* 35(1): 39–46.

Costa, W.J.E.M. 2015. Comparative morphology, phylogeny, and classification of West African callopancha-cine killifishes (Teleostei: Cyprinodontiformes: Nothobranchiidae). *Zoological Journal of the Linnean Zociety.* doi: 10.1111/zoj.12270

D'Angelo, I. 2013. Brain atlas of an emerging teleostean model: *Nothobranchius furzeri. The Anatomical Record* 296: 681–691. doi:10.1002/ar.22668

Dorn, A., Z. Musilová, M. Platzer, K. Reichwald, and A. Cellerino. 2014. The strange case of East African annual fishes: Aridification correlates with diversification for a savannah aquatic group? *BMC Evolutionary Biology* (4)14: 210. doi:10.1186/s12862–014–0210–3

Elder, J.F. Jr., B.J. Tuyrner, J.E. Thomerson, and D.C. Taphorn. 1991. Chromosomal divergence and heterogameity in two annual killifishes of the genus. *Pterolebias. Genome* 34: 674–676.

Elliott, T.A., and T.R. Gregory. 2015. Do larger genomes contain more diverse transposable elements? *BMC Evolutionary Biology* 15: 69. doi:10.1186/s12862-015-0339-8

Ewulonu, U.K., R. Haas, and B.J. Turner. 1985. A multiple sex chromosome system in the annual killifish, *Nothobranchius guentheri. Copeia* 1985: 503–508.

Fernández, A.S., J.C. Rosillo, G. Casanova, and S. Olivera-Bravo. 2011. Proliferation zones in the brain of adult fish *Austrolebias* (Cyprinodontiform: Rivulidae): A comparative study. *Neuroscience* 189: 12–24. doi:19.1016/j.neuroscience.2011.o5.063

Furness, A.I. 2015. The evolution of an annual life cycle in killifish: Adaptation to ephemeral aquatic environments through embryonic diapause. *Biological Reviews of the Cambridge Philosophical Society.* doi: 10.1111/brv.12194

Furness, A.I., D.N. Reznick, M.S. Springer, and R.W. Meredith. 2015. Convergent evolution of alternative developmental trajectories associated with diapauses in African and South American killifish. *Proceedings of the Royal Society of London B* 282: 20142189. doi: 10.1098/rspb.2014.2189

García, G., V. Gutiérrez, N. Ríos et al. 2014. Burst speciation processes and genomic expansion in the neotropical annual killifish genus *Austrolebias* (Cyprinodontiformes, Rivulidae). *Genetica* 142: 87–98. doi:10.1007/s10709–014–9756–7

García, G., M. Loureiro, N. Berois et al. 2009. Pattern of differentiation in the annual killifish genus *Austrolebias* (Cyprinodontiformes: Rivulidae) from a biosphere reserve site in South America: A multi-disciplinary approach. *Biological Journal of the Linnean Society, London* 98: 620–635.

García, G., N. Ríos, and V. Gutiérrez. 2015. Next-generation sequencing detects repetitive elements expansion in giant genomes of annual killifish genus Austrolebias (Cyprinodontiformes, Rivulidae). *Genetica,* 143: 353–360. doi: 10.1007/s10709-015-9834-5

Genade, T., M. Benedetti, E. Terzibasi, P. Roncaglia, D.R. Valenzano, A. Cattaneo, and A. Cellerino. 2005. Annual fishes of the genus *Nothobranchius* as a model system for aging research. *Aging Cell* 2005: 1–11. doi: 10.1111/j.14749726.2005.00165.x

Harel, I., B.A. Benayoun, B. Machado et al. 2015. A platform for rapid exploration of aging and diseases in a naturally short-lived vertebrate. *Cell* 160: 1013–1026. http://dx.doi.org/10.1016/j.cell.2015.01.038

Hendry, A.P., Y.E. Morbley, O.K. Berg, and J.K. Wenburg. 2004. Adaptive variation in senescence: Reproductive lifespan in a wild salmon population. *Proceeding of the Royal Society of London, Series B* 271: 259–266.

Herrera, M. and P. Jagadeeswaran. 2004. Annual fish: As a genetic model for aging. *Journal of Gerontology* 59A(2): 101–107. http://dx.doi.org/10.1093/gerona/59.2.B101

Hrbek, T. and A. Larson. 1999. The evolution of diapause in the killifish family Rivulidae (Atherinomorpha Atherinomorpha, Cyprinodontiformes): A molecular phylogenetic and biogeographic perspective. *Evolution* 53: 1200–1216. doi.org/10.2307/2640823

Krasue Lanés, L.E., A.C. Gonçalves, and M.V. Volcan. 2014. Discovery of endangered annual killifish Austrolebias cheradophilus (Aplocheiloidei: Rivulidae) in Brazil, with comments on habitat, population structure and conservation status. *Neotropical Ichthyology* 12(1): 117–124. http://dx.doi.org/10.1590/S1679–62252014000100012

Loureiro, M., and R.O. de Sá. 1998. Osteological analysis of the killifish genus *Cynolebias* (Cyprinodontiformes: Rivulidae). *Journal of Morphology* 238: 245–262. http://dx.doi.org/10.1002/(SICI)1097–4687(199811)238:2%3C245::AID–JMOR5%3E3.3.CO;2–6

Mable, B.K., M.A. Alexandrou, and M.I. Taylor. 2011. Genome duplication in amphibians and fish: An extended synthesis. *Journal of Zoology* 284: 151–182. doi: 10.1111/j.1469–7998.2011.00829.x

Mank, J.E., and J.C. Avise. 2006. Cladogenetic correlates of genomic expansion in the recent evolution of actinopterygiian fishes. *Proceedings of the Royal Society B* 273: 33–38. doi: 10.1098/rspb.2005.3295

Markofsky, J., and J.R. Matias. 1977. The effects of temperature and season of collection on the onset and duration of diapause in embryos of the annual fish *Nothobranchius guentheri*. *Journal of Experimental Zoology* 202: 49–56. doi: 10.1002/jez.1402020107

Murphy, W.J., and G.E. Collier. 1997. A molecular phylogeny for aplocheiloid fishes (Atherinomorpha, Cyprinodontiformes): The role of vicariance and the origins of annualism. *Molecular Biology and Evolution* 14: 790–799. doi: 10.1093/oxfordjournals.molbev.a025819

Murphy, W.J., and G.E. Collier. 1999a. Phylogenetic relationships of African killifishes in the genera *Aphyosemion* and *Fundulopanchax* inferred from mitochondrial DNA sequences. *Molecular Phylogenetics and Evolution* 11: 351–360.

Murphy, W.J., N.P. Nguyen, E.B. Taylor, and G.E. Collier. 1999b. Mitochondrial DNA phylogeny of West African aplocheiloid killifishes (Cyprinodontiformes, Aplochechilidae). *Molecular Phylogenetics and Evolution* 11: 343–350.

Murphy, W.J., J.E. Thomerson, and G.E. Collier. 1999a. Phylogeny of the neotropical killifish family Rivulidae (Cyprinodontiformes, Aplocheiloidei) inferred from mitochondrial DNA sequences. *Molecular Phylogenetics and Evolution* 13: 289–301. doi:10.1006/mpev.1999.0656

Myers, G.S. 1952. Annual fishes. *Aquarium Journal* 23: 125–141.

Ord, T.J., and T.C. Summers. 2015. Repeated evolution and the impact of evolutionary history on adaptation. *BMC Evolutionary Biology* 15:137. doi 10.1186/s12862-015-0424-z

Parenti, L.R. 1981. A phylogenetic and biogeographic analysis of cyprinodontiform fishes (Teleostei: Atherinomorpha). *Bulletin of the American Museum of Natural History* 168(4): 341–557.

Pennisi, E. 2013. The CRISPR craze. *Science* 341(6148): 833–836. doi: 10.1126/science.341.6148.833

Podrabsky, J.E., I.D. Garrett, and Z.F. Kohl. 2010a. Alternative developmental pathways associated with diapause regulated by temperature and maternal influences in embryos of the annual killifish *Austrofundulus limnaeus*. *Journal of Experimental Biology* 213: 3280–3288. doi: 10.1242/jeb.045906

Podrabsky, J.E., A. Tingaud-Sequeira, and J. Cerdà. 2010b. Metabolic dormancy and responses to environmental desiccation in fish embryos. In: *Topics in Current Genetics. Dormancy and Resistance in Harsh Environments*, eds. E. Lubzens, J. Cerdà, and M. Clark, pp. 203–226. Springer, The Netherlands.

Reichwald, K., C. Lauber, I. Nanda et al. 2009. High tandem repeat content in the genome of the short-lived annual fish *Nothobranchius furzeri*: A new vertebrate model for aging research. *Genome Biology* 10: R16.1. doi: 10.1186/gb-2009-10-2-r16

Reznick, D.B., M.J. Bryant, D. Roff, C.K. Ghalambor, and D.E. Ghalambor. 2004. Effect of extrinsic mortality on the evolution of senescence in guppies. *Nature*, 431(7012): 1095–1099. http://dx.doi.org/10.1038/nature02936

Stearns, S. 1989. The evolutionary significance of phenotypic plasticity: Phenotypic sources of variation among organisms can be described by developmental switches and reaction norms. *BioScience* 39(7): 436–46.

Suffia, D. 2004. Breeding and husbandry—*Fundulopanchax puerzli*. *Journal of the Amercian Killifish Association* 37: 150–152.

Terzibasi Tozzini, E., A. Dorn, E. Ng 'Oma, M. Polačik, R. Blažek, K. Reichwald, A. Petzold, B. Watters, M. Reichard, and A. Cellerino. 2013. Parallel evolution of senescence in annual fishes in response to extrinsic mortality, *BMC Evolutionary Biology* 13: 77. doi: 10.1186/1471-2148-13-77

Valenzano, D.R., J. Kirschner, R. Kamber, E. Zhang, D. Weber, A. Cellerino, C. Englert, M. Platzer, K. Reichwald, and A. Brunet. 2009. Mapping loci associated with tail color and sex determination in the short-lived fish *Nothobranchius furzeri*. Genetics 183: 1385–1395. doi:10.1534/genetics.109.108670

Valenzano, D.R., S. Sharp, and A. Brunet. 2011. Transposon-mediated transgenesis in the short-lived African killifish *Nothobranchius furzeri*, a vertebrate model for aging. *G3 Genes genomes genetics* 1: 531–538. doi: 10.1534/g3.111.001271

Vermeulen, F.B.M., and T. Hrbek. 2005. *Kryptolebias sepia* n.sp. (Actinopterygii: Cyprinodontiformes: Rivulidae), a new killifish from the Tapanahony River drainage in southeast Surinam. *Zootaxa* 928: 1–20.

Wourms, J.P. 1972a. The developmental biology of annual fishes. I. Stages in the normal development of *Austrofundulus myersi* Dahl. *Journal of Experimental Zoology* 182: 143–168. doi: 10.1002/jez.1401820202

Wourms, J.P. 1972b. The developmental biology of annual fishes. II. Naturally occurring dispersion and reaggregation of blastomers during the development of annual fish eggs. *Journal of Experimental Zoology*
182: 169–200. doi: 10.1002/jez.1401820203

Wourms, J.P. 1972c. The developmental biology of annual fishes. III. Pre-embryonic and embryonic diapause of variable duration in the eggs of annual fishes. *Journal of Experimental Zoology* 182: 389–414.
doi: 10.1002/jez.1401820310

Yokoya, S. 1966. Cell dissociation and reaggregation in early stage embryo of a teleost *Oryzias latipes*. *The
Science Reports of the Tohoku University Ser* 32: 229–236.

Index

T - #0178 - 111024 - C344 - 254/178/16 - PB - 9780367575311 - Gloss Lamination